无 网 格 法
理论与算法

**Meshless Method
Theories and Approaches**

杨建军　文丕华　著

科学出版社
北京

内 容 简 介

本书以简明易懂的方式，系统地介绍了无网格法的基本理论及各种代表性算法，使初学者很容易掌握这一计算方法的原理和知识。在内容组织上，以固体力学作为应用背景，以无网格法"介点原理"为主线，较为全面地介绍了无网格全局弱式法、局部弱式法、配点类方法、边界型方法和结合式方法等各类离散方法的基本原理及其算法。此外，对移动最小二乘近似法(MLS)的简化和稳定化、介点原理的应用，以及对配点类方法的完善和发展，是本书重点阐述的内容。

本书可作为应用数学、力学、土木、机械、航空航天等专业高年级本科生、研究生的教材，也可供相关专业教师、工程技术人员和科研人员参考使用。

图书在版编目(CIP)数据

无网格法：理论与算法/杨建军，文丕华著.—北京：科学出版社，2018.12
ISBN 978-7-03-060300-5

Ⅰ.①无… Ⅱ.①杨… ②文… Ⅲ.①网格计算–研究 Ⅳ.①TP393.028

中国版本图书馆 CIP 数据核字(2018) 第 297154 号

责任编辑：刘信力 郭学雯 / 责任校对：杨聪敏
责任印制：徐晓晨 / 封面设计：无极书装

科学出版社 出版
北京东黄城根北街16号
邮政编码：100717
http://www.sciencep.com

北京虎彩文化传播有限公司 印刷
科学出版社发行 各地新华书店经销
*
2018年12月第 一 版 开本：720×1000 1/16
2020年 1 月第二次印刷 印张：16
字数：312 000
定价：128.00元
(如有印装质量问题，我社负责调换)

前　　言

　　无网格法从 20 世纪末开始兴起，是 21 世纪初以来计算力学领域研究最为活跃、进展最为显著的计算方法分支。英国物理学家牛顿曾经讲过："要想探求自然界的奥秘，在于解微分方程。"法国数学家拉普拉斯也有类似的观点："只要能解微分方程，我就能预测宇宙的过去和将来。"由此说明，求解微分方程对人类深刻理解自然客观规律是何其重要。

　　或许超乎想象，我们周围的自然世界和人工设备，其状态和运行只是被非常有限的几个微分方程所统治。比如，建筑物的安定与失稳，机车或飞行器平稳运行或破坏，江河的奔流与变迁，气候的变幻莫测，星系的诞生和演化等，这些现象都可以被微分方程所定义。在这些微分方程中，有三类是最主要和最基本的：一是描述振动和波动特征的波动方程；二是反映输运过程的扩散（热传导）方程；三是描绘稳定过程的泊松方程。这些方程在形式上都很简单，但对于实际的问题而言，由于几何结构的复杂性，或者是边界条件的复杂性，求解这些方程却并不总是一个简单任务。

　　无网格法正是一种借助散点信息，应用计算机技术求解微分方程的计算科学方法。该方法学的研究需要集合数学、物理学和计算机工程学等多种学科的知识体系，并发展对应的理论、算法和计算程序等成果。单纯以力学的观点而言，无网格法可被应用于固体力学、流体力学、热力学、电磁力学、生物力学、天体力学、爆炸力学、微观力学等分支领域。

　　无网格法以散点信息作为计算要素，在计算模型构建时无须构造复杂的网格信息。相比于以有限元法为代表的传统网格类方法，无网格法具有诸多竞争优势。一是具有数值实施上的便利性，用散点进行离散要容易得多，尤其对三维问题而言，散点离散具有明显的便捷性；二是无网格法的近似函数通常是高阶连续的，保证了应力结果在全局的光滑性，无须进行额外的光顺化后处理；三是容易实现自适应分析，散点的局部增删和全局重构很容易实现，在计算收敛性校验和移动边界问题中具有明显优势；四是具有求解的灵活性，无网格法避免了对网格的依赖，也就无须担心网格畸变效应，因此容易处理大变形、断裂、冲击与爆炸等一些特殊问题；五是对物质对象描述的普适性，"点"是最基本的几何元素，容易实现对天文星系、原子晶体、生物细胞等物质结构的直接描述。因此，无网格法作为一种易于实施、具有更广泛适用性的数值求解技术，被学者誉为"新一代计算方法"。

　　近年来随着无网格法的快速发展，新的观点、新的理论、新的算法层出不穷，相关研究呈现"百家争鸣、千花齐放"的繁荣景象。国内外一些学者也先后出版了无网格法的专著，比如 S. N. Atluri、G. R. Liu、T. Belytschko、张雄、陈文、程玉民等陆续有专著面世。即便如此，作者还是希望借助本书的出版，能够推介基于作者认知和理解的独立学术观点，更全面地梳理无网格法的主要发展成就，介绍新近的无网格法研究成果。即使难辞挂一漏万之嫌，本书作为一家之言或可有些许参考之益。

时光如流水，飘忽不相待。本书是作者 10 余年来在此研究专题上的一个阶段性总结。同时，本书能得以完成，也有赖于更广泛的同行和学者在此研究专题上的贡献。本书在内容的取舍上，虽然希望尽可能不遗漏一些重要的研究工作，但在浩如烟海的研究文献面前，达成这样的目标显然并不现实。在推荐和评论研究工作时，虽然希望尽可能保持公允而客观的态度，事实上却难以避免作者学术观点的偏好和视野的局限性。作者虽然尽可能地努力来完善本书的内容，但很多方面依然难尽如人意，希望读者批评指正。

作者特别感谢国家自然科学基金 (No.51478053) 对本书研究工作的资助和支持！

作　者

2018 年 5 月于长沙

目 录

前言
第1章 绪论 ··· 1
1.1 传统计算力学方法概述 ··· 1
1.1.1 有限元法 ··· 1
1.1.2 有限差分法 ·· 2
1.1.3 边界元法 ··· 2
1.2 无网格法简介 ··· 3
1.3 无网格法研究简史 ·· 5
1.3.1 无网格法的早期研究历史 ··· 5
1.3.2 近代无网格法研究进展 ·· 6
1.3.3 国内无网格法研究状况 ·· 9
1.4 无网格法研究展望 ·· 11
参考文献 ··· 12
第2章 无网格法基础理论 ·· 25
2.1 固体力学基本理论 ·· 25
2.2 加权残值法 ·· 28
2.3 弹性力学变分原理 ·· 33
2.3.1 虚功原理 ··· 33
2.3.2 最小势能原理 ·· 34
2.3.3 最小余能原理 ·· 34
2.3.4 Hellinger-Reissner 变分原理 ····································· 35
2.3.5 胡海昌–鹫津变分原理 ··· 36
2.4 边界积分方程法 ·· 36
2.5 无网格法介点原理 ·· 39
2.6 本章小结 ·· 43
参考文献 ··· 44
第3章 无网格近似法 ·· 45
3.1 无网格近似函数的性质 ·· 45
3.2 移动最小二乘法 ·· 48
3.3 光滑粒子流体动力学法 ·· 50

3.4　重构核粒子法 ··· 52
　　3.5　点插值法 ··· 54
　　　　3.5.1　多项式基点插值法 ··· 54
　　　　3.5.2　径向基点插值法 ··· 55
　　　　3.5.3　多项式基与径向基耦合的点插值法 ······················· 58
　　3.6　单位分解法 ·· 59
　　3.7　自然邻接点插值法 ·· 60
　　3.8　Kriging 插值法 ··· 62
　　3.9　广义有限差分法 ·· 63
　　3.10　本章小结 ··· 65
　　参考文献 ·· 67

第 4 章　MLS 稳定性及其导数近似 ·· 71
　　4.1　MLS 的构造思想 ··· 71
　　4.2　MLS 的权函数 ··· 73
　　4.3　MLS 稳定近似的几何条件 ·· 77
　　4.4　MLS 核近似法 ··· 82
　　4.5　改进的 MLS 近似法 ·· 89
　　　　4.5.1　改进的 MLS 法 ·· 89
　　　　4.5.2　复变量 MLS 法 ·· 91
　　4.6　MLS 导数近似的讨论 ·· 92
　　4.7　本章小结 ··· 103
　　参考文献 ·· 103

第 5 章　无网格全局弱式法 ··· 105
　　5.1　强式和弱式 ·· 105
　　5.2　Galerkin 弱式 ·· 106
　　5.3　位移边界条件的施加 ··· 111
　　　　5.3.1　形函数具有插值特性 ·· 111
　　　　5.3.2　形函数不具有插值特性 ······································ 113
　　5.4　数值积分方法 ·· 118
　　　　5.4.1　背景网格积分法 ·· 118
　　　　5.4.2　有限元积分法 ··· 119
　　　　5.4.3　节点积分法 ··· 120
　　　　5.4.4　介点积分法 ··· 121
　　5.5　XFEM 法及其对有限元法的改进 ································ 122
　　5.6　本章小结 ··· 124

参考文献 ········· 124

第 6 章　无网格局部弱式法 ········· 127
6.1　Petrov-Galerkin 局部弱式 ········· 127
6.2　无网格局部 Petrov-Galerkin 法 ········· 129
6.3　阶跃检验函数 MLPG 法 ········· 131
6.4　局部边界积分方程法 ········· 132
6.5　其他局部弱式离散法 ········· 133
6.5.1　Galerkin 型 MLPG 法 ········· 133
6.5.2　最小二乘 MLPG 法 ········· 134
6.5.3　配点法 ········· 135
6.6　本章小结 ········· 135
参考文献 ········· 136

第 7 章　配点类无网格法 ········· 138
7.1　有限点法 ········· 138
7.2　双网格扩散配点法 ········· 140
7.3　最小二乘配点法 ········· 143
7.4　无网格介点法 ········· 146
7.5　无网格全局介点法 ········· 162
7.6　本章小结 ········· 168
参考文献 ········· 168

第 8 章　边界型无网格法 ········· 171
8.1　边界节点法 ········· 171
8.2　杂交边界节点法 ········· 172
8.3　基本解方法 ········· 176
8.4　边界点法 ········· 179
8.5　奇异边界法 ········· 180
8.6　边界分布源方法 ········· 181
8.7　广义基本解法 ········· 183
8.8　本章小结 ········· 196
参考文献 ········· 197

第 9 章　结合式无网格法 ········· 202
9.1　无网格强弱式法 ········· 202
9.2　杂交有限差分法 ········· 203
9.3　无限元无网格法 ········· 204
9.4　最小二乘序列函数法 ········· 208

9.5 无网格局部强弱法 ··· 215
9.6 本章小结 ··· 221
参考文献 ·· 221

第 10 章　无网格法应用 ·· 224
10.1 大变形问题中的应用 ·· 224
10.2 断裂与破坏问题中的应用 ·· 226
10.3 冲击与爆炸问题中的应用 ·· 229
10.4 微细观力学问题中的应用 ·· 231
10.5 流体力学问题中的应用 ··· 233
10.6 生物医学中的应用 ··· 236
10.7 无网格法商业软件开发 ··· 238
10.8 本章小结 ·· 239
参考文献 ··· 240

第 1 章 绪 论

计算方法和理论分析、实验统计并列为科学研究的三大支柱[1]。我们面临的力学问题中，许多实际问题通常很难通过解析方法得到问题的精确解，需要借助数值方法求解。除了少数情况，要得到微分方程的解析解是很困难的。而对微分方程进行数值求解，必须将问题域的几何结构进行离散化，将包含无限个自由度的问题域转换为有限个自由度的计算单元。传统上，这些单元是由网格来描述的。这样的离散化处理，可以将一个连续的微分方程求解问题转换为离散化的代数方程问题进行求解，从而使人工计算转化为计算机计算成为可能。当今，借助计算机执行的数值计算方法已经成为求解各种复杂力学问题的重要手段，数值计算方法已经成为工程科学的基本工具之一[2]。我们面临的实际问题通常可进行必要的简化和假设后建立起数学模型，这些数学模型一般是用包含边界条件和初始条件的偏微分方程(或者是常微分方程)描述。

1.1 传统计算力学方法概述

1.1.1 有限元法

有限元法 (finite element method, FEM) 是基于网格离散思想的一种主流计算方法，是计算力学领域最重要的分支之一，并具有良好的通用性。FEM 作为一种独立的计算方法，其名称在 1960 年由 Clough[3] 提出，并在此后得到迅速发展[4-6]，被誉为 "20 世纪工程科学领域最为重要的成就之一"[7]。该法基于 Lagrange 描述法，即用网格直接描述求解对象，而且网格同时描述力学场，整个计算过程中网格是附着于物质结构的，力学场上的点会追随物质结构上的点运动，即网格会随着结构的运动而运动。

FEM 基于 Lagrange 描述法的优点为：其一，力学场与物质结构实现统一描述，计算实施上相对简单而高效；其二，物质点上的所有场变量的整个时间历程可以很容易地追踪；其三，结构的物质特性容易描述，比如，对于非均匀材料问题，可在材料交界面处设置网格节点，其边界条件可通过这些节点自动施加和追踪；其四，网格只需在实体几何域内布置，网格数使用量相对较少，所以计算效率高。

然而，FEM 也有一定的局限性，主要表现在：

其一，难以应对结构大变形的情况。由于力学场与物质结构的结合性，所以节点的相对位移会导致网格单元的扭曲变形。因为场变量的近似函数是以网格为基础的，当网格变形太大的时候，近似精度将受到影响，从而导致求解结果失真，甚至出现计算终止的情况。

其二，实现自适应分析比较困难。克服网格畸变的方案就是在计算过程中自动识别，当网格变形超过设定的允许范围时，重新划分网格，我们通常将这种技术称为自适应分析。然而，重新划分网格通常有较大难度，尤其是结构较为复杂时。此外，自适应分析将很多计算时间用于网格划分，往往导致计算低效。

其三，建模成本较大。由于 FEM 使用网格来描述对象，对于结构复杂的问题，尤其是三维问题，当几何结构比较复杂时，对结构的网格划分，并得到协调优质的网格，是一个非常烦琐的工作，需要投入大量的人力成本。

其四，对移动边界问题难以处理。FEM 中的网格需要严格的连通性要求，在分析一些特殊问题，如断裂、结构破坏和爆炸等移动边界问题时，通常很难处理。

其五，对粒子类物质体系不能直观描述。如果物质体系是以粒子来表现的，比如微、细观上的原子晶体、分子团，或者超宏观上的宇宙星系。用网格只能来描述宏观的力学场，与真实的物质结构体系将很难对应，网格的空间特性与物质的结构性将不能统一。

1.1.2 有限差分法

有限差分法 (finite difference method, FDM) 是基于网格离散思想的另外一种计算方法，其发展历史差不多与 FEM 同时开始[8-10]。该法基于 Euler 描述法，即网格张在一个规划区域上，并用网格来描述此区域的力学场，而且该区域应覆盖描述对象的运动区域。当物质结构穿过网格时，网格单元是固定的，不随物质的运动而运动，在整个计算过程中网格单元的形状都保持不变。因此物质结构的大变形不会引起网格本身的变形，所以该法能很好地适应大变形问题。

但 Euler 法也有一些缺点，例如，其一，由于网格只能等待物质运动引起的响应，对物质运动不能追踪，因此对物质点上的场变量时间历程很难分析；其二，很难处理复杂的几何结构问题，对物质界面的力学描述不够精确；其三，网格划分区域通常要大于物质结构几何区域，这就意味着需要使用更多的网格进行离散，所以计算效率通常低于有限元法；其四，也是最为重要的一点，Euler 描述法所建立的求解方程通常要复杂得多，远不及 Lagrange 描述法那么简单直接。因此，在实际应用中，FDM 与 FEM 相比，应用的广度和范围不具有竞争优势。

1.1.3 边界元法

边界元法 (boundary element method, BEM) 是稍晚于 FEM 而发展起来的，早期比较有影响力的是 Rizzo 和 Cruse 的研究工作[11,12]。而使用"边界元法"作为此类方法的专有名称，则直到 1977 年由 Brebbia，Dominguez，Banerjee，Butterfield 等 [13] 共同商定，此后相关的专著陆续得到出版 [14,15]。与区域解法不同，BEM 是一种求解边值问题的数值方法。该法以边界积分方程为数学基础，同时采用了与 FEM 相似的网格单元离散技术。该法借助特定问题解析得到的基本解函数，在域内已满足泛定的方程，并可用有限个边界上节点的解的叠加来满足给定的边界条件。BEM 的最大特点是降低了问题域的网格离散维数，问题域内不需要划分网格，只需在边界上设置网格，从而大幅度节约了网格离散的工作量。它只以边界变量为基本变量，域内未知量可以在需要时根据边界变量求得。

BEM 的优点可以归纳为 [13]：其一，BEM 通常可获得高精度的数值解。由于在边界积分方程中采用了解析基本解，因此相应的数值方法实际上是一种半解析、半数值方法，故通常具有比一般数值方法更高的计算精度。其二，因为 BEM 仅需在边界上使用网格离散，单元数相对于 FEM 和 FDM 要少得多，因此具有更高的计算效率。其三，BEM 比区域解法降低了离散维度，其直接好处就是离散的便捷，很容易处理复杂的几何结构问题；此外，BEM 可以借助无限域的基本解函数，很容易处理无限域问题。其四，BEM 仅需边界离散的特点，使其特别适用于仅部分边界条件可探测的工程反问题。

但是，BEM 也有其固有局限性。其一，BEM 在应用范围上受到限制。该法是以存在解析基本解为前提的，对于特定的一类问题均需求寻其解析基本解，对于较为复杂的问题，寻找这样的解通常是极具挑战性的一项工作。而有些问题，比如非均匀介质问题，解析基本解是不存在的，该法在此类问题上也就无能为力了。其二，BEM 不宜用于大规模求解问题。由于该法的系统方程组系数矩阵是满阵，而且通常为非对称阵，这一弱点限制了其求解规模。

1.2　无网格法简介

无网格法 (meshless method) 是一种基于散点近似，用于求解偏微分方程问题的数值计算方法。不同于基于网格的传统数值计算方法，无网格法的计算框架是基于散点的，即无网格法使用的基本单元是离散的场节点，场变量的近似函数是通过预定义的散点信息进行描述的。无网格法作为一种独立数值方法的专有名称直至 1996 年才出现，在 Belytschko 等 [16] 的一篇综述性文献中，用 meshless method 作为标题来评述此类方法，这个命名随后得到学者的认可，逐渐成为一种较为通用的名称。为了进一步了解无网格法的特点，我们将其与其他方法进行一个简单的比较，其比较信息列于表 1-1 中。

基于网格的传统数值方法已经非常成熟，其中 FEM 是应用最为广泛的一种计算方法。然而这些使用网格框架求解的方法也面临许多挑战，比如，对一些复

表 1-1 无网格法与其他数值方法的比较

Table 1-1 Meshless method in comparison with other numerical method

比较内容	无网格法	有限元法	有限差分法	边界元法
求解对象	偏微分方程	偏微分方程	偏微分方程	偏微分方程
单元类型	节点	网格	网格	网格
基本单元信息	位置信息	位置及连通信息	位置及连通信息	位置及连通信息
离散方案	散点	全局网格	全局网格	边界网格
形函数构造	散点近似	网格插值	差分插值	网格插值
形函数连续性	全局光滑	分片连续	分片连续	分片连续
理论基础	加权残值法	加权残值法 变分原理	加权残值法	基本解方法 边界积分方程法

杂的几何结构，尤其是三维问题中，形成网格通常需要花费大量的人力成本，所以在数值实施上用于建模的时间通常要高于计算机求解时间。随着计算机技术的发展，FEM 的人工费用将比计算费用更高。对于一些实际问题，比如大变形问题等，基于网格求解的方法通常较难处理，因此在应用上受到诸多限制，数值方法的灵活性及适用性还不够理想。无网格法作为一种求解偏微分方程的方法，相比于其他基于网格的传统数值方法 (FEM，FDM，BEM 等) 具有如下优势:

其一，具有数值实施上的便利性。相比于用网格离散问题域，无网格法用散点进行离散要容易得多，尤其对三维问题而言，散点离散具有明显的便捷性。在计算成本越来越低，而人力成本越来越高的发展趋势下，无网格法的总体成本优势非常明显。无网格法的近似函数可以是全局高阶连续的，保证了应力结果在全局的光滑性，无须像 FEM 那样进行额外的光顺化后处理。

其二，容易实现自适应分析。为了保证数值解的精确性，或者特殊问题的需要 (如网格类方法消除网格的畸变)，数值方法须采用自适应分析技术。无网格法通常都是 h 自适应的，重新布设场节点或加密场节点要比网格类方法重新生成网格容易得多，因此无网格法处理自适应分析很容易实现。

其三，具有求解的灵活性。网格类方法处理一些特殊问题时，由于受网格的束缚，具有诸多的不适应性。而无网格法因为不使用网格，无须担心网格畸变，在问题域或移动边界上可以自由地增设场节点。因此，无网格法在处理大变形、断裂、流动、冲击与爆炸等问题时表现出更好的适应性。

其四，对求解对象具有描述的灵活性。自然界万物都能以"点"为基础描述，由点连线，由线造面，由面构体。对一些不适宜用网格描述的对象，无网格法也具有明显优势。比如，天文力学中在宇宙尺度下对星系的描述，微观物理学中对原子晶体的描述，生物力学中对分子或细胞的描述等。

此外，无网格法采用的散点近似法通常至少是二阶连续的，有关研究结果表明，无网格法的求解精度通常要高于 FEM。因此，无网格法作为一种易于实施、具有更广泛适用性的计算方法，被一些学者誉为"新一代计算方法"。其应用前景值得期待，并受到学界广泛关注，已经成为近 20 年来国际计算力学领域最为活跃的研究专题之一。

1.3 无网格法研究简史

了解无网格法的发展历史可查阅 Belytschko 等[16]、Li 和 Liu[17]、张雄等[18,19]、Gu 和丁桦[20,21]等的综述性文献。此外 Liu 和 Gu[22,23]、Atluri 和 Shen[24,25]、张雄和刘岩[26]、Belytschko 和 Chen[27]、刘更等[28]、赵光明[29]、刘欣[30]、龙述尧[31]、程玉民[32]等也先后出版了一些系统介绍无网格法研究成果的学术专著，参阅这些文献更有利于全面地了解无网格的研究状况。接下来，通过梳理一些比较重要的基本方法，来反映无网格法的研究进展。

1.3.1 无网格法的早期研究历史

无网格法的研究历史，甚至可以追溯到差不多与有限元法发展的同时代，不过早期的研究是零散的，不成体系的。研究的局部进展通常归类于传统方法中，或者将其列为一些具体的独特方法，一些文献也将早期这类方法统称为粒子法或自由网格法[17]。

能检索到的最早的粒子类方法，可能是由 Harlow 等提出的一种用于模拟流体动力学行为的胞含粒子 (PIC) 法[33-37]，PIC 法在 Euler 网格内设置一组粒子，粒子携带质量和位置信息，便于追踪物质的运动。而在背景网格上构造差分近似，并计算空间导数。因此，该法兼具 Lagrange 和 Euler 特性。PIC 法的弱式离散形式及其算法优化后又衍生出两种方法，分别称为流体隐含粒子 (FLIP) 法[38,39] 和物质点法 (MPM)[40-44]。

在 FDM 的计算框架下，Giraul[45]，Perrone 和 Kao[46]，Liszka 和 Orkisz[47-49]等发展了广义有限差分法 (GFDM)，此类方法在差分网格的基础上，使用距离函数抓取必要的节点执行计算，由于该法已经具备散点近似的一些特征，故可将其视为早期无网格法的一种表现形式。

为了模拟宇宙星系的演化，Gingold 和 Monaghan[50,51]，Lucy[52]等提出了光滑粒子流体动力学方法 (SPH)，该法使用 SPH 近似与配点离散技术结合，是一种真正意义上的无网格方法。SPH 法有两个特点：其一是近似的自适应性，即场变量近似基于当前时刻的粒子分布信息，因此很容易处理一些极大变形问题；其二是粒子的材料性，即用于域离散的粒子兼具空间点和物质点两种功能，因此在处理液体流动、冲击与爆炸等一些特殊问题时更具灵活性，是一种应用非常广泛的方法[53-57]。

专门用于流体计算的还有另外一种粒子方法，其出现略早于 SPH 法，即 Chorin, Leonard 等 [58-62] 提出的涡流法 (vortex method)。该方法通过求解流速-涡量方程来分析流体涡流现象，是一种专门化的数值方法，其应用范围不及 SPH 法广泛。

此外，描述分子运动特性的分子动力学 (MD) 方法 [63-68]，是早期发展的一种非常重要的粒子类方法，该法主要在物理学领域用于计算物质体系的热力学量和其他物理性质。由 MD 方法逐步衍生出的格子玻尔兹曼法 (LBM)[69-71]，在 20 世纪末期得到迅速发展，并被广泛用于流体动力学分析 [72-75]。

1.3.2 近代无网格法研究进展

无网格法真正得到迅速发展，并成为一种明确的研究趋势，则是从 20 世纪末全局弱式法的提出开始的。1992 年，Nayroles 等 [76] 为了克服有限元法的一些固有局限性，提出了采用扩散近似的广义有限元法，因为这种方法已经具备了无网格法的基本特点，后来的学者为了使其区别于有限元法，倾向于用扩散元法 (diffuse element method, DEM) 指代之。此外，DEM 所采用的扩散近似 [76,77]，实际上就是后来被广泛使用的移动最小二乘近似。DEM 提出后，引起了著名计算力学科学家、美国西北大学教授 Belytschko 的高度关注，并在 DEM 的基础上，对导数近似算法和本质边界条件施加方法进行了调整，提出了无单元伽辽金法 (EFG)[78]。此后 EFG 被广泛应用于模拟材料的断裂破坏，展现了该方法在特殊问题中优于传统 FEM 法的求解能力 [79-83]。1996 年，Belytschko 等发表了第一篇系统介绍无网格法的综述文献 [16]，并受到计算力学领域研究同行的广泛关注，这可以视为无网格法成为计算力学一个独立研究分支的开端。

弱式类无网格法的另一个重要进展就是局部弱式法的提出。1998 年，Atluri 和 Zhu[84] 提出了无网格局部彼得罗夫-伽辽金法 (MLPG)，该法积分时不需要背景网格，而仅在节点定义的局部积分域上完成，其离散系统方程是基于节点组装的，在数值实施上更加简洁，在多类问题中表现出广泛的适应性，是一种极具竞争力的无网格法 [85-95]。Zhu 等将边界解法的概念引入局部弱式，提出了局部边界积分方程法 (LBIE)[85,86] 和无网格正则局部边界积分方程法 (MRLBIE)[87]，Atluri 和 Shen[24,25] 在 MLPG 的基础上，通过调整检验函数，又构造出多种不同类型的 MLPGx 法，使得 MLPG 的研究内容非常丰富，其中 MLPG5 凭借其执行简单和计算效率较高的特点，在同系列方法中更具有优势 [94]。

无网格法研究的繁荣还体现在散点近似法的发展上，这些近似方法与不同的离散技术结合，又构成多种各具特点的方法。1995 年，Liu 等 [96] 提出了重构核粒子法 (RKPM) 近似，该近似法应用于伽辽金全局弱式离散，对应的无网格方法直接称为重构核粒子法 (RKPM)[96-99]。RKPM 近似应用于配点离散方法，被称为有限云法

(FCM)[100,101]，而在 RKPM 近似中增加 Hermite 插值特性的配点方法，又被称为 Hermite 云法 (HCM)[102,104]。1995 年，Braun 等[105,106] 和 Traversoni[107] 提出采用自然邻接插值来构造近似的方法，此后，Sukumar 等拓展了该近似技术的应用，与伽辽金法离散技术结合的方法通常称为自然单元法 (NEM)[108] 和自然邻接伽辽金法 (NNGM)[109]，与局部弱式离散技术结合的方法被称为自然邻接彼得罗夫–伽辽金法 (NNPG)[110] 和彼得罗夫–伽辽金自然单元法 (PGNE)[111]。2001 年，Liu 和 Gu[112] 提出采用多项式基的点插值近似方法，其应用于伽辽金弱式离散的数值方法，即被称为点插值法 (PIM)，而应用于配点离散的数值方法，被称为点插值配点法 (PICM)[113]。1990 年，Kansa[114,115] 将复合二次函数 (multiquadrics (MQ)) 近似应用于数值计算方法，而 MQ 法是径向基近似的一种形式，Franke 等[116,117] 则进一步发展了径向基近似与配点技术结合的方法 (RBC)。随即，Wendland[118] 提出了径向基近似与伽辽金弱式离散相结合的方法 (MGM)。Wang 和 Liu[119] 则提出采用多项式基与径向基耦合的点插值法构造近似函数，其与伽辽金弱式离散相结合的方法即被称为无网格径向点插值法 (RPIM)[120]，其与局部弱式离散相结合的方法即被称为局部径向点插值法 (LRPIM)[121,122]。1996 年，Duarte 和 Oden，Melenk 和 Babuška 等两个组分别独立地提出单位分解近似的概念，Duarte 和 Oden[123,124] 将发展的方法称为 Hp 云 (Hp-clouds) 法，而 Melenk 和 Babuška[125] 将发展的方法称为单位分解有限元法 (PUFEM)。单位分解近似与局部弱式离散结合的方法，又被称为有限球法 (FSM)[126]。2004 年，Lam 等[127] 将 Kriging 插值法用于形函数构造，并将其与局部弱式离散技术结合，提出了局部 Kriging 法 (LoKriging)。2007 年，Shaw 和 Roy[128] 提出一种采用非均匀有理 B 样条 (NURBS) 基函数的误差核重构近似法 (ERKM)，其对应的全局弱式方法被称为误差重构无网格法 (ERMF)[129]。

配点离散技术的方法，通常具有纯无网格特性、执行简单、计算效率高的优点。随着无网格研究的深入，配点技术方法也得到进一步发展。1996 年，Onate 等[130,131] 将 MLS 近似与配点离散技术结合，提出了有限点法 (FPM)，类似的还有 Lee 和 Yoon[132] 于 2004 年提出的无网格配点法 (MPCM)。2000 年，Breitkopf 等将 MLS 与配点技术结合，并引入估值点进行两阶段近似，提出了双网格扩散配点法 (DGDC)[133]。为了减轻传统配点法计算不稳定的问题，还发展了采用最小二乘离散技术的配点型方法，例如，Zhang 等[134] 于 2001 年提出的最小二乘配点法 (LSCM)，Park 和 Yoon 于 2001 年提出的最小二乘无网格法 (LSMM)[135]，张雄等[136] 于 2003 年提出的加权最小二乘无网格法 (WLSM)，Liu 等[137] 于 2006 年提出的最小二乘径向基配点法 (LS-RPCM)，Kee 等[138] 于 2007 年提出的规则化最小二乘径向基配点法 (RLS-RPCM)。为了消除配点法不稳定计算的问题，杨建军和郑健龙采用局部介点近似技术，于 2013 年提出了一种具有 h-p-d 适应性的无网格介点法 (MIP)[139,140]，于 2017 年提出采用有限点变分法导出的无网格全局介

点法 (MGIP)[141]。

为了克服单一一种离散技术方法的某些缺憾,或者是为了提高计算稳定性,或者是为了提高求解效率,抑或是为了便于施加边界条件,一类耦合离散技术的方法,或者是结合式方法也相继被提出。比如,Belytschko 等提出的有限元与 EFG 的结合法 (FE-EFG)[142],Liu 和 Gu 提出的 EFG 与 BEM 的结合法 (EFG-HBEM)[143]、MLPG 与 FEM 的结合法 (MLPG-FE)[144]、配点与 MLPG 结合的弱强式法 (MWS)[145],Pan 等提出的最小二乘配点与 EFG 相结合的无网格伽辽金最小二乘法 (MGLS) [146],de Vuyst 等提出的 SPH 与 FEM 的结合法 (SPH-FE)[147],Atluri 等提出的 MLPG 与 FDM 的结合法 (MLPG-FD)[148]、MLPG 与配点结合法 (MLPG-CM)[149],Zhang 等提出的 EFG 和 BEM 结合的方法 (EFG-BEM)[150],杨建军和郑健龙 [151] 提出的 MLPG 与 MIP 结合的无网格局部强弱法 (MLSW) 等。

借助无网格法的求解理念和研究成果,传统 FEM 也得到进一步发展。在有限元网格求解框架下,在局部采用单位分解技术,可以更为灵活地处理裂纹扩展问题,赋予 FEM 新的能力。此类方法的代表性成果有 Sukumar 等提出的扩展有限元法 (XFEM)[152-156],Duarte 等提出的广义有限元法 (GFEM)[157,158],一些学者也用 XFEM 统一指代此类方法 [159−161]。严格来说,此类方法应属于 FEM,但其最显著的新特性就是在特殊区域运用了无网格法的技术,因此可将此类方法列为无网格法研究的衍生性成果之一。

此外,将无网格近似技术应用于边界元法 (BEM),便可消除其对网格的依赖性,进而发展出一类边界型无网格法。例如,Mukherjee 等提出的边界节点法 (BNM)[162,163],Li 和 Aluru 提出的边界云法 (BCM)[164],Zhang 等提出的杂交边界点法 (HBNM)[165],Chen 等提出的边界配点法 (BCoM)[166],Gu 和 Liu 提出的边界径向基点插值法 (BRPIM)[167],程玉民和陈美娟提出的边界无单元法 (BEFM)[168],Chen 提出的边界粒子法 (BPM)[169]、边界点法 (BKM)[170,171] 和奇异边界法 (SBM)[172,173],Ren 等 [174] 提出的插值型边界无单元法 (IBEFM) 等。Yang 等 [175] 基于介点原理对 MFS 改进后提出 GMFS。

以上是对无网格法发展历史及目前已经发展出的无网格方法的一个简要总结。综上,可以将目前发展出的主要的一些无网格方法,按所使用的近似技术和离散方程导出规则列于表 1-2 中,这张表基本能够反映当前无网格法的研究,尤其是基本方法的发展现状。以上对无网格法研究进展的介绍是非常粗略与概括性的,具体方法的多样性和应用研究的繁荣已很难进行准确而细致的介绍。目前关于无网格法的应用研究方兴未艾,并已成功应用于天体物理学、微观粒子力学、流体力学、固体力学、断裂力学、爆炸力学、热力学、生物力学、电磁力学、拓扑优化等领域。

表 1-2 主要无网格方法汇总
Table 1-2 A summary of meshless methods

	配点类方法	全局弱式	局部弱式	边界法	结合式法
SPH 近似	SPH[50-52]				SPH-FE[147]
MLS 近似	FPM[130,131]	DEM[76]	MLPG[84]	BNM[162,163]	FE-EFG[142]
	MPCM[132]	EFG[78]	LBIE[85,86]	BCM[164]	EFG-HBEM[143]
	DGDC[133]		MRLBIE[87]	HBNM[165]	MLPG-FE[144]
	LSCM[134]			BFM[183]	MWS[145]
	LSMM[135]			BEFM[168]	MGLS[146]
	WLSM[136]			IBEFM[174]	MLPG-FD[148]
	MIP[139,140]				MLPG-CM[149]
	MGIP[141]				EFG-BEM[150]
	Hp-MC[181]				MLSW[151]
RKPM 近似	FCM[100,101]	RKPM[95]	MLPGx[94]		
	HCM[102-104]				
自然邻接法		NEM[108]	NNPG[110]		
		NNGM[109]	PGNE[111]		
点插值法	PICM[113]	PIM[112]	LRPIM[121,122]	MFS[176-178]	
	RBC[116,117]	MGM[118]		GMFS[175]	
	LS-RPCM[137]	RPIM[120]		BPM[169]	
	RLS-RPCM[138]			BKM[170,17]	
				MMFS[179]	
				BCoM[166]	
				BRPIM[167]	
				SBM[172,173]	
				BDS[180]	
单位分解法		Hp-clouds[123,124]	FSM[125]		
		PUFEM[125]			
		XFEM[152-156]			
		GFEM[157,158]			
Kriging 法		EFG-MK[223]	LoKriging[126]		
ERKM 近似		ERMF[128,129]			
差分法	PIC[33-37]	FLIP[38,39]			HFDM[182]
	Vortex[58-62]	MPM[40-44]			
	GFDM[45-49]				
其他	MD[63-68]				
	LBM[69-71]				

1.3.3 国内无网格法研究状况

随着国际性的无网格法研究兴起，国内学者也开始了追踪研究。我国的无网格法研究大约起步于 2000 年以后，在时间上比国外要晚。但随着越来越多的国内学者参与到网格法研究当中，也逐步取得一些比较有影响力的研究成果。

清华大学张雄等应当是国内最早开始追踪无网格法研究的学者，2003 年，张雄研究组在国内发表了第一篇介绍无网格法研究进展的综述文献[184]，2004 年在国内出版了第一本系统介绍无网格法理论与方法的专著[26]，对推动国内学者关注无网格法研究发挥了积极作用。该组在具体的数值方法发展上也有一定贡献，提出使用最小二乘离散原理来提高配点法求解稳定性的思想，发展了最小二乘型的无网格方法，代表性工作有，提出最小二乘配点无网格法 (LSCM)[134]，提出加权最小二乘无网格法 (WLSM)[136]。

河海大学陈文研究组则在边界型无网格方面做出一系列重要的工作，发展了多种边界无网格法，先后提出完全不需要内部离散点的边界粒子法 (BPM)[169,170]、边界点法 (BKM)[171] 和奇异边界法 (SBM)[172,173] 等。

上海大学程玉民研究组对移动最小二乘近似法的改进有一系列重要的研究工作，通过采用带权正交化的基函数，提出了改进的移动最小二乘法[185,186]；将复变量理论引入移动最小二乘法，提出了复变量移动最小二乘法 (CVMLS)[186]。这两类对 MLS 的改进方法，使用正交化的基函数，可以避免对矩阵 A 进行求逆运算，使得近似运算相容性更好，计算精度更高；而使用复变量执行近似运算，其主要优点是使用更少的支撑点执行近似，提高了近似的相容性，同时也显著提高了计算效率。该研究组将这两类 MLS 改进方法应用于多种无网格离散方法[187-190]，其详细的研究成果可参考程玉民的专著[32]。

复旦大学吴宗敏则在早期开展了散点插值的理论和方法研究[191]，给出了一类紧支柱正定径向函数[192-195]，并被广泛应用于无网格径向基近似法。

本书作者所在的研究组通过跟踪研究，也取得一些进展，主要的工作是提出区域解法的收敛性观点，即"介点原理"，并发展了一类介点型方法[139,196]。这类方法均使用一种改进的 MLS，即移动最小二乘核近似 (MLSc)[197]，这种通过简单改进的方法，显著提高了 MLS 近似的精确性和稳定性。研究组先后提出了无网格介点法 (MIP)[139,140]，无网格全局介点法 (MGIP)[141]，无网格局部强弱法 (MLSW)[151]，以及属于边界求解类的广义基本解法 (GMFS)[175]。其中，MIP 法具有数值执行简单，计算精确稳定和广适应性计算特性，是一种有竞争力的无网格方法。而 GMFS 是对经典 MFS[176-178] 的一种完善和发展。

此外，湖南大学龙述尧研究组也是国内较早开展无网格法研究的团队之一，该组主要对 MLPG 类方法进行了系统的应用性研究[31,198-200]。张见明等基于修正变分原理和无网格近似，提出了一种边界型局部弱式法——杂交边界点法 (HBNM)[165,201,202] 和边界面法 (BEM)[183]。同济大学蔡永昌等提出了基于 Voronoi 结构的 MLPG 法[203]。李小林和 Zhu[204] 提出了伽辽金边界节点法 (GBN) 等。

1.4 无网格法研究展望

无网格法被确立为一个计算力学的独立研究分支以来,到现在已有二十余年的研究历史。这二十多年来,研究内容主要集中在具体方法的发展和探索,但其理论体系还很不完备,工程应用还比较薄弱,仍然面临一些尚未解决的问题。

其一是算法方面的研究。虽然,目前提出的无网格方法已经非常丰富,这种算法的丰富性一方面来源于无网格近似函数构造的多样性,不同的近似函数构造方法,通常可形成一类特征性的无网格方法;另一方面来源于离散技术的多样性,一种离散技术的方法,通常可形成一类具有明显特征的无网格方法。这两种多样性的结合,便又可形成不同的方法,再加以离散技术的耦合算法,进一步丰富了无网格方法的研究内容。当前,弱式方法凭借其计算稳定而精确的特征,仍然在无网格法的方法研究中占据优势地位,需提及的是,弱式类方法也有其固有局限性,因为弱式方法需要引入网格或胞元进行积分运算,这通常会使求解灵活性削弱或计算耗时增加。基于配点技术的网格法是纯无网格法,具有形式简单、易于实现、应用灵活和计算高效的优点。然而配点型无网格法通常具有计算精度低、计算稳定性差、适应性差的缺点,这些缺点都限制了配点型无网格法的应用范围和应用效果。今后,需要特别重视配点型方法的发展和完善。

其二是无网格法数学理论方面的研究。在保证无网格法求解精度和稳定性上,即数值方法的收敛性,还没有形成统一的认识和理论观点。无网格法要想成为一种强有力的计算工具,急需寻找一个较为统一的收敛性解释机制或者是理论支撑体系,这有助于无网格法研究的进一步完善和发展。当前无网格法的数学理论研究还很薄弱,尤其是收敛性的数学理论急需发展。像有限元法、边界元法,都有发展成熟的收敛性数学理论,比如,变分原理可视为有限元法的收敛性数学理论,边界积分方程法可视为边界元法的收敛性数学理论,而无网格法还没有类似的数学理论作为支撑,这种现状会严重阻碍无网格法的进一步发展,需要同行学者予以高度重视。

其三是无网格法计算效率方面的研究。无网格法以散点作为计算框架,为了保证对散点上场变量的完整描述,需有限个近似域对全局域形成完整覆盖,这就导致在近似计算中包含较多冗余部分,从而对计算效率带来明显影响。实际上,无网格法的计算耗时主要花费在近似计算部分,因此无网格法计算效率的提升也要特别注重从近似计算的效率提高来入手。对一种特定的近似方法,在保证近似计算精确稳定的前提下,如何用最少的支撑点来实现,是需要特别予以关注的研究方向和内容。

其四是无网格法应用软件的开发和研究。计算力学是解决实际工程问题的重要工具，而进一步推广其应用价值则需要通用的计算软件来达成。目前，无网格法还没有商业化的通用软件，在工程一线的技术人员还无法使用，这种局面已经限制了无网格法的广泛应用。因此，科研工作者应加强无网格法软件实现方面的研究，一方面要解决通用计算方面的算法优化工作，另一方面要解决对复杂实际工程对象的适应性问题。

参 考 文 献

[1] 钟万勰, 程耿东. 跨世纪的中国计算力学// 周光召. 科技进步与学科发展 [M]. 北京: 中国科学技术出版社, 1998: 24-28.

[2] 郑哲敏, 周恒, 张涵信, 等. 21 世纪初的力学发展趋势 [J]. 力学进展, 1995, 25(4): 433-441.

[3] Clough R W. The finite element method in plane stress analysis[C]. Proc. 2nd ASCE Conf. on Electronic Computation. Pittsburgh, Pa. Sept., 1960.

[4] Zienkiewicz O C, Taylor R L, Taylor R L. The Finite Element Method[M]. London: McGraw-Hill, 1977.

[5] 冯康, 石钟慈. 弹性结构数学理论 [M]. 北京: 科学出版社, 1981.

[6] 石钟慈, 王鸣. 有限元方法 [M]. 北京: 科学出版社, 2010.

[7] Owen D R J, Feng Y T. Fifty years of finite elements—a solid mechanics perspective[J]. Theoretical and Applied Mechanics Letters, 2012, 2(5): 051001.

[8] 冯康. 基于变分原理的差分格式 [J]. 应用数学与计算数学, 1965, 2(4): 238-262.

[9] Richtmyer R D. Difference Methods for Initial-Value Problems[M]. New York: Interscience Pub., 1957.

[10] Richtmyer R D, Morton K W. Difference Methods for Initial-value Problems[M]. 2nd ed. Malabar, Fla.: Krieger Publishing Co., 1994.

[11] Rizzo F J. An integral equation approach to boundary value problems of classical elastostatics[J]. Quarterly of Applied Mathematics, 1967, 25(1): 83-95.

[12] Cruse T A, Rizzo F J. A direct formulation and numerical solution of the general transient elastodynamic problem[J]. Journal of Mathematical Analysis and Applications, 1968, 22(1): 244-259.

[13] 姚振汉, 王海涛. 边界元法 [M]. 北京: 高等教育出版社, 2010.

[14] Brebbia C A. The Boundary Element Method for Engineers[M]. London: Pentech Press, 1978.

[15] Banerjee P K, Butterfield R. Boundary element methods in engineering science[M]. London: McGraw-Hill, 1981.

[16] Belytschko T, Krougauz Y, Organ D, et al. Meshless method: an overview and recent developments[J]. Comput. Methods Appl. Mech. Engrg., 1996, 139: 3-47.

[17] Li S, Liu W K. Meshfree and particle methods and their applications[J]. Applied Mechanics Reviews, 2002, 55(1): 1-34.

[18] 张雄, 宋康祖, 陆明万. 无网格法研究进展及其应用 [J]. 计算力学学报, 2003, 20(6): 730-742.

[19] 张雄, 刘岩, 马上. 无网格法的理论及应用 [J]. 力学进展, 2009, 39(1): 1-36.

[20] Gu Y T. Meshfree methods and their comparisons[J]. International Journal of Computational Methods, 2005, 2(4): 477-515.

[21] 顾元通, 丁桦. 无网格法及其最新进展 [J]. 力学进展, 2005, 35(3): 323-337.

[22] Liu G R. Mesh Free Methods: Moving Beyond the Finite Element Method [M]. New York: CRC Press, 2002.

[23] Liu G R, Gu Y T. An Introduction to Meshfree Methods and Their Programming[M]. Netherlands: Springer Science & Business Media, 2005.

[24] Atluri S N, Shen S. The Meshless Local Petrov-Galerkin (MLPG) Method[M]. Forsyth: Tech Science Press (CREST), 2002.

[25] Atluri S N. The Meshless Method (MLPG) for Domain & BIE Discretizations[M]. Forsyth: Tech Science Press, 2004.

[26] 张雄, 刘岩. 无网格法 [M]. 北京: 清华大学出版社, 2004.

[27] Belytschko T, Chen J S. Meshfree and Particle Methods[M]. Hoboken: Wiley, 2009.

[28] 刘更, 刘天祥, 谢琴. 无网格法及其应用 [M]. 西安: 西北工业大学出版社, 2005.

[29] 赵光明. 无单元法理论与应用 [M]. 合肥: 中国科学技术大学出版社, 2009.

[30] 刘欣. 无网格方法 [M]. 北京: 科学出版社, 2011.

[31] 龙述尧. 无网格方法及其在固体力学中的应用 [M]. 北京: 科学出版社, 2014.

[32] 程玉民. 无网格方法 [M]. 北京: 科学出版社, 2015.

[33] Evans M W, Harlow F H, Bromberg E. The particle-in-cell method for hydrodynamic calculations[R]. Los Alamos National Lab NM, 1957.

[34] Harlow F H. Hydrodynamic problems involving large fluid distortions[J]. Journal of the ACM (JACM), 1957, 4(2): 137-142.

[35] Harlow F H. The particle-in-cell computing method for fluid dynamics[J]. Methods in Computational Physics, 1964, 3(3): 319-343.

[36] Harlow F H, Shannon J P, Welch J E. Liquid waves by computer[J]. Science, 1965, 149(3688): 1092, 1093.

[37] Harlow F H, Welch J E. Numerical calculation of time-dependent viscous incompressible flow of fluid with free surface[J]. The Physics of Fluids, 1965, 8(12): 2182-2189.

[38] Brackbill J U, Ruppel H M. FLIP: A method for adaptively zoned, particle-in-cell calculations of fluid flows in two dimensions[J]. Journal of Computational Physics, 1986, 65(2): 314-343.

[39] Brackbill J U, Kothe D B, Ruppel H M. FLIP: A low-dissipation, particle-in-cell method for fluid flow[J]. Computer Physics Communications, 1988, 48(1): 25-38.

[40] Sulsky D, Zhou S J, Schreyer H L. Application of a particle-in-cell method to solid mechanics[J]. Computer Physics Communications, 1995, 87(1-2): 236-252.

[41] Bardenhagen S G, Brackbill J U, Sulsky D. The material-point method for granular materials[J]. Computer Methods in Applied Mechanics and Engineering, 2000, 187(3-4): 529-541.

[42] Bardenhagen S G, Kober E M. The generalized interpolation material point method[J]. Computer Modeling in Engineering and Sciences, 2004, 5(6): 477-496.

[43] 廉艳平, 张帆, 刘岩, 等. 物质点法的理论和应用 [J]. 力学进展, 2013, 43(2): 237-264.

[44] 张雄, 廉艳平, 刘岩, 等. 物质点法 [M]. 北京: 清华大学出版社, 2013.

[45] Girault V. Theory of a GDM on irregular net works[J]. SIAM J. Num. Anal., 1974, 11: 260-282.

[46] Perrone N, Kao R. A general finite difference method for arbitrary meshes[J]. Computers & Structures, 1975, 5(1): 45-57.

[47] Liszka T, Orkisz J. Finite Difference Method for Arbitrary Irregular Meshes in Nonlinear Problems of Applied Mechanics[M]. San Francisco: IV SMiRt, 1977.

[48] Liszka T, Orkisz J. The finite difference method at arbitrary irregular grids and its application in applied mechanics[J]. Comput. Struct., 1980, 11: 83-95.

[49] Liszka T. An interpolation method for an irregular net of nodes[J]. International Journal for Numerical Methods in Engineering, 1984, 20(9): 1599-1612.

[50] Gingold R A, Monaghan J J. Smoothed particle hydrodynamics: Theory and application to non-spherical stars[J]. Monthly Notices of the Royal Astronomical Society, 1977, 181(3): 375-389.

[51] Monaghan J J. Smoothed particle hydrodynamics[J]. Annual Review of Astronomy and Astrophysics, 1992, 30(1): 543-574.

[52] Lucy L B. A numerical approach to the testing of the fission hypothesis [J]. The Astron J, 1977, 8(12): 1013-1024.

[53] Morris J P, Fox P J, Zhu Y. Modeling low Reynolds number incompressible flows using SPH[J]. Journal of Computational Physics, 1997, 136(1): 214-226.

[54] Colagrossi A, Landrini M. Numerical simulation of interfacial flows by smoothed particle hydrodynamics[J]. Journal of Computational Physics, 2003, 191(2): 448-475.

[55] Randles P W, Libersky L D. Smoothed particle hydrodynamics: some recent improvements and applications[J]. Computer Methods in Applied Mechanics and Engineering, 1996, 139(1-4): 375-408.

[56] Liu M B, Liu G R. Smoothed particle hydrodynamics (SPH): An overview and recent developments[J]. Archives of Computational Methods in Engineering, 2010, 17(1): 25-76.

[57] Liu G R, Liu M B. Smoothed Particle Hydrodynamics: A Meshfree Particle Method[M]. Singapore: World Scientific, 2003.

[58] Chorin A J. Numerical study of slightly viscous flow[J]. Journal of Fluid Mechanics, 1973, 57(04): 785-796.

[59] Chorin A J. Vortex sheet approximation of boundary layers[J]. Journal of Computational Physics, 1978, 27(3): 428-442.

[60] Leonard A. Vortex methods for flow simulation[J]. Journal of Computational Physics, 1980, 37(3): 289-335.

[61] Leonard A. Computing three-dimensional incompressible flows with vortex elements[J]. Annual Review of Fluid Mechanics, 1985, 17(1): 523-559.

[62] Anderson C, Greengard C. On vortex methods[J]. SIAM Journal on Numerical Analysis, 1985, 22(3): 413-440.

[63] Alder B J, Wainwright T E. Studies in molecular dynamics. I. General method[J]. The Journal of Chemical Physics, 1959, 31(2): 459-466.

[64] Alder B J, Wainwright T E. Studies in molecular dynamics. II. Behavior of a small number of elastic spheres[J]. The Journal of Chemical Physics, 1960, 33(5): 1439-1451.

[65] Alder B J. Studies in molecular dynamics. III. A mixture of hard spheres[J]. The Journal of Chemical Physics, 1964, 40(9): 2724-2730.

[66] Car R, Parrinello M. Unified approach for molecular dynamics and density-functional theory[J]. Physical Review Letters, 1985, 55(22): 2471.

[67] Kresse G, Hafner J. Ab initio molecular dynamics for liquid metals[J]. Physical Review B, 1993, 47(1): 558-561.

[68] Berendsen H J C, Postma J P M, van Gunsteren W F, et al. Molecular dynamics with coupling to an external bath[J]. The Journal of Chemical Physics, 1984, 81(8): 3684-3690.

[69] Rapaport D C. The Art of Molecular Dynamics Simulation[M]. London: Cambridge University Press, 2004.

[70] Frisch U, Hasslacher B, Pomeau Y. Lattice-gas automata for the Navier-Stokes equation[J]. Physical review letters, 1986, 56(14): 1505-1508.

[71] Chen H, Chen S, Matthaeus W H. Recovery of the Navier-Stokes equations using a lattice-gas Boltzmann method[J]. Physical Review A, 1992, 45(8): R5339.

[72] Chen S, Doolen G D. Lattice Boltzmann method for fluid flows[J]. Annual Review of Fluid Mechanics, 1998, 30(1): 329-364.

[73] Wolf-Gladrow D A. Lattice-gas cellular automata and lattice Boltzmann models: An introduction[M]. Berlin: Springer, 2004.

[74] He X, Chen S, Doolen G D. A novel thermal model for the lattice Boltzmann method in incompressible limit[J]. Journal of Computational Physics, 1998, 146(1): 282-300.

[75] He X, Luo L S. Theory of the lattice Boltzmann method: From the Boltzmann

equation to the lattice Boltzmann equation[J]. Physical Review E, 1997, 56(6): 6811-6817.

[76] Nayroles B, Touzot G, Villon P. Generalizing the finite element method: Diffuse approximation and diffuse elements[J]. Computational Mechanics, 1992, 10(5): 307-318.

[77] Nayroles B, Touzot G, Villon P. The diffuse approximation[J]. Comptes Rendus De L Academie Des Sciences Serie II, 1991, 313(3): 293-296.

[78] Belytschko T, Lu Y Y, Gu L. Element-free Galerkin methods[J]. International Journal for Numerical Methods in Engineering, 1994, 37(2): 229-256.

[79] Belytschko T, Gu L, Lu Y Y. Fracture and crack growth by element free Galerkin methods[J]. Modelling and Simulation in Materials Science and Engineering, 1994, 2(3A): 519-534.

[80] Belytschko T, Lu Y Y, Gu L. Crack propagation by element-free Galerkin methods[J]. Engineering Fracture Mechanics, 1995, 51(2): 295-315.

[81] Belytschko T, Lu Y Y, Gu L, et al. Element-free Galerkin methods for static and dynamic fracture[J]. International Journal of Solids and Structures, 1995, 32(17-18): 2547-2570.

[82] Belytschko T, Tabbara M. Dynamic Fracture using element-free Galerkin methods[J]. International Journal for Numerical Methods in Engineering, 1996, 39(6): 923-938.

[83] Fleming M, Chu Y A, Moran B, et al. Enriched element-free Galerkin methods for crack tip fields[J]. International Journal for Numerical Methods in Engineering, 1997, 40(8): 1483-1504.

[84] Atluri S N, Zhu T. A new meshless local Petrov-Galerkin (MLPG) approach in computational mechanics[J]. Computational Mechanics, 1998, 22(2): 117-127.

[85] Zhu T, Zhang J D, Atluri S N. A local boundary integral equation (LBIE) method in computational mechanics, and a meshless discretization approach[J]. Computational Mechanics, 1998, 21(3): 223-235.

[86] Zhu T, Zhang J, Atluri S N. A meshless local boundary integral equation (LBIE) method for solving nonlinear problems[J]. Computational Mechanics, 1998, 22(2): 174-186.

[87] Zhu T. A new meshless regular local boundary integral equation (MRLBIE) approach[J]. International Journal for Numerical Methods in Engineering, 1999, 46(8): 1237-1252.

[88] Atluri S N, Kim H G, Cho J Y. A critical assessment of the truly meshless local Petrov-Galerkin (MLPG), and local boundary integral equation (LBIE) methods[J]. Computational Mechanics, 1999, 24(5): 348-372.

[89] Atluri S N, Cho J Y, Kim H G. Analysis of thin beams, using the meshless local Petrov-Galerkin method, with generalized moving least squares interpolations[J].

Computational Mechanics, 1999, 24(5): 334-347.

[90] Atluri S N, Zhu T L. The meshless local Petrov-Galerkin (MLPG) approach for solving problems in elasto-statics[J]. Computational Mechanics, 2000, 25(2): 169-179.

[91] Atluri S N, Zhu T. New concepts in meshless methods[J]. International Journal for Numerical Methods in Engineering, 2000, 47(1-3): 537-556.

[92] Lin H, Atluri S N. Meshless local Petrov-Galerkin(MLPG) method for convection diffusion problems[J]. Computer Modelling in Engineering & Sciences, 2000, 1(2): 45-60.

[93] Lin H, Atluri S N. The meshless local Petrov-Galerkin (MLPG) method for solving incompressible Navier-Stokes equations[J]. Computer Modeling in Engineering and Sciences, 2001, 2(2): 117-142.

[94] Atluri S N, Shen S. The basis of meshless domain discretization: The meshless local Petrov-Galerkin (MLPG) method[J]. Advances in Computational Mathematics, 2005, 23(1-2): 73-93.

[95] Sladek J, Stanak P, Han Z D, et al. Applications of the MLPG method in engineering & sciences: a review[J]. Comput. Model. Eng. Sci., 2013, 92(5): 423-475.

[96] Liu W K, Jun S, Zhang Y F. Reproducing kernel particle methods[J]. International Journal for Numerical Methods in Fluids, 1995, 20(8-9): 1081-1106.

[97] Liu W K, Jun S, Li S, et al. Reproducing kernel particle methods for structural dynamics[J]. International Journal for Numerical Methods in Engineering, 1995, 38(10): 1655-1679.

[98] Chen J S, Pan C, Wu C T, et al. Reproducing kernel particle methods for large deformation analysis of non-linear structures[J]. Computer Methods in Applied Mechanics and Engineering, 1996, 139(1-4): 195-227.

[99] Liu W K, Chen Y, Jun S, et al. Overview and applications of the reproducing kernel particle methods[J]. Archives of Computational Methods in Engineering, 1996, 3(1): 3-80.

[100] Aluru N R. A point collocation method based on reproducing kernel approximations[J]. International Journal for Numerical Methods in Engineering, 2000, 47(6): 1083-1121.

[101] Aluru N R, Li G. Finite cloud method: A true meshless technique based on a fixed reproducing kernel approximation[J]. International Journal for Numerical Methods in Engineering, 2001, 50(10): 2373-2410.

[102] Li H, Ng T Y, Cheng J Q, et al. Hermite-Cloud: A novel true meshless method[J]. Computational Mechanics, 2003, 33(1): 30-41.

[103] Li H, Cheng J Q, Ng T Y, et al. A meshless Hermite-Cloud method for nonlinear fluid-structure analysis of near-bed submarine pipelines under current[J]. Engineering Structures, 2004, 26(4): 531-542.

[104] Lam K Y, Li H, Yew Y K, et al. Development of the meshless Hermite-Cloud method for structural mechanics applications[J]. International Journal of Mechanical Sciences, 2006, 48(4): 440-450.

[105] Braun J, Sambridge M. A numerical method for solving partial differential equations on highly irregular evolving grids[J]. Nature, 1995, 376(6542): 655.

[106] Sambridge M, Braun J, McQueen H. Geophysical parametrization and interpolation of irregular data using natural neighbours[J]. Geophysical Journal International, 1995, 122(3): 837-857.

[107] Traversoni L. Natural neighbor finite elements[C]. Int Confon Hydraulic Engineering Software Hydrosoft Proc, Vol 2, Computational Mechanics Publ, 1994: 291-297.

[108] Sukumar N, Moran B, Belytschko T. The natural element method in solid mechanics[J]. International Journal for Numerical Methods in Engineering, 1998, 43: 839-887.

[109] Sukumar N, Moran B, Yu Semenov A, et al. Natural neighbour Galerkin methods[J]. International Journal for Numerical Methods in Engineering, 2001, 50(1): 1-27.

[110] Wang K, Zhou S, Shan G. The natural neighbour Petrov-Galerkin method for elastostatics[J]. International Journal for Numerical Methods in Engineering, 2005, 63(8): 1126-1145.

[111] Cho J R, Lee H W. A Petrov-Galerkin natural element method securing the numerical integration accuracy[J]. Journal of Mechanical Science and Technology, 2006, 20(1): 94-109.

[112] Liu G R, Gu Y T. A point interpolation method for two-dimensional solids[J]. International Journal for Numerical Methods in Engineering, 2001, 50(4): 937-951.

[113] Liu X, Tai K. Point interpolation collocation method for the solution of partial differential equations[J]. Engineering Analysis with Boundary Elements, 2006, 30(7): 598-609.

[114] Kansa E J. Multiquadrics—A scattered data approximation scheme with applications to computational fluid-dynamics—I surface approximations and partial derivative estimates[J]. Computers & Mathematics with Applications, 1990, 19(8-9): 127-145.

[115] Kansa E J. Multiquadrics—A scattered data approximation scheme with applications to computational fluid-dynamics—II solutions to parabolic, hyperbolic and elliptic partial differential equations[J]. Computers & Mathematics with Applications, 1990, 19(8): 147-161.

[116] Franke C, Schaback R. Solving partial differential equations by collocation using radial basis functions[J]. Applied Mathematics and Computation, 1998, 93(1): 73-82.

[117] Zhang X, Song K Z, Lu M W, et al. Meshless methods based on collocation with radial basis functions[J]. Computational Mechanics, 2000, 26(4): 333-343.

[118] Wendland H. Meshless Galerkin methods using radial basis functions[J]. Mathematics of Computation of the American Mathematical Society, 1999, 68(228): 1521-1531.

[119] Wang J G, Liu G R. A point interpolation meshless method based on radial basis functions[J]. International Journal for Numerical Methods in Engineering, 2002, 54(11): 1623-1648.

[120] Liu G R, Zhang G Y, Gu Y T, et al. A meshfree radial point interpolation method (RPIM) for three-dimensional solids[J]. Computational Mechanics, 2005, 36(6): 421-430.

[121] Liu G R, Gu Y T. A local radial point interpolation method (LRPIM) for free vibration analyses of 2-D solids[J]. Journal of Sound and Vibration, 2001, 246(1): 29-46.

[122] Wu Y L, Liu G R. A meshfree formulation of local radial point interpolation method (LRPIM) for incompressible flow simulation[J]. Computational Mechanics, 2003, 30(5-6): 355-365.

[123] Duarte C A, Oden J T. An hp adaptive method using clouds[J]. Computer Methods in Applied Mechanics and Engineering, 1996, 139(1-4): 237-262.

[124] Duarte C A, Oden J T. Hp clouds-an hp meshless method[J]. Numerical Methods for Partial Differential Equations, 1996, 12(6): 673-706.

[125] Melenk J M, Babuška I. The partition of unity finite element method: Basic theory and applications[J]. Computer Methods in Applied Mechanics and Engineering, 1996, 139(1-4): 289-314.

[126] De S, Bathe K J. The method of finite spheres[J]. Computational Mechanics, 2000, 25(4): 329-345.

[127] Lam K Y, Wang Q X, Li H. A novel meshless approach-Local Kriging (LoKriging) method with two-dimensional structural analysis[J]. Computational Mechanics, 2004, 33(3): 235-244.

[128] Shaw A, Roy D. A NURBS-based error reproducing kernel method with applications in solid mechanics[J]. Computational Mechanics, 2007, 40(1): 127-148.

[129] Shaw A, Roy D. Analyses of wrinkled and slack membranes through an error reproducing mesh-free method[J]. International Journal of Solids and Structures, 2007, 44(11): 3939-3972.

[130] Onate E, Idelsohn S, Zienkiewicz O C, et al. A finite point method in computational mechanics. Applications to convective transport and fluid flow[J]. International Journal for Numerical Methods in Engineering, 1996, 39(22): 3839-3866.

[131] Onate E, Idelsohn S, Zienkiewicz O C, et al. A stabilized finite point method for analysis of fluid mechanics problems[J]. Computer Methods in Applied Mechanics and Engineering, 1996, 139(1-4): 315-346.

[132] Lee S H, Yoon Y C. Meshfree point collocation method for elasticity and crack problems[J]. International Journal for Numerical Methods in Engineering, 2004, 61(1):

22-48.

[133] Breitkopf P, Touzot G, Villon P. Double grid diffuse collocation method[J]. Computational Mechanics, 2000, 25(2): 199-206.

[134] Zhang X, Liu X H, Song K Z, et al. Least-squares collocation meshless method[J]. International Journal for Numerical Methods in Engineering, 2001, 51(9): 1089-1100.

[135] Park S H, Youn S K. The least-squares meshfree method[J]. International Journal for Numerical Methods in Engineering, 2001, 52(9): 997-1012.

[136] 张雄, 胡炜, 潘小飞, 等. 加权最小二乘无网格法 [J]. 力学学报, 2003, 35(4): 425-430.

[137] Liu G R, Kee B B T, Chun L. A stabilized least-squares radial point collocation method (LS-RPCM) for adaptive analysis[J]. Computer Methods in Applied Mechanics and Engineering, 2006, 195(37): 4843-4861.

[138] Kee B B T, Liu G R, Lu C. A regularized least-squares radial point collocation method (RLS-RPCM) for adaptive analysis[J]. Computational Mechanics, 2007, 40(5): 837-853.

[139] Yang J J, Zheng J L. Intervention-point principle of meshless method[J]. Chinese Science Bulletin, 2013, 58(4-5): 478-485.

[140] 杨建军, 郑健龙. 无网格介点法: 一种具有 h, p, d 适应性的无网格法 [J]. 应用数学和力学, 2016, 37(10): 1013-1025.

[141] 杨建军, 郑健龙. 无网格全局介点法 [J]. 应用力学学报, 2017, 34(5): 956-962.

[142] Belytschko T, Organ D, Krongauz Y. A coupled finite element-element-free Galerkin method[J]. Computational Mechanics, 1995, 17(3): 186-195.

[143] Liu G R, Gu Y T. Coupling of element free Galerkin and hybrid boundary element methods using modified variational formulation[J]. Computational Mechanics, 2000, 26(2): 166-173.

[144] Liu G R, Gu Y T. Meshless local Petrov-Galerkin (MLPG) method in combination with finite element and boundary element approaches[J]. Computational Mechanics, 2000, 26(6): 536-546.

[145] Liu G R, Gu Y T. A meshfree method: meshfree weak-strong (MWS) form method, for 2-D solids[J]. Computational Mechanics, 2003, 33(1): 2-14.

[146] Pan X F, Zhang X, Lu M W. Meshless Galerkin least-squares method[J]. Computational Mechanics, 2005, 35(3): 182-189.

[147] de Vuyst T, Vignjevic R, Campbell J C. Coupling between meshless and finite element methods[J]. International Journal of Impact Engineering, 2005, 31(8): 1054-1064.

[148] Atluri S N, Liu H T, Han Z D. Meshless local Petrov-Galerkin (MLPG) mixed finite difference method for solid mechanics[J]. Computer Modeling in Engineering & Sciences, 2006, 15(1): 1-16.

[149] Atluri S N, Liu H T, Han Z D. Meshless local Petrov-Galerkin (MLPG) mixed collocation method for elasticity problems[J]. Computer Modeling in Engineering & Sciences,

2006, 4(3): 141-152.

[150] Zhang Z, Liew K M, Cheng Y. Coupling of the improved element-free Galerkin and boundary element methods for two-dimensional elasticity problems[J]. Engineering Analysis with Boundary Elements, 2008, 32(2): 100-107.

[151] 杨建军, 郑健龙. 无网格局部强弱 (MLSW) 法求解不规则域问题 [J]. 力学学报, 2017, 49(3): 659-666.

[152] Sukumar N, Moës N, Moran B, et al. Extended finite element method for three-dimensional crack modelling[J]. International Journal for Numerical Methods in Engineering, 2000, 48(11): 1549-1570.

[153] Sukumar N, Chopp D L, Moës N, et al. Modeling holes and inclusions by level sets in the extended finite-element method[J]. Computer Methods in Applied Mechanics and Engineering, 2001, 190(46): 6183-6200.

[154] Sukumar N, Belytschko T. Arbitrary branched and intersecting cracks with the extended finite element method[J]. Int. J. Numer. Meth. Eng., 2000, 48: 1741-1760.

[155] Moës N, Belytschko T. Extended finite element method for cohesive crack growth[J]. Engineering Fracture Mechanics, 2002, 69(7): 813-833.

[156] Sukumar N, Prévost J H. Modeling quasi-static crack growth with the extended finite element method Part I: Computer implementation[J]. International Journal of Solids and Structures, 2003, 40(26): 7513-7537.

[157] Duarte C A, Babuška I, Oden J T. Generalized finite element methods for three-dimensional structural mechanics problems[J]. Computers & Structures, 2000, 77(2): 215-232.

[158] Duarte C A, Hamzeh O N, Liszka T J, et al. A generalized finite element method for the simulation of three-dimensional dynamic crack propagation[J]. Computer Methods in Applied Mechanics and Engineering, 2001, 190(15): 2227-2262.

[159] Zi G, Belytschko T. New crack-tip elements for XFEM and applications to cohesive cracks[J]. International Journal for Numerical Methods in Engineering, 2003, 57(15): 2221-2240.

[160] Belytschko T, Gracie R, Ventura G. A review of extended/generalized finite element methods for material modeling[J]. Modelling and Simulation in Materials Science and Engineering, 2009, 17(4): 043001.

[161] Fries T P, Belytschko T. The extended/generalized finite element method: An overview of the method and its applications[J]. International Journal for Numerical Methods in Engineering, 2010, 84(3): 253-304.

[162] Mukherjee Y X, Mukherjee S. The boundary node method for potential problems[J]. International Journal for Numerical Methods in Engineering, 1997, 40(5): 797-815.

[163] Kothnur V S, Mukherjee S, Mukherjee Y X. Two-dimensional linear elasticity by the boundary node method[J]. International Journal of solids and Structures, 1999, 36(8):

1129-1147.

[164] Li G, Aluru N R. Boundary cloud method: A combined scattered point/boundary integral approach for boundary-only analysis[J]. Computer Methods in Applied Mechanics and Engineering, 2002, 191(21): 2337-2370.

[165] Zhang J, Yao Z, Li H. A hybrid boundary node method[J]. International Journal for Numerical Methods in Engineering, 2002, 53(4): 751-763.

[166] Chen J T, Chang M H, Chen K H, et al. The boundary collocation method with meshless concept for acoustic eigenanalysis of two-dimensional cavities using radial basis function[J]. Journal of Sound and Vibration, 2002, 257(4): 667-711.

[167] Gu Y T, Liu G R. A boundary radial point interpolation method (BRPIM) for 2-D structural analyses[J]. Structural Engineering and Mechanics, 2003, 15(5): 535-550.

[168] 程玉民, 陈美娟. 弹性力学的一种边界无单元法 [J]. 力学学报, 2003, 35(2): 181-186.

[169] Chen W. Meshfree boundary particle method applied to Helmholtz problems[J]. Engineering Analysis with Boundary Elements, 2002, 26(7): 577-581.

[170] Chen W, Tanaka M. A meshless, integration-free, and boundary-only RBF technique[J]. Computers & Mathematics with Applications, 2002, 43(3): 379-391.

[171] Chen W, Hon Y C. Numerical investigation on convergence of boundary knot method in the analysis of homogeneous Helmholtz, modified Helmholtz, and convection-diffusion problems[J]. Computer Methods in Applied Mechanics and Engineering, 2003, 192(15): 1859-1875.

[172] 陈文. 奇异边界法: 一个新的、简单、无网格、边界配点数值方法 [J]. 固体力学学报, 2009, 30(6): 592-599.

[173] Chen W, Fu Z J. A novel numerical method for infinite domain potential problems[J]. Chinese Sci Bull, 2010, 55: 1598-1603.

[174] Ren H P, Cheng Y M, Zhang W. An interpolating boundary element-free method (IBEFM) for elasticity problems[J]. Science China Physics, Mechanics and Astronomy, 2010, 53(4): 758-766.

[175] Yang J J, Zheng J L, Wen P H. Generalized method of fundamental solution (GMFS) for boundary value problems[J]. Engineering Analysis with Boundary Elements, 2018, 94: 25-33.

[176] Kupradze V D, Aleksidze M A. The method of functional equations for the approximate solution of certain boundary value problems[J]. USSR Computational Mathematics & Mathematical Physics, 1964, 4(4): 82-126.

[177] Kupradze V D. A method for the approximate solution of limiting problems in mathematical physics[J]. USSR Computational Mathematics and Mathematical Physics, 1964, 4(6): 199-205.

[178] Kupradze V D. On the approximate solution of problems in mathematical physics[J]. Russian Mathematical Surveys, 1967, 22(2): 58-108.

[179] Young D L, Chen K H, Chen J T, et al. A modified method of fundamental solutions with source on the boundary for solving Laplace equations with circular and arbitrary domains[J]. Computer Modeling in Engineering and Sciences, 2007, 19(3): 197-221.

[180] Liu Y J. A new boundary meshfree method with distributed sources[J]. Engineering Analysis with Boundary Elements, 2010, 34(11): 914-919.

[181] Liszka T J, Duarte C A M, Tworzydlo W W. Hp-Meshless cloud method[J]. Computer Methods in Applied Mechanics & Engineering, 1996, 139(1): 263-288.

[182] Wen P H, Aliabadi M H. A hybrid finite difference and moving least square method for elasticity problems[J]. Engineering Analysis With Boundary Elements, 2012, 36(4): 600-605.

[183] Zhang J, Qin X, Han X, et al. A boundary face method for potential problems in three dimensions[J]. International Journal for Numerical Methods in Engineering, 2009, 80(3): 320-337.

[184] 张雄, 宋康祖, 陆明万. 无网格法研究进展及其应用 [J]. 计算力学学报, 2003, 20(6): 730-742.

[185] 李九红, 程玉民. 无网格方法的研究进展与展望 [J]. 力学季刊, 2006, 27(1): 143-152.

[186] 程玉民, 彭妙娟, 李九红. 复变量移动最小二乘法及其应用 [J]. 力学学报. 2005, 37(6): 719-723.

[187] Cheng Y M, Peng M J. Boundary element-free method for elastodynamics[J]. Science in China Ser. G Physics, Mechanics & Astronomy, 2005, 48(6): 641-657.

[188] 程玉民, 李九红. 断裂力学的复变量无网格方法 [J]. 中国科学: 物理学 力学 天文学, 2005, 35(5): 548-560.

[189] Liew K M, Cheng Y M, Kitipornchai S. Boundary element-free method (BEFM) and its application to two-dimensional elasticity problems[J]. International Journal for Numerical Methods in Engineering, 2006, 65(8): 1310-1332.

[190] Cheng Y M, Li J H. A complex variable meshless method for fracture problems[J]. Science in China Ser. G Physics, Mechanics & Astronomy, 2006, 49(1): 46-59.

[191] 吴宗敏. 散乱数据拟合的模型、方法和理论 [M]. 北京: 科学出版社, 2007.

[192] Wu Z M. Hermite-Birkhoff interpolation of scattered data by radial basis function[J]. Approx. Theory & Its Appl., 1992, 8: 1-10.

[193] Wu Z M, Schaback R. Local errorestimates for radial basis function interpolation of scattered data[J]. IMA journal of Numerical Analysis, 1993, 13: 13-27.

[194] Wu Z M. Compactly supported positive definite radial functions [J]. Adv. Comput. Math., 1995, 4: 283-292.

[195] Wu Z M. Multivariate compactly supported positive definite radial functions [J]. Advances in Computational Mathematics, 1995, 4: 283-292.

[196] 杨建军, 郑健龙. 无网格法介点原理 [J]. 科学通报, 2012, 57(26): 2456-2462.

[197] 杨建军, 郑健龙. 移动最小二乘法的近似稳定性 [J]. 应用数学学报, 2012, 35(4): 637-648.

[198] 龙述尧, 许敬晓. 弹性力学问题的局部边界积分方程方法 [J]. 力学学报, 2000, 32(5): 566-578.

[199] 龙述尧. 弹性力学问题的局部 Petrov-Galerkin 方法 [J]. 力学学报, 2001, 33(4): 508-517.

[200] Long S Y, Atluri S N. A meshless local Petrov-Galerkin method for solving the bending problem of a thin plate[J]. Computer Modeling in Engineering & Sciences, 2002, 3(1): 53-63.

[201] 张见明, 姚振汉. 一种新型无网格法——杂交边界点法 [A]. 广州: 中国计算力学大会 2001 会议论文集, 2001: 339-343.

[202] Zhang J M, Yao Z H. A new regular meshless hybrid boundary method[C]. Proceedings of the First MIT Conference on Computational Fluid and Solid Mechanics, Amsterdam: Elsevier, 2001.

[203] 蔡永昌, 朱合华, 王建华. 基于 Voronoi 结构的无网格局部 Petrov-Galerkin 方法 [J]. 力学学报, 2003, 35(2): 187-193.

[204] Li X L, Zhu J L. Galerkin boundary node method for exterior Neumann problems[J]. Journal of Computational Mathematics, 2011, 29(3): 243-260.

第 2 章 无网格法基础理论

对于一个特定的偏微分方程 (partial differential equations, PDEs) 问题，用无网格求解需要面临的一个重要任务就是构建其求解系统方程，即建立约束散点自由度的方程组。我们通常把这一构建求解系统方程的过程称为数值离散。使用不同的数值离散方法，将形成风格迥异的求解系统方程，其求解效果也可能不尽相同。本章将重点介绍数值离散方法的一些基础知识和基本理论。

2.1 固体力学基本理论

作为预备知识，先简要介绍固体力学的一些基本理论和基本方程，以方便后续章节论述中查阅引用。

为表述的简单起见，仅考虑弹性静力学问题。一个问题域为 Ω，边界为 Γ 的弹性体，其静力学平衡方程用矩阵形式写为

$$\boldsymbol{L}^{\mathrm{T}}\boldsymbol{\sigma} + \boldsymbol{b} = 0, \quad \text{在 } \Omega \text{ 内} \tag{2-1}$$

对该方程求解需要给定边界条件，定义如下：

$$\boldsymbol{n} \cdot \boldsymbol{\sigma} = \bar{\boldsymbol{t}}, \quad \text{在 } \Gamma_t \text{ 上} \tag{2-2}$$

$$\boldsymbol{u} = \bar{\boldsymbol{u}}, \quad \text{在 } \Gamma_u \text{ 上} \tag{2-3}$$

式 (2-2) 称为自然边界条件 (natural boundary condition, NBC)，力学问题中也被称为应力边界条件；式 (2-3) 称为本质边界条件 (essential boundary condition, EBC)，力学问题中也被称为位移边界条件。Γ_t 为自然边界，Γ_u 为本质边界，并有 $\Gamma_u \cup \Gamma_t = \Gamma$，$\boldsymbol{b}$ 为弹性体所受的外体力向量，$\bar{\boldsymbol{t}}$ 和 $\bar{\boldsymbol{u}}$ 分别为弹性被施加的给定表面力和位移，$\boldsymbol{\sigma}$ 表示固体的应力向量。一个三维立方单元的表面应力分量如图 2-1 所示。由 $\tau_{ij} = \tau_{ji}$，所以对三维固体上的任意一点处，有 6 个独立应力分量，而对二维固体上的任意一点，则有 3 个独立应力分量，即

$$\boldsymbol{\sigma}^{\mathrm{T}} = \begin{cases} [\sigma_x, \sigma_y, \sigma_z, \tau_{yz}, \tau_{xz}, \tau_{xy}], & \text{3D} \\ [\sigma_x, \sigma_y, \tau_{xy}], & \text{2D} \end{cases} \tag{2-4}$$

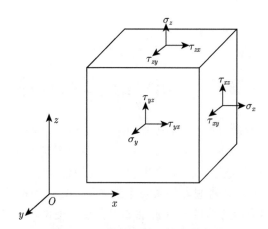

图 2-1 三维立方单元的应力分量

Fig. 2-1 Stress components of a 3D-cube element

与应力向量相对应,应变向量 ε 也有相同的应变分量,即

$$\boldsymbol{\varepsilon}^{\mathrm{T}} = \begin{cases} [\varepsilon_x, \varepsilon_y, \varepsilon_z, \gamma_{yz}, \gamma_{xz}, \gamma_{xy}], & \text{3D} \\ [\varepsilon_x, \varepsilon_y, \gamma_{xy}], & \text{2D} \end{cases} \quad (2\text{-}5)$$

式 (2-3) 中,u 为位移向量,其包含的分量表示为

$$\boldsymbol{u}^{\mathrm{T}} = \begin{cases} [u_x, u_y, u_z], & \text{3D} \\ [u_x, u_y], & \text{2D} \end{cases} \quad (2\text{-}6)$$

相应地,有

$$\boldsymbol{b}^{\mathrm{T}} = \begin{cases} [b_x, b_y, b_z], & \text{3D} \\ [b_x, b_y], & \text{2D} \end{cases} \quad (2\text{-}7)$$

$$\bar{\boldsymbol{t}}^{\mathrm{T}} = \begin{cases} [\bar{t}_x, \bar{t}_y, \bar{t}_z], & \text{3D} \\ [\bar{t}_x, \bar{t}_y], & \text{2D} \end{cases} \quad (2\text{-}8)$$

$$\bar{\boldsymbol{u}}^{\mathrm{T}} = \begin{cases} [\bar{u}_x, \bar{u}_y, \bar{u}_z], & \text{3D} \\ [\bar{u}_x, \bar{u}_y], & \text{2D} \end{cases} \quad (2\text{-}9)$$

而应变向量 ε 和位移向量 u 存在如下计算关系:

$$\boldsymbol{\varepsilon} = \boldsymbol{L}\boldsymbol{u} \quad (2\text{-}10)$$

上式称为弹性固体的应变-位移方程,也被称为几何方程。其中 L 表示一个微分算子,与式 (2-1) 中的符号定义一致,并定义如下:

对三维问题，有

$$^{3D}\boldsymbol{L} = \begin{bmatrix} \partial/\partial x & 0 & 0 \\ 0 & \partial/\partial y & 0 \\ 0 & 0 & \partial/\partial z \\ 0 & \partial/\partial z & \partial/\partial y \\ \partial/\partial z & 0 & \partial/\partial x \\ \partial/\partial y & \partial/\partial x & 0 \end{bmatrix} \tag{2-11}$$

对二维问题，则有

$$^{2D}\boldsymbol{L} = \begin{bmatrix} \partial/\partial x & 0 \\ 0 & \partial/\partial y \\ \partial/\partial y & \partial/\partial x \end{bmatrix} \tag{2-12}$$

弹性体的应力和应变服从广义胡克定律，定义为

$$\boldsymbol{\sigma} = \boldsymbol{D}\boldsymbol{\varepsilon} \tag{2-13}$$

上式被称为弹性材料的本构方程，$\boldsymbol{D}$ 是材料常数矩阵，对各向同性材料，三维问题可写为

$$^{3D}\boldsymbol{D} = \begin{bmatrix} D_{11} & D_{12} & D_{12} & 0 & 0 & 0 \\ & D_{11} & D_{12} & 0 & 0 & 0 \\ & & D_{11} & 0 & 0 & 0 \\ & & & (D_{11}-D_{12})/2 & 0 & 0 \\ & \text{对称} & & & (D_{11}-D_{12})/2 & 0 \\ & & & & & (D_{11}-D_{12})/2 \end{bmatrix} \tag{2-14a}$$

其中，

$$\left. \begin{array}{l} D_{11} = \dfrac{E(1-\mu)}{(1-2\mu)(1+\mu)} \\ D_{12} = \dfrac{E\mu}{(1-2\mu)(1+\mu)} \\ \dfrac{D_{11}-D_{12}}{2} = \dfrac{E}{2(1+\mu)} \end{array} \right\} \tag{2-14b}$$

式中，E 和 μ 分别为材料的弹性模量和泊松比。

对于二维问题，材料常数矩阵有平面应力和平面应变两种形式：

$$^{2D}\boldsymbol{D} = \frac{E}{1-\mu^2} \begin{bmatrix} 1 & \mu & 0 \\ \mu & 1 & 0 \\ 0 & 0 & (1-\mu)/2 \end{bmatrix}, \quad \text{平面应力} \tag{2-15a}$$

$$^{2D}\boldsymbol{D} = \frac{E(1-\mu)}{(1-2\mu)(1+\mu)} \begin{bmatrix} 1 & \dfrac{\mu}{1-\mu} & 0 \\ \dfrac{\mu}{1-\mu} & 1 & 0 \\ 0 & 0 & \dfrac{1-2\mu}{2(1-\mu)} \end{bmatrix}, \quad \text{平面应变} \quad (2\text{-}15\text{b})$$

更一般地讲，平面应变固体的材料常数矩阵可由平面应力固体的材料常数矩阵中的 E 和 μ 分别换成 $E/(1-\mu^2)$ 和 $\mu/(1-\mu)$ 而得到。

式 (2-2) 中的 $\boldsymbol{n}$ 表示边界外法向向量，一个三维固体的局部边界面上一点处的外法向向量如图 2-2 所示，其分量向量组写为 $\boldsymbol{n}:(n_x, n_y, n_z)$。对于二维固体，其分量向量组对应写为 $\boldsymbol{n}:(n_x, n_y)$。$\boldsymbol{n}$ 的矩阵表示形式参照微分算子 $\boldsymbol{L}$ 的转置结构表示如下：

三维问题：

$$^{3D}\boldsymbol{n} = \begin{bmatrix} n_x & 0 & 0 & 0 & n_z & n_y \\ 0 & n_y & 0 & n_z & 0 & n_x \\ 0 & 0 & n_z & n_y & n_x & 0 \end{bmatrix} \quad (2\text{-}16)$$

二维问题：

$$^{2D}\boldsymbol{n} = \begin{bmatrix} n_x & 0 & n_y \\ 0 & n_y & n_x \end{bmatrix} \quad (2\text{-}17)$$

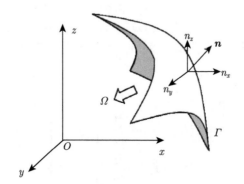

图 2-2 边界外法向向量

Fig. 2-2 Outer normal vector of a boundary

2.2 加权残值法

加权残值法 (method of weighted residual, MWR) 是一种数学方法，可以直接从微分方程中得出近似解，是数值求解的理论基础。用这种方法解微分方程，首先

2.2 加权残值法

要假设一个试函数 (trial function) 作为满足微分方程及其边界条件的近似解，这个近似解中有已经确定的试函数项，也有待定的系数。将试函数代入控制微分方程式和边界条件，一般不能严格满足，便会出现残值。而数值方法的任务就是构建一系列的代数方程组，并尽可能地消除这些残值，或者使残值减到最小。联立求解这些方程组，便可得到待求的系数。这些系数一旦确定，试函数也就确定，这个试函数，或者是试函数的组合，便成为满足控制微分方程及其边界条件的近似解[1,2]。

设某一给定边值常微分方程 (ordinary differential equation, ODE) 或偏微分方程 (partial differential equation, PDE) 问题，其求解域为 Ω，边界为 Γ，并有如下定义：

$$F(u) - f = 0, \quad 在 \Omega 内 \tag{2-18}$$

$$G(u) - g = 0, \quad 在 \Gamma 上 \tag{2-19}$$

式中，u 为待求函数；F 和 G 为微分算子；f 和 g 是不含 u 的项。其中，式 (2-18) 被定义为控制微分方程，式 (2-19) 被定义为边界条件方程。如果 G 定义为 0 阶微分算子，则式 (2-19) 变成一个一元方程，因此该式表征包含本质边界条件的广义微分方程。

实际问题中，满足式 (2-18) 并同时满足式 (2-19) 的函数 u 通常很难解析得到。因此，我们假设一个试函数为

$$u^h = \sum_{t=1}^{m} C_t N_t \tag{2-20}$$

式中，N_t 为试函数的项；C_t 为其对应的待定系数。将式 (2-20) 代入式 (2-18) 和式 (2-19)，一般不会满足，于是出现了内部残值 R_I 及边界残值 R_B：

$$R_\mathrm{I} = F(u^h) - f \neq 0 \tag{2-21}$$

$$R_\mathrm{B} = G(u^h) - g \neq 0 \tag{2-22}$$

为了消除残值，需要引入一个权函数 w，内部权函数和边界权函数可以采用不同的形式，并分别用 w_I 和 w_B 表示，则消除内部残值和边界残值的加权残值方程定义为

$$\int_\Omega w_\mathrm{I} \cdot R_\mathrm{I} \mathrm{d}\Omega = 0 \tag{2-23}$$

$$\int_\Gamma w_\mathrm{B} \cdot R_\mathrm{B} \mathrm{d}\Gamma = 0 \tag{2-24}$$

上式所定义的加权残值方程可以视为数值方法离散的最基本形式，由此可确定试函数的表达形式，即式 (2-18) 和式 (2-19) 的近似解。将试函数代入上述两个公式，

其残值将是最小的。此外,需注意到式 (2-23) 和式 (2-24) 定义的加权残值方程是积分形式,因此,加权残值法提供了一种将微分方程转化成积分方程的方法。采用积分形式的好处之一就是可通过分部积分以降低求导阶数,从而降低对近似函数连续性阶数的要求;另外一个好处就是积分通常是以积分点来构造离散方程的,其数学意义就是在包含稠密位置信息的区域上来消除残值,通常会加大对求解未知量自由度的约束,故可改善解的稳定性。

由式 (2-20) 定义的试函数,可能形式是比较复杂的,见后续第 3 章将要介绍的各类无网格试函数。推及广泛的数值方法,试函数的表现形式会足够丰富。但不管怎么假设,试函数只可能有三类:

其一,边界型试函数。试函数满足式 (2-19) 定义的边界条件,但不满足式 (2-18) 定义的控制微分方程。对应的数值方法只需要消除内部残值即可,该类数值方法可称为内部解法。这类解法通常在求解无限域问题中较多采用。

其二,内部型试函数。试函数满足控制微分方程,但不满足边界条件。对应的数值方法只需要消除边界残值即可,该类数值方法可称为"边界解法",例如,边界元法即这类方法的代表。

其三,混合型试函数。试函数既不满足控制微分方程,也不满足边界条件。对应的数值方法不仅需要消除区域内部的残值,也需要消除边界上的残值。该类数值方法被称为"混合解法",有时也与边界解法相对应,被称为"区域解法"。这种解法对试函数的要求较低,求解通用性更好,是目前数值方法中较为普遍采用的求解形式。

通过上述分析,可说明加权残值法是求解 ODE 或 PDE 问题的通用的、强有力的工具。可以说,现有的主流数值方法,比如有限元法、差分法、边界元法、无网格法等都是建立在加权残值法的基础之上的。

加权残值法中的权函数选取,将赋予数值方法不同的离散形式,而且也影响着数值方法的求解效果。按使用的权函数类型,加权残值法可分为五种基本的方法。

1. 配点法 (collocation method)

配点法采用 Dirac δ 函数作为权函数,即

$$w_i = \delta(x - x_i), \quad i = 1, 2, \cdots, n \tag{2-25}$$

Dirac δ 函数具有如下性质:

$$\delta(x - x_i) = \begin{cases} \infty, & x = x_i \\ 0, & x \neq x_i \end{cases} \tag{2-26}$$

$$\int_{-\infty}^{\infty} \delta(x - x_i) \mathrm{d}x = 1 \tag{2-27}$$

2.2 加权残值法

$$\int_{x_i-c}^{x_i+c} \delta(x-x_i)\mathrm{d}x = 1 \quad (c>0,\ c\to 0) \tag{2-28}$$

则式 (2-21) 和式 (2-22) 定义的加权残值方程变为

$$\int_{\Omega} w_i \cdot R_{\mathrm{I}}\mathrm{d}\Omega = R_{\mathrm{I}} = 0 \tag{2-29}$$

$$\int_{\Gamma} w_i \cdot R_{\mathrm{B}}\mathrm{d}\Gamma = R_{\mathrm{B}} = 0 \tag{2-30}$$

配点法的优点是形式非常简单，数值实施便捷，计算效率高。然而配点法也有明显的不足，主要表现在计算不稳定，求解精度较低，收敛速度慢。

2. **最小二乘法** (least squares method)

最小二乘法采用式 (2-21) 和式 (2-22) 定义的残值函数本身作为权函数，即

$$w_i = R_i \tag{2-31}$$

则式 (2-21) 和式 (2-22) 定义的加权残值方程变为

$$\int_{\Omega} R_{\mathrm{I}} \cdot R_{\mathrm{I}}\mathrm{d}\Omega = 0 \tag{2-32}$$

$$\int_{\Gamma} R_{\mathrm{B}} \cdot R_{\mathrm{B}}\mathrm{d}\Gamma = 0 \tag{2-33}$$

以上两式不能直接求解，需要采用特别的求解技术。以式 (2-32) 为例，需首先定义一个泛函：

$$J_{\mathrm{I}}(C_t) = \int_{\Omega} R_{\mathrm{I}} \cdot R_{\mathrm{I}}\mathrm{d}\Omega \tag{2-34}$$

寻求该泛函的最小值，要求

$$\frac{\partial J_{\mathrm{I}}}{\partial C_t} = \int_{\Omega} \frac{\partial (R_{\mathrm{I}} \cdot R_{\mathrm{I}})}{\partial C_t}\mathrm{d}\Omega = 2\int_{\Omega} \frac{\partial (R_{\mathrm{I}})}{\partial C_t} R_{\mathrm{I}}\mathrm{d}\Omega = 0 \tag{2-35}$$

或

$$\int_{\Omega} \frac{\partial (R_{\mathrm{I}})}{\partial C_t} R_{\mathrm{I}}\mathrm{d}\Omega = 0 \tag{2-36}$$

类似可得到

$$\int_{\Gamma} \frac{\partial (R_{\mathrm{B}})}{\partial C_t} R_{\mathrm{B}}\mathrm{d}\Gamma = 0 \tag{2-37}$$

在混合解法中，通常使用控制方程和边界方程的加权残值方程叠加的形式，即

$$\int_{\Omega} \frac{\partial (R_{\mathrm{I}})}{\partial C_t} R_{\mathrm{I}}\mathrm{d}\Omega + \int_{\Gamma} \frac{\partial (R_{\mathrm{B}})}{\partial C_t} R_{\mathrm{B}}\mathrm{d}\Gamma = 0 \tag{2-38}$$

在数值方法中为了便于数值实施，通常使用式 (2-38) 的强形式。

3. 矩量法 (moment method)

矩量法采用单项式序列函数作为权函数，即

$$w_i = x_i^{i-1}, \quad i = 1, 2, \cdots, n \tag{2-39}$$

则式 (2-21) 和式 (2-22) 定义的加权残值方程变为

$$\int_\Omega x_i^{i-1} \cdot R_\mathrm{I} \mathrm{d}\Omega = 0 \tag{2-40}$$

$$\int_\Gamma x_i^{i-1} \cdot R_\mathrm{B} \mathrm{d}\Gamma = 0 \tag{2-41}$$

矩量法所求解的方程通常为病态条件类型，一种变形形式是使用 Chebyshev 多项式代替单项式。

4. 子域法 (subdomain method)

子域法采用 Heaviside 阶跃函数作为权函数，即

$$w_i = H(x - x_i), \quad i = 1, 2, \cdots, n \tag{2-42}$$

Heaviside 阶跃函数有如下性质：

$$H(x - x_i) = 1, \quad 在 \ \Omega_i \ 内 \tag{2-43}$$

$$H(x - x_i) = 0, \quad 在 \ \Omega_i \ 外 \tag{2-44}$$

式中，Ω_i 表示伴随 x_i 定义的一个局部子域，则式 (2-21) 定义的加权残值方程变为

$$\int_\Omega H_i \cdot R_\mathrm{I} \mathrm{d}\Omega = \int_{\Omega_i} R_\mathrm{I} \mathrm{d}\Omega = 0 \tag{2-45}$$

在上式中，局部域 Ω_i 上再引入一个权函数 w_{is}，等价于初始权函数变为一个复权函数，即

$$w_0 = w_i \cdot w_{is} \tag{2-46}$$

则式 (2-45) 可改写为

$$\int_{\Omega_i} w_{is} \cdot R_\mathrm{I} \mathrm{d}\Omega = 0 \tag{2-47}$$

上式即彼得罗夫–伽辽金 (Petrov-Galerkin) 局部弱式法的基本形式。对其进行分部积分，并应用散度定理，则边界条件会被自动引入。相比于伽辽金 (Galerkin) 法，Petrov-Galerkin 局部弱式法的数值实施较为简单直接。

5. 伽辽金法 (Galerkin method)

伽辽金法采用场函数的试函数作为其权函数，即

$$w_i = N_i \tag{2-48}$$

则伽辽金法的加权残值方程写为

$$\int_\Omega N_i \cdot R_{\mathrm{I}} \mathrm{d}\Omega - \int_\Gamma N_i \cdot R_{\mathrm{B}} \mathrm{d}\Gamma = 0 \tag{2-49}$$

式中，N_i 通常取其变分形式，即 δN_i。伽辽金法所获得的系统矩阵通常是对称阵，而且可得到与变分原理的能量法相同的公式形式，有特定的物理意义。因此，伽辽金法被视为加权残值法中非常有效的一种形式，在数值方法中被广泛采用，特别是在有限元法中，伽辽金法是非常普遍的一种计算格式。

实际上，已经发展出的一些具体的数值方法中，有一部分也很难归类于上述五种基本方法的某一个类别。因此，可说这五种基本方法并不是加权残值法的所有基本形式，尚有继续完善和扩充的空间。

2.3 弹性力学变分原理

变分原理 (variational principle) 是求解泛函极值问题的一种有效方法，是弹性力学的重要组成部分，并在有限元法的数值离散中得到广泛应用 [3,4]。在无网格法中，全局弱式法的基本方法，即 Galerkin 法，也采用变分的离散形式。因此，有必要将变分原理的基础知识作一个简单的介绍。

2.3.1 虚功原理

变分原理的一大特点就是将微分方程求解的数学问题，引入了"能量"的概念，使其具有了明确的物理意义。根据能量守恒原理，由式 (2-1)～式 (2-3) 定义的一般线弹性力学中，外力在可能位移上所做之功，等于可能应力在相应的可能应变上所做之功，即

$$\int_\Omega \boldsymbol{b}^{\mathrm{T}} \boldsymbol{u} \mathrm{d}\Omega + \int_\Gamma \boldsymbol{t}^{\mathrm{T}} \boldsymbol{u} \mathrm{d}\Gamma = \int_\Omega \boldsymbol{\sigma}^{\mathrm{T}} \boldsymbol{\varepsilon} \mathrm{d}\Omega \tag{2-50}$$

这个关系式即被称为"虚功原理"，有时也被称为"虚位移原理"。式中，t 表示全部可能的表面力，而不仅仅是给定的表面力 $\bar{t}$。利用已知的关系式，替换 b、t 和 ε，则可导出

$$\int_\Omega \left\{ \left(\boldsymbol{L}^{\mathrm{T}} \boldsymbol{\sigma}\right)^{\mathrm{T}} \boldsymbol{u} + \boldsymbol{\sigma}^{\mathrm{T}} \left(\boldsymbol{L} \boldsymbol{u}\right) \right\} \mathrm{d}\Omega = \int_\Gamma \left(\boldsymbol{n} \boldsymbol{\sigma}\right)^{\mathrm{T}} \boldsymbol{u} \mathrm{d}\Gamma \tag{2-51}$$

这个关系式表示 $\boldsymbol{\sigma}$ 与 $\boldsymbol{u}$ 的恒等关系。

假设 u_1，ε_1 和 σ_1 分别表示弹性结构的第一种位移、应变和应力状态，并用 u_2，ε_2 和 σ_2 表示对应的第二种状态，由式 (2-50) 可推得

$$\sigma_2^{\mathrm{T}}\varepsilon_1 = \sigma_2^{\mathrm{T}}\boldsymbol{D}^{-1}\sigma_1 = \sigma_1^{\mathrm{T}}\boldsymbol{D}^{-1}\sigma_2 = \sigma_1^{\mathrm{T}}\varepsilon_2 \tag{2-52}$$

$$\int_\Omega \boldsymbol{b}_2^{\mathrm{T}}\boldsymbol{u}_1 \mathrm{d}\Omega + \int_\Gamma \boldsymbol{t}_2^{\mathrm{T}}\boldsymbol{u}_1 \mathrm{d}\Gamma = \int_\Omega \boldsymbol{b}_1^{\mathrm{T}}\boldsymbol{u}_2 \mathrm{d}\Omega + \int_\Gamma \boldsymbol{t}_1^{\mathrm{T}}\boldsymbol{u}_2 \mathrm{d}\Gamma \tag{2-53}$$

这两个关系式被称为"功的互等定理"，其中式 (2-52) 是内功的互等定理，式 (2-53) 是外功的互等定理。

2.3.2　最小势能原理

弹性系统的势能包含两部分，其一是弹性体的应变能：

$$\Pi_2^{\mathrm{PE}} = \int_\Omega \frac{1}{2}\varepsilon^{\mathrm{T}}\boldsymbol{D}\varepsilon \mathrm{d}\Omega \tag{2-54}$$

其二是外荷载的势能：

$$\Pi_1^{\mathrm{PE}} = -\int_\Omega \boldsymbol{b}^{\mathrm{T}}\boldsymbol{u} \mathrm{d}\Omega - \int_\Gamma \bar{\boldsymbol{t}}^{\mathrm{T}}\boldsymbol{u} \mathrm{d}\Gamma \tag{2-55}$$

所以整个系统的势能为

$$\Pi^{\mathrm{PE}} = \Pi_1^{\mathrm{PE}} + \Pi_2^{\mathrm{PE}} = \int_\Omega \left(\frac{1}{2}\varepsilon^{\mathrm{T}}\boldsymbol{D}\varepsilon - \boldsymbol{b}^{\mathrm{T}}\boldsymbol{u}\right)\mathrm{d}\Omega - \int_\Gamma \bar{\boldsymbol{t}}^{\mathrm{T}}\boldsymbol{u} \mathrm{d}\Gamma \tag{2-56}$$

对上式求总势能的变分，并令此泛函值为 0，得到

$$\delta \Pi^{\mathrm{PE}} = -\int_\Omega \left(\boldsymbol{L}^{\mathrm{T}}\sigma + \boldsymbol{b}\right)^{\mathrm{T}}\delta \boldsymbol{u} \mathrm{d}\Omega + \int_\Gamma \left(\boldsymbol{n}\cdot\sigma - \bar{\boldsymbol{t}}\right)^{\mathrm{T}}\delta \boldsymbol{u} \mathrm{d}\Gamma = 0 \tag{2-57}$$

这个关系式被称为"最小势能原理"，显然，其满足平衡方程和外力边界条件，并要求自变函数 u 和 ε 满足几何方程和位移边界条件。如果将最小势能原理推广到动力学系统，则被称为 Hamilton 原理。

2.3.3　最小余能原理

弹性系统的余能也包含两部分[5]，其一是弹性体的余应变能：

$$\Pi_2^{\mathrm{CE}} = \frac{1}{2}\int_\Omega \sigma^{\mathrm{T}}\boldsymbol{D}^{-1}\sigma \mathrm{d}\Omega \tag{2-58}$$

其二是已知边界位移的余能：

$$\Pi_1^{\mathrm{CE}} = -\int_{\Gamma_u} \boldsymbol{t}^{\mathrm{T}}\bar{\boldsymbol{u}} \mathrm{d}\Gamma \tag{2-59}$$

2.3 弹性力学变分原理

因此整个系统的余能为

$$\Pi^{\mathrm{CE}} = \Pi_1^{\mathrm{CE}} + \Pi_2^{\mathrm{CE}} = \frac{1}{2}\int_\Omega \boldsymbol{\sigma}^{\mathrm{T}}\boldsymbol{D}^{-1}\boldsymbol{\sigma}\mathrm{d}\Omega - \int_{\Gamma_u} \boldsymbol{t}^{\mathrm{T}}\bar{\boldsymbol{u}}\mathrm{d}\Gamma \tag{2-60}$$

对上式求总余能的变分，并令此泛函值为 0，得到

$$\delta \Pi^{\mathrm{CE}} = \delta\left\{\frac{1}{2}\int_\Omega \boldsymbol{\sigma}^{\mathrm{T}}\boldsymbol{D}^{-1}\boldsymbol{\sigma}\mathrm{d}\Omega - \int_{\Gamma_u} \boldsymbol{t}^{\mathrm{T}}\bar{\boldsymbol{u}}\mathrm{d}\Gamma\right\} = 0 \tag{2-61}$$

这个关系式被称为"最小余能原理"。其中，要求自变函数 $\boldsymbol{\sigma}$ 满足平衡方程和应力边界条件。显然，这是一个条件极值原理。

2.3.4 Hellinger-Reissner 变分原理

为了将式 (2-60) 变为无条件的极值问题，可使用 Lagrange 乘子法将式 (2-1) 定义的平衡方程和式 (2-2) 定义的应力边界条件强行引入，以保证自变函数 $\boldsymbol{\sigma}$ 满足平衡方程和应力边界条件，即

$$\Pi^{\mathrm{HR}} = \int_\Omega \left\{\frac{1}{2}\boldsymbol{\sigma}^{\mathrm{T}}\boldsymbol{D}^{-1}\boldsymbol{\sigma} + \left(\boldsymbol{L}^{\mathrm{T}}\boldsymbol{\sigma} + \boldsymbol{b}\right)^{\mathrm{T}}\boldsymbol{\lambda}\right\}\mathrm{d}\Omega - \int_{\Gamma_u} \boldsymbol{t}^{\mathrm{T}}\bar{\boldsymbol{u}}\mathrm{d}\Gamma - \int_{\Gamma_t} (\boldsymbol{t}-\bar{\boldsymbol{t}})^{\mathrm{T}}\boldsymbol{\lambda}\mathrm{d}\Gamma \tag{2-62}$$

式中，$\boldsymbol{\lambda}$ 为 Lagrange 乘子，令 $\boldsymbol{\lambda} = \boldsymbol{u}$，则有

$$\Pi^{\mathrm{HR}} = \int_\Omega \left\{\frac{1}{2}\boldsymbol{\sigma}^{\mathrm{T}}\boldsymbol{D}^{-1}\boldsymbol{\sigma} + \left(\boldsymbol{L}^{\mathrm{T}}\boldsymbol{\sigma} + \boldsymbol{b}\right)^{\mathrm{T}}\boldsymbol{u}\right\}\mathrm{d}\Omega - \int_{\Gamma_u} \boldsymbol{t}^{\mathrm{T}}\bar{\boldsymbol{u}}\mathrm{d}\Gamma - \int_{\Gamma_t} (\boldsymbol{t}-\bar{\boldsymbol{t}})^{\mathrm{T}}\boldsymbol{u}\mathrm{d}\Gamma \tag{2-63}$$

这个关系式包含应力函数 $\boldsymbol{\sigma}$ 和位移函数 $\boldsymbol{u}$ 这两个自变量，可称为二类变量广义余能。利用式 (2-51)，上式还可改写为另外一种形式：

$$\Pi^{\mathrm{HR}} = \int_\Omega \left\{\boldsymbol{\sigma}^{\mathrm{T}}\boldsymbol{L}\boldsymbol{u} - \frac{1}{2}\boldsymbol{\sigma}^{\mathrm{T}}\boldsymbol{D}^{-1}\boldsymbol{\sigma} - \boldsymbol{b}^{\mathrm{T}}\boldsymbol{u}\right\}\mathrm{d}\Omega - \int_{\Gamma_u} \boldsymbol{t}^{\mathrm{T}}(\boldsymbol{u}-\bar{\boldsymbol{u}})\mathrm{d}\Gamma - \int_{\Gamma_t} \bar{\boldsymbol{t}}^{\mathrm{T}}\boldsymbol{u}\mathrm{d}\Gamma \tag{2-64}$$

这个关系式可称为二类变量广义势能。为了得到泛函极值，对式 (2-63) 或式 (2-64) 进行变分，然后利用式 (2-51) 消去自变函数变分的导数，得到

$$\delta \Pi^{\mathrm{HR}} = \int_\Omega \left\{\delta\boldsymbol{\sigma}^{\mathrm{T}}\left(\boldsymbol{L}\boldsymbol{u} - \boldsymbol{D}^{-1}\boldsymbol{\sigma}\right) - \left(\boldsymbol{L}^{\mathrm{T}}\boldsymbol{\sigma} + \boldsymbol{b}\right)^{\mathrm{T}}\delta\boldsymbol{u}\right\}\mathrm{d}\Omega$$
$$- \int_{\Gamma_u} \delta\boldsymbol{t}^{\mathrm{T}}(\boldsymbol{u}-\bar{\boldsymbol{u}})\mathrm{d}\Gamma + \int_{\Gamma_t} \left(\boldsymbol{t}^{\mathrm{T}}-\bar{\boldsymbol{t}}^{\mathrm{T}}\right)\delta\boldsymbol{u}\mathrm{d}\Gamma = 0 \tag{2-65}$$

这个变分形式是以应力和位移为自变函数的无条件变分原理，由 Hellinger[5] 和 Reissner[6] 分别独立提出，因此通常被称为 Hellinger-Reissner 二类变量广义变分原理。

2.3.5 胡海昌-鹫津变分原理

为了使势能方程式 (2-56) 变为无条件成立，需要将几何方程和位移边界条件引入，则有

$$\Pi^{\mathrm{HW}} = \int_{\Omega} \left[\frac{1}{2} \boldsymbol{\varepsilon}^{\mathrm{T}} \boldsymbol{D} \boldsymbol{\varepsilon} - \boldsymbol{b}^{\mathrm{T}} \boldsymbol{u} - \boldsymbol{\sigma}^{\mathrm{T}} (\boldsymbol{\varepsilon} - \boldsymbol{L}\boldsymbol{u}) \right] \mathrm{d}\Omega - \int_{\Gamma_u} \boldsymbol{t}^{\mathrm{T}} (\boldsymbol{u} - \bar{\boldsymbol{u}}) \mathrm{d}\Gamma - \int_{\Gamma_t} \bar{\boldsymbol{t}}^{\mathrm{T}} \boldsymbol{u} \mathrm{d}\Gamma \tag{2-66}$$

上式包含位移、应变、应力等三类自变函数，可称为三类变量的广义势能。利用式 (2-51)，可得到另外一种形式：

$$\Pi^{\mathrm{HW}} = \int_{\Omega} \left[\boldsymbol{\sigma}^{\mathrm{T}} \boldsymbol{\varepsilon} - \frac{1}{2} \boldsymbol{\varepsilon}^{\mathrm{T}} \boldsymbol{D} \boldsymbol{\varepsilon} + \left(\boldsymbol{L}^{\mathrm{T}} \boldsymbol{\sigma} + \boldsymbol{b} \right)^{\mathrm{T}} \boldsymbol{u} \right] \mathrm{d}\Omega - \int_{\Gamma_u} \boldsymbol{t}^{\mathrm{T}} \bar{\boldsymbol{u}} \mathrm{d}\Gamma - \int_{\Gamma_t} (\boldsymbol{t} - \bar{\boldsymbol{t}})^{\mathrm{T}} \boldsymbol{u} \mathrm{d}\Gamma \tag{2-67}$$

这个公式可称为三类变量的广义余能。对这两个公式的任意一个用变分的方法寻求极值，则得到

$$\begin{aligned} \delta \Pi^{\mathrm{HW}} = & \int_{\Omega} \left[\left(\boldsymbol{L}^{\mathrm{T}} \boldsymbol{\sigma} + \boldsymbol{b} \right)^{\mathrm{T}} \delta \boldsymbol{u} + (\boldsymbol{\sigma} - \boldsymbol{D}\boldsymbol{\varepsilon})^{\mathrm{T}} \delta \boldsymbol{\varepsilon} + \delta \boldsymbol{\sigma}^{\mathrm{T}} (\boldsymbol{\varepsilon} - \boldsymbol{L}\boldsymbol{u}) \right] \mathrm{d}\Omega \\ & + \int_{\Gamma_u} \delta \boldsymbol{t}^{\mathrm{T}} (\boldsymbol{u} - \bar{\boldsymbol{u}}) \mathrm{d}\Gamma - \int_{\Gamma_t} \left(\boldsymbol{t}^{\mathrm{T}} - \bar{\boldsymbol{t}}^{\mathrm{T}} \right) \delta \boldsymbol{u} \mathrm{d}\Gamma \\ = & 0 \end{aligned} \tag{2-68}$$

这种变分形式由胡海昌[7]和 Washizu[8] 分别独立提出，因此被称为胡海昌-鹫津 (Hu-Washizu) 三类变量广义变分原理。Hu-Washizu 变分原理包含了弹性力学的全部三类基本变量，满足弹性力学的所有方程和所有边界条件，因此被认为是弹性力学中最一般的变分原理，是弹性力学问题的单纯变分法的提法。

2.4 边界积分方程法

边界积分方程法是数值边界解法的数学基础，边界元法就是这种数学方法的数值运用。在无网格方法的发展中，弱式边界型无网格法也同样基于这一理论体系。

由式 (2-1)~式 (2-3) 表示的矩阵或向量形式的弹性力学基本方程，可改写为指标符号公式表示的形式，即

$$\sigma_{ij,j} + b_i = 0, \quad \text{在 } \Omega \text{ 内} \tag{2-69}$$

$$t_i = n_j \sigma_{ij} = \bar{t}_i, \quad \text{在 } \Gamma_t \text{ 上} \tag{2-70}$$

$$u_i = \bar{u}_i, \quad \text{在 } \Gamma_u \text{ 上} \tag{2-71}$$

2.4 边界积分方程法

则几何方程写为

$$\varepsilon_{ij} = \varepsilon_{ji} = \frac{1}{2}\left(u_{i,j} + u_{j,i}\right) \tag{2-72}$$

本构方程写为

$$\sigma_{ij} = E_{ijkl}\varepsilon_{kl} \tag{2-73}$$

设平面闭域 Ω 由光滑或分段光滑的曲线边界 Γ 围成，如图 2-3 所示。对于具有连续一阶偏导数的任意矢量函数 v，具有如下恒等式：

$$\int_{\Omega} v_{i,j}\mathrm{d}\Omega = \int_{\Gamma} n_j v_i \mathrm{d}\Gamma \tag{2-74}$$

上式即格林 (Green) 公式，它对任意的可微场函数，建立了边界上函数本身与区域内函数微分的积分等价关系，是边界积分方程法的数学基本原则。将格林公式推广到三维域上，即被称为高斯 (Gauss) 公式。

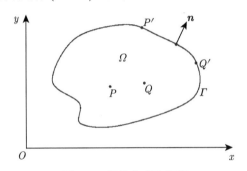

图 2-3 格林公式定义图

Fig. 2-3 Definition schematics of Green's formula

若不考虑体力，二维弹性力学的边值问题可以表示为

$$\lambda u_{k,ki} + G\left(u_{i,jj} + u_{j,ij}\right) = 0, \quad \text{在 } \Omega \text{ 内} \tag{2-75}$$

$$t_i = \lambda u_{k,k}n_i + G\left(u_{i,j}n_j + u_{j,i}n_j\right) = \bar{t}_i, \quad \text{在 } \Gamma_t \text{ 上} \tag{2-76}$$

$$u_i = \bar{u}_i, \quad \text{在 } \Gamma_u \text{ 上} \tag{2-77}$$

对式 (2-75) 左边乘以基本解 u_{li}^*，并进行区域积分：

$$\int_{\Omega} \left[\lambda u_{k,ki} + G\left(u_{i,jj} + u_{j,ij}\right)\right] u_{li}^* \mathrm{d}\Omega = 0 \tag{2-78}$$

式中，基本解 u_{li}^* 需要解析得到。对于域内的任意两个点 P 和 Q，如果在源点 P 施加一个 l 方向的集中力，则 u_{li}^* 表示由此集中力在场点 Q 的 i 方向所产生的位移。平面应变问题的基本解为 Kelvin 解：

$$u_{li}^* = \frac{1}{8\pi G\left(1-\mu\right)}\left[(3-4\mu)\ln\frac{1}{r}\delta_{li} + r_{,l}r_{,i}\right] \tag{2-79}$$

$$t_{li}^* = \frac{-1}{4\pi(1-\mu)r}\{r_{,n}[(1-2\mu)\delta_{li} + 2r_{,l}r_{,i}] - (1-2\mu)(r_{,l}n_i - r_{,i}n_l)\} \quad (2\text{-}80)$$

式中，下标中 "," 表示求导运算；$r = \|\boldsymbol{x}_P - \boldsymbol{x}_Q\|$ 表示源点 P 到目标点 Q 的距离。基本解满足表面力边界条件：

$$\lambda u_{lk,k}^* n_i + G(u_{li,j}^* n_j + u_{lj,i}^* n_j) = t_{li}^* \quad (2\text{-}81)$$

以及满足方程：

$$\lambda u_{lk,ki}^* + G(u_{li,jj}^* + u_{lj,ij}^*) + \delta_{li}\delta(P,Q) = 0 \quad (2\text{-}82)$$

则有

$$\int_\Omega u_i[\lambda u_{lk,ki}^* + G(u_{li,jj}^* + u_{lj,ij}^*)]\,\mathrm{d}\Omega = \int_\Omega u_i[-\delta_{li}\delta(P,Q)]\,\mathrm{d}\Omega = -u_l(P) \quad (2\text{-}83)$$

运用格林公式，由式 (2-78) 可得

$$\int_\Omega [\lambda u_{k,ki} + G(u_{i,jj} + u_{j,ij})]u_{li}^*\,\mathrm{d}\Omega = \int_\Omega u_i[\lambda u_{lk,ki}^* + G(u_{li,jj}^* + u_{lj,ij}^*)]\,\mathrm{d}\Omega$$

$$+ \int_\Gamma u_{li}^*[\lambda u_{k,k}n_i + G(u_{i,j}n_j + u_{j,i}n_j)]\,\mathrm{d}\Gamma$$

$$- \int_\Gamma u_i[\lambda u_{lk,k}^* n_i + G(u_{li,jj}^* + u_{lj,ij}^*)]\,\mathrm{d}\Gamma$$

$$= 0 \quad (2\text{-}84)$$

将式 (2-76)、式 (2-81)、式 (2-83) 代入，并将下标 i 替换成 k，可得

$$u_l = \int_\Gamma (u_{lk}^* t_k - t_{lk}^* u_k)\,\mathrm{d}\Gamma \quad (2\text{-}85)$$

这个公式在弹性理论中被称为 Somigliana 恒等式。它建立了区域内任意点位移 u_l 与边界上位移 u_k、面力 t_k 的转换关系。如果边界上的位移和面力能够得到，则场内任意点的位移即可由式 (2-85) 得到。而通常面临的实际问题中，边界上的位移和面力都是给定的。

当源点 P 和场点 Q 均位于边界上时，分别用 P' 和 Q' 表示，且 P' 向 Q' 趋近时，基本解会产生奇异性。因此，式 (2-85) 只能用于区域内，如果将该公式推广到在边界上也成立，需要特别的处理。常用的方法就是积分运算采用虚拟路径的技术避开 Q' 点，则式 (2-85) 将改写为

$$\frac{u_l}{2} = \int_{\hat{\Gamma}} (u_{lk}^* t_k - t_{lk}^* u_k)\,\mathrm{d}\Gamma \quad (2\text{-}86)$$

式中，$\int_{\hat{\Gamma}}(*)\,\mathrm{d}\Gamma$ 表示 Cauchy 主值积分。上式可通过边界上的已知表面力和位移来

求得域内任意点的位移,因此被称为边界积分方程的位移法,是 Rizzo[9] 与 Cruse 和 Rizzo[10] 最初提出的边界积分方程法。式 (2-86) 的等号两边对源点 P 的坐标求导,并利用几何方程和本构方程,即可得到域内任意点的应力。

为了在已知位移的边界上取得较好的数值结果,胡海昌[11] 提出了另一种边界积分方程法,它将弹性体中任意一点的应力用边界上的位移和面力直接表达出来,这种边界积分方程法正好与边界积分方程的位移法互为补充,故被称为边界积分方程的应力法。

边界积分方程法是用一种基本解的叠加来近似满足边界值要求的方法,基本解在域内已满足泛定的方程,这种基本解就是加权残值法的内部型试函数。某类问题的基本解存在,是边界积分方程法求解的前提,这个基本解通常是带奇异性的某个特殊问题的解析解[12]。例如,在无限大的弹性介质中某一点受集中力作用时,能解析得到 Kelvin 解;在无限大的弹性介质中沿某个带状平面的两侧有一定量的相对位移时,能解析得到 Crouch 解;由这两类基础性的基本解,还可以扩展出一些其他类型的基本解。此外,需特别注意的是,基本解并不具有通用性。适用于某一类问题的基本解,可能并不适用于另外一类问题。因此,数值求解需要预先了解基本解的特性,以便针对问题来选用。实际上,每选用一种基本解,就可以构成一种边界积分方程法,所以,边界积分方程法的发展有赖于基本解的发展。

2.5 无网格法介点原理

为了更为统一地描述无网格法的收敛性,从而进一步完善无网格法的基础理论,本书作者尝试性地提出了一种无网格法的收敛性观点,即无网格法介点原理[13-15]。

首先给出介点的定义[13]:无网格法组建离散系统方程时,可能需要借助某种不同于场节点的另外一种点,这种与问题域离散无关的点,或者说最终得到的求解结果不通过这些点信息来表达的点,被定义为介点 (intervention point)。使用"介点"这个名称,有几个方面的考虑:其一,有"介入"的含义,无网格法的基本元素是节点,而介点是从系统外引入的;其二,有"调解和干涉"的含义,可借助介点的调解来帮助无网格法实现快速收敛;其三,有"媒介"的含义,指介点在近似中可能起到中间介质的作用。实际上,弱式方法使用的积分点,就属于介点的定义。因此,可以把支持积分运算的介点称为"弱介点",而把不支持积分运算的介点称为"强介点"。

实践表明,基于不同离散方法 (加权残值法) 的无网格法,其收敛速率可能有较大差别,对于稀疏场节点分布情况,或散乱布点情况,一些方法可能表现为很差的收敛性。而对于一些特定问题,一些方法是不能有效收敛的。接下来,我们从理论

上讨论无网格法的收敛性, 首先提出一个命题:

命题 1　介点是无网格法保证收敛性的一个必要条件。

这一命题被称为 "介点原理"[14,15]。对式 (2-18) 和式 (2-19) 定义的边值问题, 将其转换成一个包含介点的加权残值泛函:

$$J = \int_\Omega w \left\{ F\left[u^h(x_n, x_p)\right] - f \right\} \mathrm{d}\Omega + \int_\Gamma w \left\{ G\left[u^h(x_n, x_p)\right] - g \right\} \mathrm{d}\Gamma \tag{2-87}$$

式中, x_n 表示场节点; x_p 表示介点, 则介点原理可被定义为

$$\left\{ R^*(x_n, x_p) : \left((u^h)^{(d)} \to u^{(d)} \right) \right\} \Rightarrow \left\{ R^*(u^h) \to 0 \right\} \tag{2-88}$$

式中, 场函数及其近似函数的上标 (d) 表示实际运算的微分阶数, 而不一定是微分算子 F 和 G 的阶数。上式表示使用介点的加权残值泛函 J, 可使得试函数能满足直到 d 阶导数的逼近, 从而使残值消除而保证数值计算收敛。简而言之, 使用介点有助于试函数逼近, 从而保证数值计算收敛。

接下来对这一命题做一个简单的证明:

$$R_u^* = \int_\Omega w\left[F(u) - f\right] \mathrm{d}\Omega + \int_\Gamma w\left[G(u) - g\right] \mathrm{d}\Gamma \equiv 0 \tag{2-89}$$

将上式与式 (2-87) 两边分别相减, 可得

$$R_{u^h}^* = \int_\Omega wF(u^h - u) \mathrm{d}\Omega + \int_\Gamma wG(u^h - u) \mathrm{d}\Gamma \tag{2-90}$$

应用 Schwarz 不等式:

$$\begin{aligned} \{R_{u^h}^*\}^2 &= \left\{ \int_\Omega wF(u^h - u) \mathrm{d}\Omega + \int_\Gamma wG(u^h - u) \mathrm{d}\Gamma \right\}^2 \\ &\leqslant \int_\Omega [wF(u^h - u)]^2 \mathrm{d}\Omega + \int_\Gamma [wG(u^h - u)]^2 \mathrm{d}\Gamma \\ &\leqslant \|w\|_\Omega^2 \cdot \|F(u^h - u)\|_\Omega^2 + \|w\|_\Gamma^2 \cdot \|F(u^h - u)\|_\Gamma^2 \end{aligned} \tag{2-91}$$

由此表明加权残值泛函 R^* 的收敛性取决于以下几点。

(1) 算子 F 和 G 的光滑性。对于特定的一类问题, F 和 G 的光滑性是确定的。

(2) 权函数 w 的光滑化。采用不同的权函数, 将对算子 F 和 G 造成不同的光滑化, 即形成不同光滑化的加权残值方法。

(3) $|(u^h)^{(k)} - u^{(k)}|_{k=0}^d$ 的有效逼近。其中, 既包含可能的场函数 u 的光滑性条件, 也包含权函数 w 及算子 F 和 G 共同对其构造的光滑性要求。该要求是对于任意的、普遍性的权函数和微分算子的一致收敛条件。令

$$\delta = |(u^h)^{(k)} - u^{(k)}|_{k=0}^d, \quad d \geqslant 1 \tag{2-92}$$

则有
$$\lim_{\delta \to 0} R^* \left(u^h\right) \to 0 \tag{2-93}$$

试函数的逼近，必然意味着残值泛函的收敛，这是显然的结论。然而，实际上试函数不总是无条件的逼近，介点原理给出的是创造条件的逼近。接下来，对介点原理给出几个推论。

推论 1 使用弱介点，利用积分转换，可以降低试函数的导数近似阶数，从而有助于数值方法的收敛。

例如，弹性力学控制方程的加权残值泛函：

$$\begin{aligned}
R^* \left(x_n, x_p\right) &= \int_{\Omega|\Omega_S} w_{\mathrm{I}} \left(\sigma_{ij,j} + b_i\right) \mathrm{d}\Omega \\
&= \int_{\Omega|\Omega_S} w_{\mathrm{I}} \sigma_{ij,j} \mathrm{d}\Omega + \int_{\Omega|\Omega_S} w_{\mathrm{I}} b_i \mathrm{d}\Omega \\
&= \int_{\Gamma|\Gamma_S} n_j w_{\mathrm{I}} \sigma_{ij} \mathrm{d}\Gamma - \int_{\Omega|\Omega_S} w_{\mathrm{I},j} \sigma_{ij} \mathrm{d}\Omega + \int_{\Omega|\Omega_S} w_{\mathrm{I}} b_i \mathrm{d}\Omega \\
&= 0
\end{aligned} \tag{2-94}$$

可见，原控制方程中 $\sigma_{ij,j}$ 包含试函数的二阶导数近似，即 $u^h_{i,ij}$，而经弱式转换后，仅留下 σ_{ij}，只包含试函数的一阶导数近似，即 $u^h_{i,j}$。这种对试函数导数近似的降阶要求，更有利于试函数的逼近，从而有助于数值方法的收敛。

此外，式 (2-94) 的积分运算必须引入积分点，即介点 x_p。可以说，积分单元或积分网格未必是严格必要的，但积分点是不可缺少的。另，式中积分符号 $\int_{\Omega|\Omega_S}$ 和 $\int_{\Gamma|\Gamma_S}$ 表示全局积分和局域积分的通用形式，因此，该推论对全局弱式方法和局部弱式方法均适用。

推论 2 使用强介点，并令控制方程或边界条件方程的残值在介点上消除，可增加对场节点上场变量的自由度约束，从而有助于数值方法的收敛。

可以写为两种形式。

其一，

$$R^* \left(x_n, x_p\right) = \sum_{\bar{x}_p} \left[R_{\mathrm{I}} \left(x_n, \bar{x}_p\right) + R_{\mathrm{B}} \left(x_n, \bar{x}_p\right)\right] = 0, \quad N_p > N_n \tag{2-95}$$

表示残值全部在介点上消除，并要求介点数量多于场节点数量，即 $N_p > N_n$。

其二，

$$R^* \left(x_n, x_p\right) = \sum_{\bar{x}_p} \left[R_{\mathrm{I}} \left(x_n, \bar{x}_p\right) + R_{\mathrm{B}} \left(x_n, \bar{x}_p\right)\right] + \sum_{\bar{x}_n} \left[R_{\mathrm{I}} \left(x_n\right) + R_{\mathrm{B}} \left(x_n\right)\right] = 0 \tag{2-96}$$

表示残值不仅在介点上消除，也在场节点上消除，并要求 $N_p \geqslant 0$。

式 (2-96) 实际上是式 (2-95) 的特例，如果介点序列包含节点序列，即 $\{x_p\} \supseteq \{x_n\}$，则两个泛函式等价。这两类泛函构造的方程组，方程数多于求解未知数，所以不能直接求解，需要变分法或最小二乘法进行求解。此外，这类泛函对试函数的逼近 $\|(u^h - u)_{wF}\|$ 是强条件的，不一定能保证严格逼近，但可以加速收敛。第 7 章中介绍到的 LSCM 和 MGIP 法均满足此推论。

推论 3 使用强介点，可以构造出两步近似的方法，是创造条件的试函数逼近，从而有助于数值方法的收敛。

这种方法构造的残值泛函更为简单：

$$R^*(x_n, x_p) = \{R_\mathrm{I}(x_n, x_p), R_\mathrm{B}(x_n, x_p)\} = 0 \tag{2-97}$$

其中，隐含的介点近似函数写为

$$u_\mathrm{I}^h(x) = \sum_{p=1}^{N_p} \sum_{J=1}^{N_J} \phi_p(x) \phi_J(x) \hat{u}_J \tag{2-98}$$

这种借助介点的两步近似方法，对试函数的逼近 $\|(u^h - u)_{wF}\|$ 是弱条件的，能有条件地保证严格逼近。第 7 章中介绍到的 DGDC 法和 MIP 法均满足此推论。

推论 4 使用弱介点，并结合问题的基本解，可以推导出基于边界积分方程的收敛性边界解法。定义为

$$\int_\Gamma w^* R_\mathrm{B}(x_n, x_p) \mathrm{d}\Gamma = 0 \tag{2-99}$$

比如式 (2-85) 表示的 Somigliana 恒等式，即可视为边界弱介点的收敛性原则。概而言之，所有的弱式边界解法均满足推论4。实际上，有些弱式边界解法，并不一定是边界积分方程的标准形式。比如，在第 8 章中将要介绍到的 BDS 法，并不是严格采用边界积分方程法求解的，却能够满足该推论。

推论 5 使用强介点，并结合问题的基本解，可以得到稳定化和保收敛性的边界解法。定义为

$$R_\mathrm{B}(x_n, x_p) = 0 \tag{2-100}$$

第 8 章中，将介绍到 GMFS。这种方法可视为满足该推论的一个示范。

以上给出的 5 个推论，是介点原理对数值方法收敛性的 5 种途径，是否还有其他的途径，尚需进一步归纳和发展。介点原理本质上是给出了无网格法收敛性的更统一的、一般性的解释。由此，无网格法的收敛性，并不是无条件的，而是需要创造条件或优化条件保证场量逼近，此处所指场量并非单指场变量本身，还包含 PDEs 的算子作用后的场量微分形式。介点原理就是阐述创造条件的场量逼近，从

而有助于无网格法的收敛,如图 2-4 所示。对于那些成熟的离散方法,比如,全局弱式、局部弱式、边界积分方程方法等,介点原理只不过是另外一种表述方式。但对于尚需进一步发展的方法,尤其是配点类方法,介点原理作为统一、概括性的理性解释就显得尤为必要。

图 2-4 无网格法收敛性机制

Fig. 2-4 The convergence mechanism of meshless method

2.6 本章小结

加权残值法可视为是以 ODE 或 PDE 问题为求解对象的数值方法进行数值离散的统一的理论基础,例如,有限元法、有限差分法、边界元法和无网格法都是基于加权残值法得到的。变分法是寻求泛函极值的一种有效方法,因此,变分原理可以视为有限元法和有限差分法数值收敛性的理论基础。那么,边界积分方程法就可视为边界元法数值收敛性的理论基础。

论及无网格法的收敛性,全局弱式方法的收敛性也可用变分原理解释,因为全局弱式方法多采用变分的形式。而边界型的方法,其收敛性或可用边界积分方程法解释。但对其他类型,或者是更广泛的方法,还很难用现有发展的理论来较为统一地解释其收敛性。这一点可以说是无网格法进一步发展的最大障碍。为了应对这一问题,本书作者尝试性地提出了一种无网格法的收敛性观点,即无网格法介点原理。

配点型方法,或者说强式类方法,是真正意义上的无网格法。其计算框架是完全基于散点信息的,不采用积分运算,不需要任何积分网格或积分单元,具有计算效率高、数值实施简单等一系列显而易见的优点。然而此类方法有明显的局限性,主要表现在数值收敛性难以保证,从而导致计算精度低、计算不稳定等一系列问题。介点原理的提出可能为配点型方法的完善提供新的思路,有助于此类方法的进一步发展。

介点原理作为一种框架性的解释机制,为建立无网格法普遍性的收敛理论提供了一种可供选择的途径。当前,介点原理作为一个唯象的观点,与理论化还差之甚远,其完整而严密的理论架构尚需进一步发展。

参 考 文 献

[1] 徐次达. 固体力学加权残值法 [M]. 上海: 同济大学出版社, 1987.
[2] 邱吉宝. 加权残值法的理论与应用 [M]. 北京: 宇航出版社, 1991.
[3] 胡海昌. 弹性力学的变分原理及其应用 [M]. 北京: 科学出版社, 1981.
[4] 钱令希. 余能理论 [J]. 中国科学, 1950, 1(2): 449-456.
[5] Hellinger E. Die Allgemeinen Ansätze der Mechanik der Kontinua[M]. Germany: Eneyelopadie der Mathematischen Wissenschaften, 1914.
[6] Reissner E. On a variational theorem in elasticity[J]. Studies in Applied Mathematics, 1950, 29(1-4): 90-95.
[7] 胡海昌. 论弹性体力学和受范性体力学中的一般变分原理 [J]. 物理学报, 1954, 10(3): 259-289.
[8] Washizu K. On the variational principles of elasticity and plasticity[R]. Boston: Massachusetts Inst of Tech, Cambridge Aeroelastic and Structures Research Lab, 1955.
[9] Rizzo F J. An integral equation approach to boundary value problems of classical elastostatics[J]. Quarterly of Applied Mathematics, 1967, 25(1): 83-95.
[10] Cruse T A, Rizzo F J. A direct formulation and numerical solution of the general transient elastodynamic problem. I[J]. Journal of Mathematical Analysis and Applications, 1968, 22(1): 244-259.
[11] 胡海昌. 弹性力学中一类新的边界积分方程 [J]. 中国科学 A 辑, 1986, 11: 1170-1174.
[12] 吴永礼. 计算固体力学方法 [M]. 北京: 科学出版社, 2003.
[13] 杨建军. 无网格法的介点原理及其应用 [D]. 长沙理工大学博士学位论文, 2012.
[14] 杨建军, 郑健龙. 无网格法介点原理 [J]. 科学通报, 2012, 57(26): 2456-2462.
[15] Yang J J, Zheng J L. Intervention-point principle of meshless method[J]. Chinese Science Bulletin, 2013, 58(4-5): 478-485.

第 3 章 无网格近似法

有限元法 (FEM) 的近似函数是通过网格单元的一组固定节点利用插值技术构造的, 故被称为基于固定单元的插值近似。而无网格法中, 问题域是由一组场节点离散的, 单元的形态为节点, 需要特殊的形函数构造技术, 以适应离散节点的场函数近似。可以说, 使用散点信息来构造近似函数是无网格法最基本的特征。

3.1 无网格近似函数的性质

对计算节点 x_I, 其场函数值 $u_I(x)$ 由邻近的一组场节点 $\{x_J\}_{J=1}^N$ 信息来表达, 如图 3-1 表示。此处, 特别约定, 本书用黑体 x 表示的点信息或变量是多维度的一般形式, 即, 对一维问题, $x = [x]$; 二维问题, $x = [x,y]^{\mathrm{T}}$; 三维问题, $x = [x,y,z]^{\mathrm{T}}$。图中 Ω_S 表示该计算节点的支撑域, 本书有时也将其表示为 Ω_S^I, 则有 $\Omega \subset \bigcup_{I=1}^n \Omega_S^I$, 即所有节点对应的支撑域将对问题域 Ω 形成完整覆盖。x_J 表示被该支撑域覆盖的任意一个场节点 (通常被称为支撑点), 则该计算节点的无网格近似函数 $u_I^h(x)$ 可写成如下一般形式:

$$u_I(x) \approx u_I^h(x) = \sum_{J=1}^N \phi_J(x) u_J = \Phi(x) u \tag{3-1}$$

$$\Phi(x) = [\phi_J(x)]_{J=1}^N = [\phi_1(x), \phi_2(x), \cdots, \phi_N(x)] \tag{3-2}$$

$$u = [u_J;]_{J=1}^N = [u_1; u_2; \cdots; u_N] = [u_1, u_2, \cdots, u_N]^{\mathrm{T}} \tag{3-3}$$

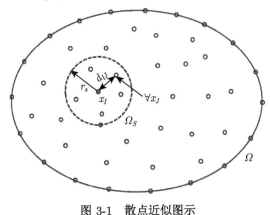

图 3-1 散点近似图示

Fig. 3-1 Schematics of scatter-point approximation

式 (3-1) 是带有计算节点信息的近似函数，并有 $u_I^h(\boldsymbol{x}) \equiv u^h(\boldsymbol{x}_I)$。本书中，通常采用上标 "$h$" 表示近似函数，有时也用顶部带 "$\sim$" 符号的形式表示，即 $u^h = \tilde{u}$。在数值计算中，近似函数通常是逐点计算的。这种表示法将计算节点信息包含其中，有表达直接、避免引起符号混淆的优点。此外，在弱式方法或一些具体的特殊方法中，计算点 $\boldsymbol{x}_I$ 有可能不是一个节点，而或许是一个积分点或"介点"。故本书有时用 $\boldsymbol{x}_*$ 一般性地表示计算点 (目标点)。其中，$\boldsymbol{\Phi}(\boldsymbol{x})$ 被称为对应于 N 个支撑点的整体无网格形函数，而 $\phi_J(\boldsymbol{x})$ 被称为支撑点 $\boldsymbol{x}_J$ 对应的形函数。

在数值方法中，计算点场函数 $u_I(\boldsymbol{x})$ 的导数通常直接由其近似函数 $u_I^h(\boldsymbol{x})$ 的导数来计算，并定义为

$$\left\{\begin{array}{l} u_{I,i}(\boldsymbol{x}) \\ u_{I,ij}(\boldsymbol{x}) \end{array}\right\} \approx \left\{\begin{array}{l} u_{I,i}^h(\boldsymbol{x}) \\ u_{I,ij}^h(\boldsymbol{x}) \end{array}\right\} = \left\{\begin{array}{l} \sum_{J=1}^{N} \phi_{J,i}(\boldsymbol{x}) u_J \\ \sum_{J=1}^{N} \phi_{J,ij}(\boldsymbol{x}) u_J \end{array}\right\} = \left\{\begin{array}{l} \boldsymbol{\Phi}_{,i}(\boldsymbol{x})\boldsymbol{u} \\ \boldsymbol{\Phi}_{,ij}(\boldsymbol{x})\boldsymbol{u} \end{array}\right\} \quad (3\text{-}4)$$

式中下标 "$,i$" 表示求函数对 i 的一阶偏导数，即 $\phi_{J,i} = \partial\phi/\partial i$；下标 "$,ij$" 表示求函数对 ij 的二阶偏导数，即 $\phi_{J,ij} = \partial^2\phi/\partial i \partial j$。其中，$i$ 和 j 表示空间坐标系的分量。在实际问题中，对场函数的求导运算可能是一个高阶导数运算，则式 (3-4) 可以改写为一个更一般的形式，即

$$u_I^{(d)}(\boldsymbol{x}) \approx \left(u_I^h\right)^{(d)}(\boldsymbol{x}) = \sum_{J=1}^{N} \phi_J^{(d)}(\boldsymbol{x}) u_J = \boldsymbol{\Phi}^{(d)}(\boldsymbol{x})\boldsymbol{u} \quad (3\text{-}5)$$

式中，右上标 "(d)" 即表示对函数的 d 阶导数运算。

需厘清一个概念，近似函数和插值函数在表示形式上是一样的，但二者有本质区别。如式 (3-1) 所示，在近似函数中，u_J 是一个待求的未知量；而在插值函数中，u_J 却是一个已知的测量值。无网格近似函数的性质由形函数决定，合理的无网格形函数须满足以下要求。

1. 紧支性

形函数的紧支性，不仅是保证数值方法计算高效的先决条件，也是保证数值方法精确性的重要要求。紧支性包含两层含义，其一是局部性，任意场节点处的场变量近似应在其局部邻域上完成，该局部邻域通常称为计算节点 (目标点) 的支撑域。其二是就近精确近似性，支撑域中越是靠近计算点的支撑点，越是倾向于取得一个更大的 0 阶形函数值。支撑域的尺度应当合理选择，其一应保证对周边场节点形成有效覆盖，从而保证近似计算的执行；其二应保证其尺度尽可能得小，从而使得其截取的场函数足够局部，一个足够局部的区域所截取场函数片段的形状才可能足够

3.1 无网格近似函数的性质

简单,从而保证有限近似能力的近似函数执行足够精确的近似。支撑域的形状选择具有一定的任意性,为简化计算,多采用规则的圆形 (椭圆形) 或矩形域。所有场节点对应局部支撑域的集合,将对整个问题域形成完整覆盖,如图 3-2 所示。图中大椭圆表示问题域,给出的散点表示场节点,以场节点为中心的小圆或矩形,即对应节点的支撑域。

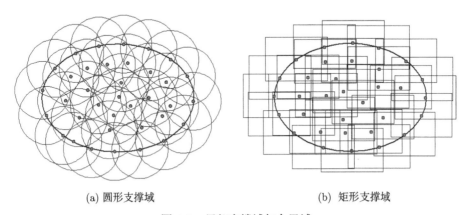

(a) 圆形支撑域   (b) 矩形支撑域

图 3-2 局部支撑域与全局域

Fig. 3-2 Local supported domain and global domain

2. 相容性 (compatibility)

相容性就是要求近似函数对支撑域中的节点 (支撑点) 数量和空间分布的敏感性足够低,从而保证支持域尺度选择以及增删节点或移动节点具有一定的自由度。而近似函数的这种性质,通常是由紧支性的权函数和容许冗余的支撑点存在来保证的。

3. 一致性 (consistency)

一致性就是要求形函数的光滑性能适应待解微分方程的求解要求,即对于 d 阶微分方程问题,如果需要计算直至 k ($k \leqslant d$) 阶的近似函数导数,则形函数至少应保证 k 阶可微 (C^k 一致性)。

以上三个性质是对无网格形函数的基本要求。实际上,一种好的无网格近似函数构造方法还需要满足更多要求。比如,适用性的要求,即方法的构造是简洁易用、便于数值实施的;效率性的要求,即执行计算所消耗的时间成本应尽可能低,无网格法的主要计算耗时是花费在近似计算中,近似计算效率的提高对特定方法求解效率的贡献是非常可观的;精确性的要求,可以确定地说,近似计算的精确性将显著影响数值方法求解的精确性。

3.2 移动最小二乘法

移动最小二乘 (moving least square, MLS) 法已经有较长的研究历史, 最初由 McLain[1] 提出用于曲面拟合。此后, Gordon 和 Wixom[2]、Lancaster 和 Salkauskas[3] 进一步推介了这种插值方法。将其应用于无网格法, 则始于 Nayroles 等 [4] 提出的扩散单元法 (diffuse element method, DEM)。此后, Belytschko 等 [5] 在提出无单元伽辽金 (element-free Galerkin, EFG) 法时, 积极推介了这种近似方法。此后, MLS 法也随着 EFG 法的成功而倍受关注 [6,7]。

因多项式函数具有计算机表示简单的优点, 所以多项式函数被广泛应用于近似函数的表示。MLS 近似函数一般采用多项式基函数的线性组合, 即

$$u(\boldsymbol{x}) \approx u^h(\boldsymbol{x}) = \sum_{t=1}^{m} p_t(\boldsymbol{x}) a_t(\boldsymbol{x}) = \boldsymbol{p}^{\mathrm{T}}(\boldsymbol{x}) \boldsymbol{a}(\boldsymbol{x}) \tag{3-6}$$

其中,

$$\boldsymbol{p}(\boldsymbol{x}) = [p_t(\boldsymbol{x})\,;]_{t=1}^{m} = [p_1(\boldsymbol{x}), p_2(\boldsymbol{x}), \cdots, p_m(\boldsymbol{x})]^{\mathrm{T}} \tag{3-7}$$

表示多项式基函数 (basis function), 向量右上标 "T" 表示矩阵的转置。如果使用二次完备多项式基函数则有

$$1D\ (m=3): \boldsymbol{p}(\boldsymbol{x}) = \begin{bmatrix} 1 & x & x^2 \end{bmatrix}^{\mathrm{T}} \tag{3-8}$$

$$2D\ (m=6): \boldsymbol{p}(\boldsymbol{x}) = \begin{bmatrix} 1 & x & y & x^2 & xy & y^2 \end{bmatrix}^{\mathrm{T}} \tag{3-9}$$

$$3D\ (m=10): \boldsymbol{p}(\boldsymbol{x}) = \begin{bmatrix} 1 & x & y & z & xy & yz & zx & x^2 & y^2 & z^2 \end{bmatrix}^{\mathrm{T}} \tag{3-10}$$

完备多项式基函数的项数 m 与其最高次数 k 及坐标维度数 D 有如下计算关系:

$$m = \frac{(k+1)(k+2)\cdots(k+D)}{1 \times 2 \times \cdots \times D} \tag{3-11}$$

而

$$\boldsymbol{a}(\boldsymbol{x}) = [a_t(\boldsymbol{x})\,;]_{t=1}^{m} = [a_1(\boldsymbol{x}), a_2(\boldsymbol{x}), \cdots, a_m(\boldsymbol{x})]^{\mathrm{T}} \tag{3-12}$$

表示近似函数的系数向量。则对计算节点 $\boldsymbol{x}_I$, 其近似函数对应地写为

$$u_I(\boldsymbol{x}) \approx u_I^h(\boldsymbol{x}) = \boldsymbol{p}_I^{\mathrm{T}}(\boldsymbol{x}) \boldsymbol{a}(\boldsymbol{x}) \tag{3-13}$$

在 MLS 近似中, 最重要的工作就是确定式 (3-13) 中的系数向量 $\boldsymbol{a}(\boldsymbol{x})$。如图 3-1 所示, 在计算节点 $\boldsymbol{x}_I$ 的局部邻域, 即支撑域 Ω_S 上, 基于散点集 $\{\boldsymbol{x}_J\}_{J=1}^{N}$

3.2 移动最小二乘法

($\forall \boldsymbol{x}_J \in \Omega_S$) 构造如下离散加权 L_2 泛函：

$$J(\boldsymbol{a}) = \sum_{J=1}^{N} w_J(\boldsymbol{x}) \left[\boldsymbol{p}_J^{\mathrm{T}}(\boldsymbol{x}) \boldsymbol{a}(\boldsymbol{x}) - \hat{u}_J \right]^2 \qquad (3\text{-}14)$$

式中，$\hat{u}_J \equiv u_J(\boldsymbol{x}) = u(\boldsymbol{x}_J)$，$w_J(\boldsymbol{x}) \equiv w_J(\boldsymbol{x} - \boldsymbol{x}_I) = w(\boldsymbol{x}_J - \boldsymbol{x}_I)$ 表示权函数 (weight function)。寻求泛函 $J(\boldsymbol{a})$ 关于 $\boldsymbol{a}(\boldsymbol{x})$ 的驻值，即令

$$\partial J / \partial \boldsymbol{a} = 0 \qquad (3\text{-}15)$$

则可解得

$$\boldsymbol{a}(\boldsymbol{x}) = \boldsymbol{A}^{-1}(\boldsymbol{x}) \boldsymbol{B}(\boldsymbol{x}) \hat{\boldsymbol{u}} \qquad (3\text{-}16)$$

其中，

$$\begin{cases} \boldsymbol{A}(\boldsymbol{x}) = \displaystyle\sum_{J=1}^{N} \boldsymbol{A}_J(\boldsymbol{x}) \\ \boldsymbol{A}_J(\boldsymbol{x}) = w_J(\boldsymbol{x}) \boldsymbol{p}_J(\boldsymbol{x}) \boldsymbol{p}_J^{\mathrm{T}}(\boldsymbol{x}) \end{cases} \qquad (3\text{-}17)$$

$$\begin{cases} \boldsymbol{B}(\boldsymbol{x}) = [\boldsymbol{B}_J(\boldsymbol{x})]_{J=1}^{N} = [\boldsymbol{B}_1(\boldsymbol{x}), \boldsymbol{B}_2(\boldsymbol{x}), \cdots, \boldsymbol{B}_N(\boldsymbol{x})] \\ \boldsymbol{B}_J(\boldsymbol{x}) = w_J(\boldsymbol{x}) \boldsymbol{p}_J(\boldsymbol{x}) \end{cases} \qquad (3\text{-}18)$$

$$\hat{\boldsymbol{u}} = [\hat{u}_J;]_{J=1}^{N} = [\hat{u}_1, \hat{u}_2, \cdots, \hat{u}_N]^{\mathrm{T}} \qquad (3\text{-}19)$$

将式 (3-16) 代入式 (3-13) 可得

$$u_I^h(\boldsymbol{x}) = \boldsymbol{p}_I^{\mathrm{T}}(\boldsymbol{x}) \boldsymbol{A}^{-1}(\boldsymbol{x}) \boldsymbol{B}(\boldsymbol{x}) \hat{\boldsymbol{u}} = \sum_{J=1}^{N} \phi_J(\boldsymbol{x}) \hat{u}_J = \boldsymbol{\Phi}(\boldsymbol{x}) \hat{\boldsymbol{u}} \qquad (3\text{-}20)$$

则 MLS 对应于 N 个节点的整体形函数 $\boldsymbol{\Phi}(\boldsymbol{x})$ 写为

$$\boldsymbol{\Phi}(\boldsymbol{x}) = [\phi_J(\boldsymbol{x})]_{J=1}^{N} = [\phi_1(\boldsymbol{x}), \phi_2(\boldsymbol{x}), \cdots, \phi_N(\boldsymbol{x})] = \boldsymbol{p}_I^{\mathrm{T}}(\boldsymbol{x}) \boldsymbol{A}^{-1}(\boldsymbol{x}) \boldsymbol{B}(\boldsymbol{x}) \qquad (3\text{-}21)$$

对应于节点 $\boldsymbol{x}_J$ 的形函数 $\phi_J(\boldsymbol{x})$，写为

$$\phi_J(\boldsymbol{x}) = \boldsymbol{p}_I^{\mathrm{T}}(\boldsymbol{x}) \left[\boldsymbol{A}^{-1}(\boldsymbol{x}) \boldsymbol{B}_J(\boldsymbol{x}) \right] \qquad (3\text{-}22)$$

对形函数 $\phi_J(\boldsymbol{x})$ 的导数运算，Nayroles 等[4] 建议为

$$\begin{cases} \phi_{J,i}(\boldsymbol{x}) = \boldsymbol{p}_{I,i}^{\mathrm{T}} \left(\boldsymbol{A}^{-1} \boldsymbol{B}_J \right) \\ \phi_{J,ij}(\boldsymbol{x}) = \boldsymbol{p}_{I,ij}^{\mathrm{T}} \left(\boldsymbol{A}^{-1} \boldsymbol{B}_J \right) \end{cases} \qquad (3\text{-}23)$$

这种导数计算方法，实际上是将系数向量 $\boldsymbol{a}(\boldsymbol{x})$ 假定为一个常数向量，而不作为函数来对待。

而 Belytschko 等[5]则认为，在形函数导数运算中，系数向量 $a(x)$ 应视为一个函数，应当参与导数运算。令 $\eta^T = p_I^T A^{-1}$，由 A 的对称性，即 $A^T = A$，可解得

$$\eta = A^{-1} p_I \tag{3-24}$$

则可得到 η 的偏导数计算公式：

$$\begin{cases} \eta_{,i} = A^{-1}(p_{I,i} - A_{,i}\eta) \\ \eta_{,ij} = A^{-1}[p_{I,ij} - (A_{,i}\eta_{,j} + A_{,j}\eta_{,i} + A_{,ij}\eta)] \end{cases} \tag{3-25}$$

则 $\phi_J(x)$ 的偏导数计算公式写为

$$\begin{cases} \phi_{J,i}(x) = \eta_{,i}^T B_J + \eta^T B_{J,i} \\ \phi_{J,ij}(x) = \eta_{,ij}^T B_J + \eta_{,i}^T B_{J,j} + \eta_{,j}^T B_{J,i} + \eta^T B_{J,ij} \end{cases} \tag{3-26}$$

由上式可以看出：$a(x)$ 参与导数运算，则 $\phi_J(x)$ 的偏导数计算公式将变得非常复杂。对于其必要性，将在第 4 章中进行特别讨论。

3.3 光滑粒子流体动力学法

光滑粒子流体动力学 (smoothed particle hydrodynamics, SPH) 法[8-15]最初发展运用于研究大质量恒星解体和天体物理学中的非对称现象，并在随后的拓展性研究中，被广泛应用于流体力学、冲击与爆炸力学等领域。SPH 法构造简单，对一些特定问题，比如以密度为场变量的问题，其数值求解具有明显的优势。

SPH 近似函数定义为

$$u_I(x) \approx u_I^h(x) = \int_\Omega u_J(x) w_J(x - x_I, h) \mathrm{d}\Omega_S \tag{3-27}$$

式中，$w_J(x - x_I, h)$ 为核 (kernel) 函数；也可将其一般性地称为权函数，该函数具有如下特性：

$$\int_\Omega w_J(x - x_I, h) \mathrm{d}\Omega_S = 1 \tag{3-28}$$

$$\lim_{h \to 0} w_J(x - x_I, h) = \delta(x - x_I) \tag{3-29}$$

$$w_J(x - x_I, h) > 0, \quad \forall x \in \Omega_S(x_I) \tag{3-30}$$

$$w_J(x - x_I, h) = 0, \quad \forall x \notin \Omega_S(x_I) \tag{3-31}$$

Monaghan[9]推荐使用高斯型核函数，其形式写为

$$w(r, h) = \left(\frac{1}{\pi h^2}\right)^{3/2} \exp\left(-\frac{r^2}{h^2}\right), \quad \text{1D} \tag{3-32}$$

$$w(\boldsymbol{r},h)=\frac{3H\left(1-|\boldsymbol{r}|/h\right)}{4\pi h^{3}},\quad \text{2D} \tag{3-33}$$

$$w(\boldsymbol{r},h)=\frac{S\left(|\boldsymbol{r}|/h\right)}{h^{3}},\quad \text{3D} \tag{3-34}$$

式中，H 函数为 Heaviside 阶跃函数；S 函数为球形 δ 样条函数。在数值运算中，式 (3-27) 的积分形式通常采用求和近似运算，即

$$u_{I}^{h}(\boldsymbol{x})=\sum_{J=1}^{N}u_{J}(\boldsymbol{x})w_{J}(\boldsymbol{x}-\boldsymbol{x}_{I},h)V_{J} \tag{3-35}$$

式中，V_J 表示 J 编号粒子 $\boldsymbol{x}_J$ 所对应的空间测度，即一维问题代表长度，二维问题代表面积，三维问题代表体积，并有

$$V_{J}=m_{J}/\rho_{J} \tag{3-36}$$

式中，m_J 和 ρ_J 分别表示粒子 $\boldsymbol{x}_J$ 的质量和密度。将式 (3-36) 代入式 (3-35)，令 $u_J(\boldsymbol{x})\equiv u_J$，可得

$$u_{I}^{h}(\boldsymbol{x})=\sum_{J=1}^{N}\frac{m_{J}}{\rho_{J}}w_{J}(\boldsymbol{x}-\boldsymbol{x}_{I},h)u_{J}(\boldsymbol{x})=\sum_{J=1}^{N}\phi_{J}(\boldsymbol{x})u_{J} \tag{3-37}$$

显然，SPH 的形函数 $\phi_J(\boldsymbol{x})$ 可定义为

$$\phi_{J}(\boldsymbol{x})=\frac{m_{J}}{\rho_{J}}w_{J}(\boldsymbol{x}-\boldsymbol{x}_{I},h) \tag{3-38}$$

如果用 $\rho(\boldsymbol{x})$ 直接替换式 (3-37) 中的 $u(\boldsymbol{x})$，则可得

$$\rho_{I}^{h}(\boldsymbol{x})=\sum_{J=1}^{N}m_{J}w_{J}(\boldsymbol{x}-\boldsymbol{x}_{I},h) \tag{3-39}$$

上式通常被称为密度求和法，是 SPH 方法中用于获取粒子密度的一个非常简单而直接的形式。

对 SPH 的形函数 $\phi_J(\boldsymbol{x})$ 的导数运算，则定义如下：

$$\left\{\begin{array}{c}\phi_{J,i}(\boldsymbol{x})\\ \phi_{J,ij}(\boldsymbol{x})\end{array}\right\}=\left\{\begin{array}{c}\dfrac{m_{J}}{\rho_{J}}w_{J,i}(\boldsymbol{x}-\boldsymbol{x}_{I},h)\\ \dfrac{m_{J}}{\rho_{J}}w_{J,ij}(\boldsymbol{x}-\boldsymbol{x}_{I},h)\end{array}\right\} \tag{3-40}$$

可见，SPH 法的导数近似依赖于核函数的导数运算。因此，SPH 法对核函数的光滑性有较高的要求。不同于样条核函数，高斯型核函数具有高阶连续性，所以选用其作为核函数有一定优势。

由 SPH 法中近似函数的构造可以看出，其近似函数中包含诸多粒子物理量，这在实际问题中通常与其物理本质成对应关系，因此，对一些实际的物理问题，特别是流体流动现象有其独特的优势。

3.4 重构核粒子法

为了克服 SPH 方法在有限域边界附近一些求解不稳定现象,Liu 等[16-18]将小波分析的理论运用于散点近似函数的构造,提出了重构核粒子法 (reproducing kernel particle method, RKPM),在式 (3-27) 中引入一个校正函数 $C(\boldsymbol{x}, \boldsymbol{x}_I)$,则有

$$u_I(\boldsymbol{x}) \approx u_I^h(\boldsymbol{x}) = \int_{\Omega_S} C_J(\boldsymbol{x}, \boldsymbol{x}_I) w_J(\boldsymbol{x} - \boldsymbol{x}_I, h) u_J(\boldsymbol{x}) \mathrm{d}\Omega \tag{3-41}$$

其中,$C_J(\boldsymbol{x}, \boldsymbol{x}_I) \equiv C(\boldsymbol{x}_J, \boldsymbol{x}_I)$ 称为校正函数 (correction function),通常将其取为多项式基函数的线性组合,即

$$C(\boldsymbol{x}, \boldsymbol{x}_I) = \sum_{t=1}^{m} p_t(\boldsymbol{x} - \boldsymbol{x}_I) b_t(\boldsymbol{x}) = \boldsymbol{p}^{\mathrm{T}}(\boldsymbol{x} - \boldsymbol{x}_I) \boldsymbol{b}(\boldsymbol{x}) \tag{3-42}$$

式中,$\boldsymbol{p}(\boldsymbol{x} - \boldsymbol{x}_I) = [p_t(\boldsymbol{x} - \boldsymbol{x}_I);]_{t=1}^{m}$ 是核基函数向量,$\boldsymbol{b}(\boldsymbol{x}) = [b_t(\boldsymbol{x});]_{t=1}^{m}$ 是系数向量,而系数向量需由重构条件求得。将近似函数 $u_I^h(\boldsymbol{x})$ 在计算点处进行 Taylor 级数展开,可得

$$u_I^h(\boldsymbol{x}) = m_0(\boldsymbol{x}) u(\boldsymbol{x}) + \sum_{i=1}^{\infty} \frac{(-1)^i}{i!} m_i(\boldsymbol{x}) u^{(i)}(\boldsymbol{x}) \tag{3-43}$$

其中,

$$m_i(\boldsymbol{x}) = \sum_{J=1}^{N} p_i(\boldsymbol{x}) C_J(\boldsymbol{x}, \boldsymbol{x}_I) w_J(\boldsymbol{x} - \boldsymbol{x}_I, h) V_J \tag{3-44}$$

由式 (3-43),取核函数的重构条件为

$$\begin{cases} m_0(\boldsymbol{x}) = 1 \\ m_i(\boldsymbol{x}) = 0, \quad i = 1, 2, \cdots \end{cases} \tag{3-45}$$

则有

$$\boldsymbol{M}(\boldsymbol{x}) \boldsymbol{b}(\boldsymbol{x}) = \boldsymbol{H} \tag{3-46}$$

其中,

$$\boldsymbol{M}(\boldsymbol{x}) = \sum_{J=1}^{N} \boldsymbol{p}_J(\boldsymbol{x} - \boldsymbol{x}_I) \boldsymbol{p}_J^{\mathrm{T}}(\boldsymbol{x} - \boldsymbol{x}_I) w_J(\boldsymbol{x} - \boldsymbol{x}_I, h) V_J \tag{3-47}$$

$$\boldsymbol{H} = [1, 0, \cdots, 0]^{\mathrm{T}} \tag{3-48}$$

可解得

$$\boldsymbol{b}(\boldsymbol{x}) = \boldsymbol{M}^{-1}(\boldsymbol{x}) \boldsymbol{H} \tag{3-49}$$

式 (3-41) 表示的积分形式，采用梯形积分法得到其离散形式为

$$u_I^h(\boldsymbol{x}) = \sum_{J=1}^{N} C_J(\boldsymbol{x}, \boldsymbol{x}_I) w_J(\boldsymbol{x} - \boldsymbol{x}_I, h) u_J V_J = \sum_{J=1}^{N} \phi_J(\boldsymbol{x}) u_J \quad (3\text{-}50)$$

则 RKPM 的形函数 $\phi_J(\boldsymbol{x})$ 定义为

$$\phi_J(\boldsymbol{x}) = C_J(\boldsymbol{x}, \boldsymbol{x}_I) w_J(\boldsymbol{x} - \boldsymbol{x}_I, h) V_J \quad (3\text{-}51)$$

其中，

$$C_J(\boldsymbol{x}, \boldsymbol{x}_I) = \boldsymbol{p}_J^{\mathrm{T}}(\boldsymbol{x} - \boldsymbol{x}_I) \boldsymbol{b}(\boldsymbol{x}) \quad (3\text{-}52)$$

式 (3-50) 改用矩阵形式表示为

$$u_I^h(\boldsymbol{x}) = \boldsymbol{C}(\boldsymbol{x}) \boldsymbol{W}(\boldsymbol{x}) \boldsymbol{V} \boldsymbol{u} = \boldsymbol{\Phi}(\boldsymbol{x}) \boldsymbol{u} \quad (3\text{-}53)$$

其中，

$$\boldsymbol{C}(\boldsymbol{x}) = [C_J(\boldsymbol{x}, \boldsymbol{x}_I)]_{J=1}^{N} = [C(\boldsymbol{x}_1, \boldsymbol{x}_I), C(\boldsymbol{x}_2, \boldsymbol{x}_I), \cdots, C(\boldsymbol{x}_N, \boldsymbol{x}_I)]$$
$$= \boldsymbol{b}^{\mathrm{T}}(\boldsymbol{x}) \boldsymbol{P}(\boldsymbol{x} - \boldsymbol{x}_I) \quad (3\text{-}54)$$

$$\boldsymbol{W}(\boldsymbol{x}) = \mathrm{diag}\,[w_J(\boldsymbol{x} - \boldsymbol{x}_I)]_{J=1}^{N} = \begin{bmatrix} w(\boldsymbol{x}_1 - \boldsymbol{x}_I) & 0 & \cdots & 0 \\ 0 & w(\boldsymbol{x}_2 - \boldsymbol{x}_I) & \cdots & 0 \\ \vdots & \vdots & \ddots & \vdots \\ 0 & 0 & \cdots & w(\boldsymbol{x}_N - \boldsymbol{x}_I) \end{bmatrix}$$
$$(3\text{-}55)$$

$$\boldsymbol{V} = \mathrm{diag}\,[V_J]_{J=1}^{N} = \begin{bmatrix} V_1 & 0 & \cdots & 0 \\ 0 & V_2 & \cdots & 0 \\ \vdots & \vdots & \ddots & \vdots \\ 0 & 0 & \cdots & V_N \end{bmatrix} \quad (3\text{-}56)$$

其中，

$$\boldsymbol{P}(\boldsymbol{x} - \boldsymbol{x}_I) = [\boldsymbol{p}_J(\boldsymbol{x} - \boldsymbol{x}_I)]_{J=1}^{N} = \begin{bmatrix} p_1(\boldsymbol{x}_1 - \boldsymbol{x}_I) & p_1(\boldsymbol{x}_2 - \boldsymbol{x}_I) & \cdots & p_1(\boldsymbol{x}_N - \boldsymbol{x}_I) \\ p_2(\boldsymbol{x}_1 - \boldsymbol{x}_I) & p_2(\boldsymbol{x}_2 - \boldsymbol{x}_I) & \cdots & p_2(\boldsymbol{x}_N - \boldsymbol{x}_I) \\ \vdots & \vdots & \ddots & \vdots \\ p_m(\boldsymbol{x}_1 - \boldsymbol{x}_I) & p_m(\boldsymbol{x}_2 - \boldsymbol{x}_I) & \cdots & p_m(\boldsymbol{x}_N - \boldsymbol{x}_I) \end{bmatrix}$$
$$(3\text{-}57)$$

另，RKPM 形函数 $\phi_J(\boldsymbol{x})$ 的导数计算公式写为

$$\begin{cases} \phi_{J,i}(\boldsymbol{x}) = (C_{J,i}w_J + C_J w_{J,i}) V_J \\ \phi_{J,ij}(\boldsymbol{x}) = (C_{J,ij}w_J + C_{J,i}w_{J,j} + C_{J,j}w_{J,i} + C_J w_{J,ij}) V_J \end{cases} \quad (3\text{-}58)$$

其中，

$$\begin{cases} C_{J,i}(\boldsymbol{x}, \boldsymbol{x}_I) = \boldsymbol{p}_{J,i}^{\mathrm{T}}\boldsymbol{b} + \boldsymbol{p}_J^{\mathrm{T}}\boldsymbol{b}_{,i} \\ C_{J,i}(\boldsymbol{x}, \boldsymbol{x}_I) = \boldsymbol{p}_{J,ij}^{\mathrm{T}}\boldsymbol{b} + \boldsymbol{p}_{J,i}^{\mathrm{T}}\boldsymbol{b}_{,j} + \boldsymbol{p}_{J,j}^{\mathrm{T}}\boldsymbol{b}_{,i} + \boldsymbol{p}_J^{\mathrm{T}}\boldsymbol{b}_{ij} \end{cases} \quad (3\text{-}59)$$

$$\begin{cases} \boldsymbol{b}_{,i}(\boldsymbol{x}) = -\boldsymbol{M}^{-1}\boldsymbol{M}_{,i}\boldsymbol{b} \\ \boldsymbol{b}_{,ij}(\boldsymbol{x}) = -\boldsymbol{M}^{-1}\left(\boldsymbol{M}_{,ij}\boldsymbol{b} + \boldsymbol{M}_{,i}\boldsymbol{b}_{,j} + \boldsymbol{M}_{,j}\boldsymbol{b}_{,i}\right) \end{cases} \quad (3\text{-}60)$$

在 RKPM 形函数中，因积分运算的转换，其表达式中包含粒子的体积 V_J，在实际近似计算中，V_J 很难精确确定，可采用简约的方法计算。如果用 $\bar{h}$ 表示粒子的平均间距，则对一维问题，$V_J = \bar{h}$；对二维问题 $V_J = \bar{h}^2$；对三维问题 $V_J = \bar{h}^3$。实际运用中，为了简单起见，通常取 $V_J = 1$，这种简化操作通常不会对 RKPM 形函数的近似精度带来直接影响。

3.5 点插值法

点插值法 (point interpolation method, PIM) 是应用最早的插值型方法之一，被广泛应用于各类数值方法。Liu 和 Gu[19,20] 建议将其应用于构造无网格形函数。

3.5.1 多项式基点插值法

多项式基点插值法近似函数采用多项式基函数的线性组合，即

$$u(\boldsymbol{x}) \approx u^h(\boldsymbol{x}) = \sum_{t=1}^{m} p_t(\boldsymbol{x}) a_t = \boldsymbol{p}^{\mathrm{T}}(\boldsymbol{x}) \boldsymbol{a} \quad (3\text{-}61)$$

式中，$\boldsymbol{p}(\boldsymbol{x})$ 为多项式基函数如式 (3-7) 所示；$\boldsymbol{a}$ 为常系数向量。在计算节点 $\boldsymbol{x}_I$ 的局部邻域即支撑域 Ω_S 上基于散点集 $\{\boldsymbol{x}_J\}_{J=1}^{N}$ ($\forall \boldsymbol{x}_J \in \Omega_S$) 构造如下泛函：

$$J = \sum_{J=1}^{N} \left[\boldsymbol{p}_J^{\mathrm{T}}(\boldsymbol{x}) \boldsymbol{a} - \hat{u}_J \right] \quad (3\text{-}62)$$

令 $J = 0$ 则可解得

$$\boldsymbol{a} = \boldsymbol{P}^{-1}(\boldsymbol{x}) \hat{\boldsymbol{u}} \quad (3\text{-}63)$$

式中，

$$\boldsymbol{a} = [a_t;]_{t=1}^{m} = [a_1, a_2, \cdots, a_m]^{\mathrm{T}} \quad (3\text{-}64)$$

3.5 点插值法

$$\hat{\boldsymbol{u}} = [\hat{u}_J;]_{J=1}^N = [\hat{u}_1, \hat{u}_2, \cdots, \hat{u}_N]^T \tag{3-65}$$

$$\boldsymbol{P}(\boldsymbol{x}) = \left[\boldsymbol{p}_J^T(\boldsymbol{x});\right]_{J=1}^N = \begin{bmatrix} p_1(\boldsymbol{x}_1) & p_2(\boldsymbol{x}_1) & \cdots & p_m(\boldsymbol{x}_1) \\ p_1(\boldsymbol{x}_2) & p_2(\boldsymbol{x}_2) & \cdots & p_m(\boldsymbol{x}_2) \\ \vdots & \vdots & \ddots & \vdots \\ p_1(\boldsymbol{x}_N) & p_2(\boldsymbol{x}_N) & \cdots & p_m(\boldsymbol{x}_N) \end{bmatrix} \tag{3-66}$$

则对计算节点 $\boldsymbol{x}_I$，其近似函数对应地写为

$$u_I^h(\boldsymbol{x}) = \boldsymbol{p}_I^T(\boldsymbol{x})\boldsymbol{a} = \boldsymbol{p}_I^T(\boldsymbol{x})\boldsymbol{P}^{-1}(\boldsymbol{x})\hat{\boldsymbol{u}} = \sum_{J=1}^N \phi_J(\boldsymbol{x})\hat{u}_J = \boldsymbol{\Phi}(\boldsymbol{x})\hat{\boldsymbol{u}} \tag{3-67}$$

则有

$$\boldsymbol{\Phi}(\boldsymbol{x}) = [\phi_J(\boldsymbol{x})]_{J=1}^N = [\phi_1(\boldsymbol{x}), \phi_2(\boldsymbol{x}), \cdots, \phi_N(\boldsymbol{x})] = \boldsymbol{p}_I^T(\boldsymbol{x})\boldsymbol{P}^{-1}(\boldsymbol{x}) \tag{3-68}$$

形函数的导数计算公式写为

$$\begin{cases} \boldsymbol{\Phi}_{,i}(\boldsymbol{x}) = \boldsymbol{p}_{I,i}^T(\boldsymbol{x})\boldsymbol{P}^{-1}(\boldsymbol{x}) \\ \boldsymbol{\Phi}_{,ij}(\boldsymbol{x}) = \boldsymbol{p}_{I,ij}^T(\boldsymbol{x})\boldsymbol{P}^{-1}(\boldsymbol{x}) \end{cases} \tag{3-69}$$

多项式基点插值法的形函数 $\phi_J(\boldsymbol{x})$ 具有 Kronecker δ 函数性质，即

$$\phi_J(\boldsymbol{x}) = \begin{cases} 1, & J = I \\ 0, & J \neq I \end{cases} \tag{3-70}$$

即有 $u_I^h(\boldsymbol{x}) = \hat{u}_I$。因此该函数性质使得其很容易施加本质边界条件。然而多项式基点插值法的近似函数也有明显不利的方面，该近似法要求支撑点数 N 与基函数向量的项数 m 必须严格相等，即 $N = m$，这是一个非常严格的条件，支撑点的任意分布很容易导致 $\boldsymbol{P}(\boldsymbol{x})$ 奇异或条件数恶化，因此，这种近似函数构造方法的相容性很差。

3.5.2 径向基点插值法

径向基函数 (radial basis function, RBF) 是一类以距离函数 $r = \|\boldsymbol{x} - \boldsymbol{x}_t\|$ 为自变量的函数，$r = \|\boldsymbol{x} - \boldsymbol{x}_t\|$ 表示点 $\boldsymbol{x}$ 到点 $\boldsymbol{x}_t$ 的距离，RBF 具有形式简单、与空间维数无关、各向同性等优点[21-25]。为了避免 PIM 法所引起的奇异性问题，一种改进的方法就是使用 RBF 代替传统多项式基函数，对应的近似函数构造方法被称为径向基点插值法 (radial point interpolation method, RPIM)[26-29]。

RPIM 近似函数采用径向基函数的线性组合，即

$$u(\boldsymbol{x}) \approx u^h(\boldsymbol{x}) = \sum_{t=1}^{m=N} R_t(\boldsymbol{x})b_t = \boldsymbol{R}^T(\boldsymbol{x})\boldsymbol{b} \tag{3-71}$$

式中，$\boldsymbol{R}(\boldsymbol{x})$ 为径向基函数向量，写为

$$\boldsymbol{R}(\boldsymbol{x}) = [R_t(\boldsymbol{x});]_{t=1}^m = [R_1(\boldsymbol{x}), R_2(\boldsymbol{x}), \cdots, R_m(\boldsymbol{x})]^{\mathrm{T}} \qquad (3\text{-}72)$$

系数向量 $\boldsymbol{b}$ 写为

$$\boldsymbol{b} = [b_t;]_{t=1}^m = [b_1, b_2, \cdots, b_m]^{\mathrm{T}} \qquad (3\text{-}73)$$

式中，项数 m 将由局部支撑域 Ω_S 上的节点数 N 来自动确定。

对第 t 项径向基函数 $R_t(\boldsymbol{x})$ 是以距离函数 r 作为变量，对 $\forall \boldsymbol{x}$，r 表示 $\boldsymbol{x}$ 到 $\boldsymbol{x}_t$ 的距离，即 $r = \|\boldsymbol{x} - \boldsymbol{x}_t\|$，对二维问题则有

$$r = \sqrt{(x - x_t)^2 + (y - y_t)^2} \qquad (3\text{-}74)$$

对支撑节点 $\boldsymbol{x}_J$，则对其径向基函数向量有如下定义：

$$\boldsymbol{R}_J(\boldsymbol{x}) = [R_{tJ}(\boldsymbol{x}) \equiv R_t(\boldsymbol{x}_J);]_{t=1}^m = [R_1(\boldsymbol{x}_J), R_2(\boldsymbol{x}_J), \cdots, R_m(\boldsymbol{x}_J)]^{\mathrm{T}} \qquad (3\text{-}75)$$

目前已发展出很多可供选择的径向基函数，一些带形状参数的典型径向基函数有

$$\text{复合二次型：} R_t(\boldsymbol{x}) = \left[r^2 + (\alpha h)^2\right]^{\beta} \qquad (3\text{-}76)$$

$$\text{高斯型：} R_t(\boldsymbol{x}) = \exp\left[-\alpha(r/h)^2\right] \qquad (3\text{-}77)$$

$$\text{薄板样条型：} R_t(\boldsymbol{x}) = r^{\alpha} \qquad (3\text{-}78)$$

$$\text{对数型：} R_t(\boldsymbol{x}) = r^{\alpha} \log r \qquad (3\text{-}79)$$

式中，h 表示平均节点间距；α 和 β 为形状参数。

此外，吴宗敏和 Schaback[21,22,30] 提出的正定紧支柱径向基函数有

$$R_t(\boldsymbol{x}) = (1-\kappa)^3 \left(8 + 9\kappa + 3\kappa^2\right) \qquad (3\text{-}80)$$

$$R_t(\boldsymbol{x}) = (1-\kappa)^4 \left(4 + 16\kappa + 12\kappa^2 + 3\kappa^3\right) \qquad (3\text{-}81)$$

$$R_t(\boldsymbol{x}) = (1-\kappa)^5 \left(1 + 5\kappa + 9\kappa^2 + 5\kappa^3 + \kappa^4\right) \qquad (3\text{-}82)$$

$$R_t(\boldsymbol{x}) = (1-\kappa)^4 \left(16 + 29\kappa + 20\kappa^2 + 5\kappa^3\right) \qquad (3\text{-}83)$$

$$R_t(\boldsymbol{x}) = (1-\kappa)^5 \left(8 + 40\kappa + 48\kappa^2 + 25\kappa^3 + 5\kappa^4\right) \qquad (3\text{-}84)$$

$$R_t(\boldsymbol{x}) = (1-\kappa)^6 \left(6 + 36\kappa + 82\kappa^2 + 72\kappa^3 + 30\kappa^4 + 5\kappa^5\right) \qquad (3\text{-}85)$$

$$R_t(\boldsymbol{x}) = (1-\kappa)^7 \left(5 + 35\kappa + 101\kappa^2 + 147\kappa^3 + 101\kappa^4 + 35\kappa^5 + 5\kappa^6\right) \qquad (3\text{-}86)$$

3.5 点插值法

Wendland[23,31] 提出的紧支径向基函数有

$$R_t(\boldsymbol{x}) = (1-\kappa)^4 (1+4\kappa) \tag{3-87}$$

$$R_t(\boldsymbol{x}) = (1-\kappa)^6 (3+18\kappa+35\kappa^2) \tag{3-88}$$

$$R_t(\boldsymbol{x}) = (1-\kappa)^8 (1+8\kappa+25\kappa^2+32\kappa^3) \tag{3-89}$$

Buhmann[32-34] 提出的紧支径向基函数有

$$R_t(\boldsymbol{x}) = \frac{1}{3} + \kappa^2 - \frac{4}{3}\kappa^3 + 2\kappa^2\ln\kappa \tag{3-90}$$

$$R_t(\boldsymbol{x}) = \frac{1}{15} + \frac{19}{6}\kappa^2 - \frac{16}{3}\kappa^3 + 3\kappa^4 - \frac{16}{15}\kappa^5 + \frac{1}{6}\kappa^6 + 2\kappa^2\ln\kappa \tag{3-91}$$

其中，

$$\kappa = r/r_S \tag{3-92}$$

式中，r_S 表示局部支撑域的半径尺度。

根据式 (3-71)，在计算节点 $\boldsymbol{x}_I$ 的局部支撑域 Ω_S 上，基于散点集 $\{\boldsymbol{x}_J\}_{J=1}^N$ ($\forall \boldsymbol{x}_J \in \Omega_S$) 构造如下泛函：

$$J = \sum_{J=1}^{N} \left[\boldsymbol{R}_J^{\mathrm{T}}(\boldsymbol{x})\boldsymbol{b} - \hat{u}_J \right] \tag{3-93}$$

令 $J = 0$，则可解得

$$\boldsymbol{b} = \boldsymbol{A}^{-1}\hat{\boldsymbol{u}} \tag{3-94}$$

其中，

$$\hat{\boldsymbol{u}} = [\hat{u}_J;]_{J=1}^{N} = [\hat{u}_1, \hat{u}_2, \cdots, \hat{u}_N]^{\mathrm{T}} \tag{3-95}$$

$$\boldsymbol{A} = \left[\boldsymbol{R}_J^{\mathrm{T}}(\boldsymbol{x});\right]_{J=1}^{N} = \begin{bmatrix} R_1(\boldsymbol{x}_1) & R_2(\boldsymbol{x}_1) & \cdots & R_m(\boldsymbol{x}_1) \\ R_1(\boldsymbol{x}_2) & R_2(\boldsymbol{x}_2) & \cdots & R_m(\boldsymbol{x}_2) \\ \vdots & \vdots & \ddots & \vdots \\ R_1(\boldsymbol{x}_N) & R_2(\boldsymbol{x}_N) & \cdots & R_m(\boldsymbol{x}_N) \end{bmatrix}_{(m=N)} \tag{3-96}$$

因距离无方向性，有 $R_t(\boldsymbol{x}_J) = R_J(\boldsymbol{x}_t)$，所以矩阵 $\boldsymbol{A}$ 是一个对称阵。对于计算节点 $\boldsymbol{x}_I$，其近似函数对应地写为

$$u_I^h(\boldsymbol{x}) = \boldsymbol{R}_I^{\mathrm{T}}(\boldsymbol{x})\boldsymbol{b} = \boldsymbol{R}_I^{\mathrm{T}}(\boldsymbol{x})\boldsymbol{A}^{-1}\hat{\boldsymbol{u}} = \sum_{J=1}^{N} \phi_J(\boldsymbol{x})\hat{u}_J = \boldsymbol{\Phi}(\boldsymbol{x})\hat{\boldsymbol{u}} \tag{3-97}$$

则形函数 $\boldsymbol{\Phi}(\boldsymbol{x})$ 写为

$$\boldsymbol{\Phi}(\boldsymbol{x}) = [\phi_J(\boldsymbol{x})]_{J=1}^{N} = [\phi_1(\boldsymbol{x}), \phi_2(\boldsymbol{x}), \cdots, \phi_N(\boldsymbol{x})] = \boldsymbol{R}_I^{\mathrm{T}}(\boldsymbol{x})\boldsymbol{A}^{-1} \tag{3-98}$$

形函数的导数计算公式写为

$$\begin{cases} \boldsymbol{\Phi}_{,i}(\boldsymbol{x}) = \boldsymbol{R}_{I,i}^{\mathrm{T}}(\boldsymbol{x})\boldsymbol{A}^{-1} \\ \boldsymbol{\Phi}_{,ij}(\boldsymbol{x}) = \boldsymbol{R}_{I,ij}^{\mathrm{T}}(\boldsymbol{x})\boldsymbol{A}^{-1} \end{cases} \quad (3\text{-}99)$$

其中，

$$\begin{cases} \boldsymbol{R}_{I,i}^{\mathrm{T}} = \boldsymbol{R}_{I,r}^{\mathrm{T}} r_{,i} \\ \boldsymbol{R}_{I,ij}^{\mathrm{T}} = \boldsymbol{R}_{I,rr}^{\mathrm{T}} r_{,i} r_{,j} + \boldsymbol{R}_{I,r}^{\mathrm{T}} r_{,ij} \end{cases} \quad (3\text{-}100)$$

与多项式基点插值法的形函数类似，使用径向基函数的点插值法构造的形函数 $\phi_J(\boldsymbol{x})$ 也具有 Kronecker δ 函数性质。此外，径向基函数向量的项数与支撑节点数是自动对应的，可以避免矩阵 $\boldsymbol{A}$ 的奇异，因此具有较好的相容性。

3.5.3 多项式基与径向基耦合的点插值法

为了避免多项式基点插值法所引起的奇异性问题，所以引入径向基来保证近似函数的相容性，多项式基与径向基耦合的点插值法近似函数写为

$$u(\boldsymbol{x}) \approx u^h(\boldsymbol{x}) = \sum_{t=1}^{m} p_t(\boldsymbol{x})a_t + \sum_{s=1}^{N} R_s(\boldsymbol{x})b_s = \boldsymbol{p}^{\mathrm{T}}(\boldsymbol{x})\boldsymbol{a} + \boldsymbol{R}^{\mathrm{T}}(\boldsymbol{x})\boldsymbol{b} \quad (3\text{-}101)$$

为了区别表示，对应于式 (3-71)，上式中径向基的项数序列符号改用 s 表示，其项数则直接用支撑点数 N 表示。在计算节点 $\boldsymbol{x}_I$ 的局部支撑域 Ω_S 上，基于散点集 $\{\boldsymbol{x}_J\}_{J=1}^{N}$ ($\forall \boldsymbol{x}_J \in \Omega_S$) 构造如下泛函：

$$J_1 = \sum_{J=1}^{N} \left[\boldsymbol{p}_J^{\mathrm{T}}(\boldsymbol{x})\boldsymbol{a} + \boldsymbol{R}_J^{\mathrm{T}}(\boldsymbol{x})\boldsymbol{b} - \hat{u}_J \right] \quad (3\text{-}102)$$

$$J_2 = \sum_{J=1}^{N} p_t(\boldsymbol{x}_J)b_J, \quad t = 1, 2, \cdots, m \quad (3\text{-}103)$$

由式 (3-101) 可知，向量 $\boldsymbol{a}$ 和 $\boldsymbol{b}$ 共有 $m+N$ 个未知数，而泛函 J_1 只能建立 N 个方程，引入 J_2 泛函可以添加另外 m 个方程。令 $J_1 = 0$, $J_2 = 0$，则可得到如下矩阵方程：

$$\begin{bmatrix} \boldsymbol{A} & \boldsymbol{P} \\ \boldsymbol{P}^{\mathrm{T}} & 0 \end{bmatrix} \begin{bmatrix} \boldsymbol{b} \\ \boldsymbol{a} \end{bmatrix} = \begin{bmatrix} \hat{\boldsymbol{u}} \\ 0 \end{bmatrix} \quad (3\text{-}104)$$

其中，矩阵 $\boldsymbol{P}$ 和 $\boldsymbol{A}$ 分别由式 (3-66) 和式 (3-96) 所示，其他矩阵与前述对应，并可解得

$$\boldsymbol{a} = \underbrace{\left(\boldsymbol{P}^{\mathrm{T}}\boldsymbol{A}^{-1}\boldsymbol{P}\right)^{-1}\boldsymbol{P}^{\mathrm{T}}\boldsymbol{A}^{-1}}_{\boldsymbol{G}_a}\hat{\boldsymbol{u}} = \boldsymbol{G}_a\hat{\boldsymbol{u}} \quad (3\text{-}105)$$

$$b = \underbrace{\left[A^{-1} - A^{-1}P\left(P^{\mathrm{T}}A^{-1}P\right)^{-1}P^{\mathrm{T}}A^{-1}\right]}_{G_b}\hat{u} = G_b\hat{u} \tag{3-106}$$

则对计算节点 x_I，其近似函数对应地写为

$$u_I^h(x) = p_I^{\mathrm{T}}(x)\,a + R_I^{\mathrm{T}}(x)\,b = \left[p_I^{\mathrm{T}}(x)\,G_a + R_I^{\mathrm{T}}(x)\,G_b\right]\hat{u}$$
$$= \sum_{J=1}^{N}\phi_J(x)\hat{u}_J = \boldsymbol{\Phi}(x)\,\hat{u} \tag{3-107}$$

则形函数 $\boldsymbol{\Phi}(x)$ 写为

$$\boldsymbol{\Phi}(x) = p_I^{\mathrm{T}}(x)\,G_a + R_I^{\mathrm{T}}(x)\,G_b \tag{3-108}$$

其导数计算公式对应地写为

$$\begin{cases} \boldsymbol{\Phi}_{,i}(x) = p_{I,i}^{\mathrm{T}}(x)\,G_a + R_{I,i}^{\mathrm{T}}(x)\,G_b \\ \boldsymbol{\Phi}_{,ij}(x) = p_{I,ij}^{\mathrm{T}}(x)\,G_a + R_{I,ij}^{\mathrm{T}}(x)\,G_b \end{cases} \tag{3-109}$$

其中，$R_I^{\mathrm{T}}(x)$ 的导数运算如式 (3-100)。

多项式基与径向基耦合的点插值法，所构造的形函数既保持了多项式基点插值法的线性再生性特点，又带有径向基点插值法的对任意数量支撑节点自动适应的相容性优势。不足的方面是，形函数构造比较复杂。有关研究表明，这种耦合式的 PIM 通常具有较高的近似精度，但在采用较密的场节点时其收敛性不够稳定[20]。

3.6 单位分解法

单位分解法 (partition of unity method, PUM) 的概念由 Duarte 和 Oden[35,36]，以及 Melenk 和 Babuška[37] 提出并加以发展，PUM 近似函数可统写为如下形式：

$$u(x) \approx u^h(x) = \sum_{I=1}^{n}\varphi_I(x)\,V_I(x) \tag{3-110}$$

式中，n 表示求解域 Ω 上的场节点数量；$V_I(x)$ 表示计算节点 x_I 的局部支撑域 Ω_S^I 上的局部近似函数；$\varphi_I(x)$ 称为单位分解函数，并满足非零和单位分解条件

$$\begin{cases} \varphi_I(x) \neq 0, \quad \varphi_I \in \Omega_S^I \\ \sum_{I=1}^{n}\varphi_I(x) = 1 \end{cases} \tag{3-111}$$

构造单位分解函数的方法有多种，较多采用的有 MLS 形函数、Shepard 函数、有限元形函数等。

Melenk 和 Babuška[37] 在 PUFEM 方法中使用 Shepard 函数构造单位分解函数 $\varphi_I(\boldsymbol{x})$，而使用 Lagrange 插值函数 $L(\boldsymbol{x})$ 来构造局部近似函数 $V_I(\boldsymbol{x})$，写为

$$V_I(\boldsymbol{x}) = \sum_{J=1}^{N} L_{JI}(\boldsymbol{x}) u_J \tag{3-112}$$

Duarte 和 Oden[35,36] 在 Hp 云法中使用 MLS 形函数作为单位分解函数 $\varphi_I(\boldsymbol{x})$，而局部近似函数 $V_I(\boldsymbol{x})$ 可写为如下形式：

$$V_I(\boldsymbol{x}) = u_I + \sum_{t=1}^{m} b_{It} q_{It}(\boldsymbol{x} - \boldsymbol{x}_I) \tag{3-113}$$

式中，$q_{It}(\boldsymbol{x} - \boldsymbol{x}_I)$ 为 Legendre 多项式基函数的第 t 项；b_{It} 为其对应的系数。

单位分解函数 $\varphi_I(\boldsymbol{x})$ 也可取为有限元形函数，求解域依然用有限元划分，只在局部区域使用局部近似函数 $V_I(\boldsymbol{x})$，并将此类方法称为广义有限元法 (GFEM) 或单位分解有限元法 (PUFEM)。而扩展有限元法 (XFEM) 与此类方法相似，其局部近似函数 $V_I(\boldsymbol{x})$ 采用了更为简单的形式：

$$V_I(\boldsymbol{x}) = u_I + a_I E_I(\boldsymbol{x}) \tag{3-114}$$

式中，$E_I(\boldsymbol{x})$ 为局部加强函数；a_I 为对应的附加节点自由度，一般只对那些裂纹贯穿了其支撑域的节点引入附加自由度。如果裂纹完全贯穿了节点 $\boldsymbol{x}_I$ 的支撑域（其邻接单元的并集），则局部加强函数 $E_I(\boldsymbol{x})$ 取为 Heaviside 阶跃函数，即 $E_I(\boldsymbol{x}) = H_I(\boldsymbol{x})$。如果裂纹只是部分贯穿了节点 $\boldsymbol{x}_I$ 的支撑域，则局部加强函数取为裂纹附近的分支函数，即 $E_I(\boldsymbol{x}) = \beta_I(\boldsymbol{x})$，而 β 函数通常采用裂纹尖端渐近解基函数：

$$\left\{ \sqrt{r}\cos\frac{\theta}{2}, \sqrt{r}\sin\frac{\theta}{2}, \sqrt{r}\sin\frac{\theta}{2}\sin\theta, \sqrt{r}\cos\frac{\theta}{2}\sin\theta \right\}$$

的线性组合。

3.7 自然邻接点插值法

自然邻接点插值 (natural neighbour interpolation, NNI) 法 [38-42] 是一种利用计算几何中的 Voronoi 图来建立近似函数的方法。以二维问题为例，求解区域 Ω 可以用 n 个与节点 $\boldsymbol{x}_J$ 相关联的凸多边形 T_J 离散。多边形 T_J 中的所有点距离节点 $\boldsymbol{x}_J$ 比距其他任何节点更近，如图 3-3 所示。

3.7 自然邻接点插值法

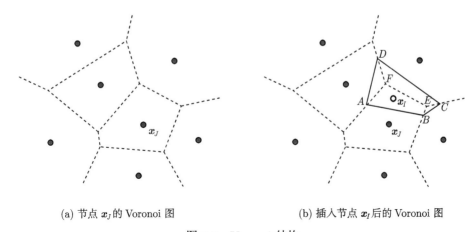

(a) 节点 x_J 的 Voronoi 图　　　　(b) 插入节点 x_I 后的 Voronoi 图

图 3-3　Voronoi 结构

Fig. 3-3　Voronoi structure

在图 3-3(a) 的基础上，插入新的计算节点 x_I，并重新规划生成新的 Voronoi 结构，如图 3-3(b) 所示。其中多边形 $ABCD$ 即为点 x_I 的一阶 Voronoi 多边形，其面积用 $A(\boldsymbol{x})$ 表示。而多边形 $ABEF$ 是点 x_I 的一阶 Voronoi 多边形 $ABCD$ 和 x_J 的 Voronoi 多边形的重叠区域，则多边形 $ABEF$ 通常被称为点 x_I 对应于节点 x_J 的二阶 Voronoi 多边形，其面积用 $A_J(\boldsymbol{x})$ 表示。则计算点 x_I 对应于节点 x_J 的自然邻接形函数可被定义为

$$\phi_J(\boldsymbol{x}) = \frac{A_J(\boldsymbol{x})}{A(\boldsymbol{x})} \tag{3-115}$$

并有

$$\boldsymbol{\Phi}_I(\boldsymbol{x}) = [\phi_J(\boldsymbol{x})]_{J=1}^N = [\phi_1(\boldsymbol{x}), \phi_2(\boldsymbol{x}), \cdots, \phi_N(\boldsymbol{x})] \tag{3-116}$$

式中，N 表示计算节点 x_I 的自然邻接点数。由式 (3-115) 微分，可得形函数的导数，即

$$\phi_{J,i}(\boldsymbol{x}) = \frac{A_{J,i}(\boldsymbol{x}) - \phi_J(\boldsymbol{x}) A_{,i}(\boldsymbol{x})}{A(\boldsymbol{x})} \tag{3-117}$$

由式 (3-115) 得到的 NNI 形函数，也被称之为 Sibson 插值形函数，是 C^0 连续的。为了应对特殊问题中二阶或者高阶导数近似的需要，Farin[43]，Sukumar 和 Moran[44] 对 Sibson 插值进行了改造，使其具有 C^1 连续性。此外，Belikov 等 [41] 发展了非 Sibson 插值方法，使用低一维的测度建立近似函数，计算量更小。而且，非 Sibson 形函数在边界上是严格线性的，可适用于凸边界和凹边界。Sibson 形函数和非 Sibson 形函数均具有 Kronecker δ 函数性质，可以很方便地施加本质边界条件。

3.8 Kriging 插值法

Kriging 插值法 (Kriging interpolation method, KIM) 最初被应用于地理信息系统的构建[45,46]，随着无网格法研究的发展，Gu 等[47-50] 将其应用于无网格形函数的构造。在计算节点 x_I 的局部邻域 Ω_S 上，对 $\forall x \in \Omega_S$，KIM 近似函数有如下定义：

$$u(x) \approx u^h(x) = \boldsymbol{p}^{\mathrm{T}}(x)\boldsymbol{a} + \boldsymbol{Z}(x) \tag{3-118}$$

式中，

$$\boldsymbol{p}^{\mathrm{T}}(x) = \begin{bmatrix} p_1(x) & p_2(x) & \cdots & p_m(x) \end{bmatrix} \tag{3-119}$$

为多项式基函数向量，而

$$\boldsymbol{a} = \begin{bmatrix} a_1 & a_2 & \cdots & a_m \end{bmatrix}^{\mathrm{T}} \tag{3-120}$$

为系数向量。$\boldsymbol{Z}(x)$ 是一个协方差矩阵，并定义如下：

$$\mathrm{Cov}\{\boldsymbol{Z}(x_i), \boldsymbol{Z}(x_j)\} = s^2 \boldsymbol{R}[r(x_i, x_j)] \tag{3-121}$$

式中，s^2 为比例参数；$\boldsymbol{R}$ 是一个关联矩阵；$r(x_i, x_j)$ 是关联函数，表示节点对 x_i 和 x_j 之间的关联性，常用的关联函数采用高斯函数，即

$$\left.\begin{array}{l} r(x_i, x_j) = \mathrm{e}^{-\alpha d_{ij}^2} \\ d_{ij} = \|x_i - x_j\| \end{array}\right\} \tag{3-122}$$

式中，$\alpha > 0$ 表示一个模型参数。

通过最佳线性无偏估计 (best linear unbiased predictor, BLUP)，基于散点集 $\{x_J\}_{J=1}^N$ ($\forall x_J \in \Omega_S$)，可给出计算节点 x_I 的 KIM 近似函数：

$$u_I^h(x) = [\boldsymbol{p}_I^{\mathrm{T}}(x)\boldsymbol{A}(x) + \boldsymbol{r}_I^{\mathrm{T}}(x)\boldsymbol{B}(x)]\hat{\boldsymbol{u}} = \sum_{J=1}^N \phi_J(x)\hat{u}_J = \boldsymbol{\Phi}(x)\hat{\boldsymbol{u}} \tag{3-123}$$

式中，

$$\boldsymbol{A}(x) = \left(\boldsymbol{P}^{\mathrm{T}}\boldsymbol{R}^{-1}\boldsymbol{P}\right)^{-1}\boldsymbol{P}^{\mathrm{T}}\boldsymbol{R}^{-1} \tag{3-124}$$

$$\boldsymbol{B}(x) = \boldsymbol{R}^{-1}(\boldsymbol{I} - \boldsymbol{P}\boldsymbol{A}) \tag{3-125}$$

$$\boldsymbol{p}_I^{\mathrm{T}}(x) = \begin{bmatrix} p_1(x_I) & p_2(x_I) & \cdots & p_m(x_I) \end{bmatrix} \tag{3-126}$$

$$\boldsymbol{r}_I^{\mathrm{T}}(x) = [r(x_I, x_J)]_{J=1}^N = \begin{bmatrix} r(x_I, x_1) & r(x_I, x_2) & \cdots & r(x_I, x_N) \end{bmatrix} \tag{3-127}$$

式中，I 表示 $N \times N$ 的单位矩阵，并对 P, R 有如下定义：

$$P = \left[p_J^{\mathrm{T}}(x); \right]_{J=1}^{N} = \begin{bmatrix} p_1(x_1) & p_2(x_1) & \cdots & p_m(x_1) \\ p_1(x_2) & p_2(x_2) & \cdots & p_m(x_2) \\ \vdots & \vdots & \ddots & \vdots \\ p_1(x_N) & p_2(x_N) & \cdots & p_m(x_N) \end{bmatrix} \tag{3-128}$$

$$R = \begin{bmatrix} 1 & r(x_1, x_2) & \cdots & r(x_1, x_N) \\ r(x_2, x_1) & 1 & \cdots & r(x_2, x_N) \\ \vdots & \vdots & \ddots & \vdots \\ r(x_N, x_1) & r(x_N, x_2) & \cdots & 1 \end{bmatrix} \tag{3-129}$$

显然，KIM 形函数 $\boldsymbol{\Phi}(x)$ 有如下计算格式：

$$\boldsymbol{\Phi}(x) = p_I^{\mathrm{T}}(x) A(x) + r_I^{\mathrm{T}}(x) B(x) \tag{3-130}$$

则其导数运算格式给出如下：

$$\left. \begin{array}{l} \boldsymbol{\Phi}_{,i}(x) = p_{I,i}^{\mathrm{T}}(x) A(x) + r_{I,i}^{\mathrm{T}}(x) B(x) \\ \boldsymbol{\Phi}_{,ij}(x) = p_{I,ij}^{\mathrm{T}}(x) A(x) + r_{I,ij}^{\mathrm{T}}(x) B(x) \end{array} \right\} \tag{3-131}$$

Gu[47] 也将这种基于紧支近似的 KIM 称为移动 Kriging 插值法 (moving Kriging interpolation method, MKIM)，就我们理解，所谓的全局近似，无非是把局部域外的散点的形函数作赋 "0" 值处理，其本质上仍然是紧支的。对任意分布的场函数，在全局域上近似，通常是不可行的。实际上，KIM 近似可以视为点插值 (PIM) 法的一种特例，如果在 KIM 近似中的关联函数或方差函数 $r(x_i, x_j)$，与多项式基及径向基耦合的 PIM 中的径向基函数取为同一形式，则两种近似方法是完全等价的 [48]。因此，PIM 形函数所具备的基本函数性质，如单位分解性、Kronecker δ 函数性等，KIM 近似函数同样具备。

3.9 广义有限差分法

早在 20 世纪 70 年代，在 FEM 得到快速发展时，为了推进经典有限差分法 (FDM) 在不规则域问题中的适应性，于是发展了对任意不规则网格适应的广义有限差分法 (general finite difference method, GFDM)[51-55]。虽然其早期的应用主要面向不规则网格，但 GFDM 本质上也属于散点近似方法的范畴，具有无网格近似的特征。但与前述几种方法不同，它不是对场函数的近似，而是对场函数的微分变量的近似。

在一个伴随目标节点 x_I 定义的局部邻域 Ω_S 上，场变量 $u(x)$ 的近似函数可以取为基于 x_I 的 Taylor 展开式。因我们实际面临的大多数问题为二阶 PDE，所以通常只需取到其二阶展开公式，即

$$u(\boldsymbol{x}) \approx u^h(\boldsymbol{x}) = u_I + h\frac{\partial u_I}{\partial x} + k\frac{\partial u_I}{\partial y} + \frac{h^2}{2}\frac{\partial^2 u_I}{\partial x^2} + \frac{k^2}{2}\frac{\partial^2 u_I}{\partial y^2} + kh\frac{\partial^2 u_I}{\partial x \partial y} + O(\Delta^3) \quad (3\text{-}132)$$

式中，

$$u_I = u(x_I, y_I) \tag{3-133}$$

$$h = x - x_I \tag{3-134}$$

$$k = y - y_I \tag{3-135}$$

$$\Delta = \sqrt{h^2 + k^2} \tag{3-136}$$

将式 (3-132) 应用于计算点 x_I 的局部支撑点集 $\{x_J\}_{J=1}^N$，需注意，此处该点集不包含计算点 x_I，则可得到如下线性方程：

$$\boldsymbol{a} \cdot \boldsymbol{d} = \boldsymbol{f} \tag{3-137}$$

式中，

$$\boldsymbol{a} = \begin{bmatrix} h_1 & k_1 & h_1^2/2 & k_1^2/2 & h_1 k_1 \\ h_2 & k_2 & h_2^2/2 & k_2^2/2 & h_2 k_2 \\ \vdots & \vdots & \vdots & \vdots & \vdots \\ h_J & k_J & h_J^2/2 & k_J^2/2 & h_J k_J \\ \vdots & \vdots & \vdots & \vdots & \vdots \\ h_N & k_N & h_N^2/2 & k_K^2/2 & h_N k_N \end{bmatrix} \tag{3-138}$$

$$\boldsymbol{d} = \begin{bmatrix} \dfrac{\partial u_I}{\partial x}, & \dfrac{\partial u_I}{\partial y}, & \dfrac{\partial^2 u_I}{\partial x^2}, & \dfrac{\partial^2 u_I}{\partial y^2}, & \dfrac{\partial^2 u_I}{\partial x \partial y} \end{bmatrix}^\mathrm{T} \tag{3-139}$$

$$\boldsymbol{f} = [(u_J - u_I);]_{J=1}^N = [(u_1 - u_I), (u_2 - u_I), \cdots, (u_J - u_I), \cdots, (u_N - u_I)]^\mathrm{T} \tag{3-140}$$

则目标点 x_I 的 5 个未知的导数变量可表示为

$$\boldsymbol{d} = \boldsymbol{a}^{-1} \boldsymbol{f} \tag{3-141}$$

这样直接求解, 需要严格的 5 个支撑节点 ($N = 5$)。而且, a 阵很容易是病态的甚至是奇异的。为了保证近似的相容性, 并避免 a 的奇异, 需要增加支撑点数量, 即 $N > 5$, 在此点集 $\{x_J\}_{J=1}^N$ 上构造如下离散二乘泛函:

$$\Pi = \sum_{J=1}^{N} \frac{1}{\Delta_J^3} (a_J d - f_J)^2, \quad N > 5 \tag{3-142}$$

式中,

$$a_J = \begin{bmatrix} h_J & k_J & h_J^2/2 & k_J^2/2 & h_J k_J \end{bmatrix} \tag{3-143}$$

$$f_J = u_J - u_I \tag{3-144}$$

由

$$\frac{\partial \Pi}{\partial d} = 0 \tag{3-145}$$

可得

$$d = A^{-1} B f \tag{3-146}$$

式中,

$$A = \sum_{J=1}^{N} \frac{1}{\Delta_J^3} a_J^{\mathrm{T}} a_J \tag{3-147}$$

$$\begin{cases} B = \begin{bmatrix} \dfrac{a_1^{\mathrm{T}}}{\Delta_1^3} & \dfrac{a_2^{\mathrm{T}}}{\Delta_2^3} & \cdots & \dfrac{a_J^{\mathrm{T}}}{\Delta_J^3} & \cdots & \dfrac{a_N^{\mathrm{T}}}{\Delta_N^3} \end{bmatrix} \\ B_J = \dfrac{a_J^{\mathrm{T}}}{\Delta_J^3} \end{cases} \tag{3-148}$$

可见, GFDM 近似得到的是计算点 x_I 的场量的微分表达式, 即

$$d_i = \sum_{J=1}^{N} \phi_{iJ} \cdot (u_J - u_I) \tag{3-149}$$

$$\phi_{iJ} = (A^{-1} B_J)_i \tag{3-150}$$

数值计算中, 只需将场量 u_I 及其微分近似 d_i 代入离散点对应的控制方程及其边界条件, 便可构造离散系统方程求解。

3.10 本章小结

本章介绍了多种无网格近似函数的构造方法, 这些近似方法大多是由早期数学研究工作中的散点插值方法演化而来的, 比如移动最小二乘 (MLS) 法、点插值

法 (PIM)、径向基点插值法 (RPIM)、自然邻接点插值 (NNI) 法、Kriging 插值法 (KIM) 等。介绍的几类形函数中，大多具有单位分解性，即

$$\sum_{J=1}^{N} \phi_J(\boldsymbol{x}) = 1 \tag{3-151}$$

比如，MLS、RKPM、PIM、RPIM、NNI、KIM 等均具有单位分解性，PUM 在近似函数中虽然包含单位分解函数，但不一定具备单位分解性，而 SPH 是不满足单位分解性条件的。

形函数的 Kronecker δ 函数性质，即

$$\phi_J(x) = \begin{cases} 1, & J = I \\ 0, & J \neq I \end{cases} \tag{3-152}$$

有利于方便地施加本质边界条件。在介绍的几种近似方法中，PIM，RPIM，NNI，KIM 等均具有该函数性质。实际上，是否方便施加本质边界条件，也与具体数值方法的离散形式密切相关，如果系统刚度矩阵是按节点顺序依次组装的，则较容易施加本质边界条件，如 MLPG 法；而如果系统刚度矩阵是整体叠加方式组装的，如 EFG 法，则精确施加本质边界条件就有一定困难。因此，形函数是否具有 Kronecker δ 函数性质，并不能作为是否方便施加本质条件的唯一判断。

虽然每一种近似方法都有其独特的发展脉络和体系，但随着研究的深入，又可发现这些近似方法之间存在密切的关联性。比如，RKPM 是基于 SPH 发展的，但是，RKPM 最终获得的计算形式与 MLS 非常接近，如果 MLS 用核基函数 $\boldsymbol{p}(\boldsymbol{x} - \boldsymbol{x}_I)$ 代替普通基函数 $\boldsymbol{p}(\boldsymbol{x})$，并选取相同的权函数，则 RKPM 形函数与 MLS 形函数完全相同，二者将是等价的[56-59]。再比如，KIM 和 RPIM，虽然二者的发展起源完全不同，但在一定约束条件下，二者也是完全等价的。

评价一种近似方法的优劣和竞争性是多方面的：一是其构造和算法是否简洁，二是对任意分布的散乱节点是否适应，三是对支撑节点的数量是否宽容，四是对任意高阶导数近似是否相容，五是近似计算是否高效，最后，至为重要的一点，近似计算是否精确而稳定。综合考虑这些要求，及其在实际应用中的检验，MLS，SPH，RKPM，RPIM 等这四种近似方法比较有竞争优势，对发展纯粹的无网格法而言，这几种方法也更具有潜力。其中，SPH 在流体力学领域的应用具有一定优势；MLS 和 RKPM 在区域求解方法中凭借其光滑性好、近似精度高等优势而被广泛采用，而 RPIM 是一种"点距"表示的近似函数，很便于在边界解法中表示源点和场点的关系，因此在边界解法中更为流行。近似函数是无网格法研究的基础问题之一，对这些具体的近似方法，进行局部的改进和完善，仍然是后续研究工作需要重点关注的内容。

参 考 文 献

[1] McLain D H. Drawing contours from arbitrary data points[J]. The Computer Journal, 1974, 17: 318-324.

[2] Gordon W J, Wixom J A. Shepard's method of "metric interpolation" to bivariate and multivariate interpolation[J]. Mathematics of Computation, 1978, 32(141): 253-264.

[3] Lancaster P, Salkauskas K. Surfaces generated by moving least squares methods[J]. Math. Comput., 1981, 37(155): 141-158.

[4] Nayroles B, Touzot G, Villon P. Generalizing the finite element method: Diffuse approximation and diffuse elements[J]. Computational Mechanics, 1992, 10(5): 307-318.

[5] Belytschko T, Lu Y Y, Gu L. Element-free Galerkin methods[J]. International Journal for Numerical Methods in Engineering, 1994, 37(2): 229-256.

[6] Belytschko T, Krougauz Y, Organ D, et al. Meshless method: An overview and recent developments[J]. Comput. Methods Appl. Mech. Engrg., 1996, 139: 3-47.

[7] 程玉民. 移动最小二乘法研究进展与述评 [J]. 计算机辅助工程, 2009, 18(2): 5-11.

[8] Gingold R A, Monaghan J J. Smoothed particle hydrodynamics: Theory and application to non-spherical stars[J]. Monthly Notices of The Royal Astronomical Society, 1977, 181(3): 375-389.

[9] Monaghan J J. Smoothed particle hydrodynamics[J]. Annual Review of Astronomy and Astrophysics, 1992, 30(1): 543-574.

[10] Lucy L B. A numerical approach to the testing of the fission hypothesis[J]. The A Stron J, 1977, 8(12): 1013-1024.

[11] Morris J P, Fox P J, Zhu Y. Modeling low Reynolds number incompressible flows using SPH[J]. Journal of Computational Physics, 1997, 136(1): 214-226.

[12] Colagrossi A, Landrini M. Numerical simulation of interfacial flows by smoothed particle hydrodynamics[J]. Journal of Computational Physics, 2003, 191(2): 448-475.

[13] Randles P W, Libersky L D. Smoothed particle hydrodynamics: Some recent improvements and applications[J]. Computer Methods in Applied Mechanics and Engineering, 1996, 139(1-4): 375-408.

[14] Liu M B, Liu G R. Smoothed particle hydrodynamics (SPH): An overview and recent developments[J]. Archives of Computational Methods in Engineering, 2010, 17(1): 25-76.

[15] Liu G R, Liu M B. Smoothed Particle Hydrodynamics: A Meshfree Particle Method[M]. Singapore: World Scientific, 2003.

[16] Liu W K, Jun S, Zhang Y F. Reproducing kernel particle methods[J]. International Journal for Numerical Methods in Fluids, 1995, 20(8-9): 1081-1106.

[17] Liu W K, Jun S, Li S, et al. Reproducing kernel particle methods for structural dynamics[J]. International Journal for Numerical Methods in Engineering, 1995, 38(10):

1655-1679.

[18] Liu W K, Chen Y, Jun S, et al. Overview and applications of the reproducing kernel particle methods[J]. Archives of Computational Methods in Engineering, 1996, 3(1): 3-80.

[19] Liu G R, Gu Y T. A point interpolation method for two-dimensional solids[J]. International Journal for Numerical Methods in Engineering, 2001, 50(4): 937-951.

[20] Liu G R, Gu Y T. An Introduction to Meshfree Methods and Their Programming[M]. Netherlands: Springer Science & Business Media, 2005.

[21] Wu Z, Schaback R. Local error estimates for radial basis function interpolation of scattered data[J]. IMA Journal of Numerical Analysis, 1993, 13(1): 13-27.

[22] Wu Z. Compactly supported positive definite radial functions[J]. Advances in Computational Mathematics, 1995, 4(1): 283-292.

[23] Wendland H. Piecewise polynomial, positive definite and compactly supported radial functions of minimal degree[J]. Advances in Computational Mathematics, 1995, 4(1): 389-396.

[24] Schaback R. Error estimates and condition numbers for radial basis function interpolation[J]. Advances in Computational Mathematics, 1995, 3(3): 251-264.

[25] Rippa S. An algorithm for selecting a good value for the parameter c in radial basis function interpolation[J]. Advances in Computational Mathematics, 1999, 11(2): 193-210.

[26] Liu G R, Gu Y T. A local radial point interpolation method (LRPIM) for free vibration analyses of 2-D solids[J]. Journal of Sound and Vibration, 2001, 246(1): 29-46.

[27] Wang J G, Liu G R. A point interpolation meshless method based on radial basis functions[J]. International Journal for Numerical Methods in Engineering, 2002, 54(11): 1623-1648.

[28] 陈文, 傅卓佳, 魏星. 科学与工程计算中的径向基函数方法 [M]. 北京: 科学出版社, 2014.

[29] Chen W, Fu Z J, Chen C S. Recent Advances in Radial Basis Function Collocation Methods[M]. Heidelberg: Springer, 2014.

[30] 吴宗敏. 散乱数据拟合的模型、方法和理论 [M]. 北京: 科学出版社, 2007.

[31] Wendland H. Error estimates for interpolation by compactly supported radial basis functions of minimal degree[J]. Journal of Approximation Theory, 1998, 93(2): 258-272.

[32] Buhmann M D. Multivariate cardinal interpolation with radial-basis functions[J]. Constructive Approximation, 1990, 6(3): 225-255.

[33] Buhmann M D. Radial functions on compact support[J]. Proceedings of the Edinburgh Mathematical Society (Series 2), 1998, 41(01): 33-46.

[34] Buhmann M D. Radial Basis Functions: Theory and Implementations[M]. Cambridge: Cambridge University Press, 2003.

[35] Duarte C A, Oden J T. An hp adaptive method using clouds[J]. Computer Methods in Applied Mechanics and Engineering, 1996, 139(1-4): 237-262.

[36] Duarte C A, Oden J T. Hp clouds-an hp meshless method[J]. Numerical Methods for Partial Differential Equations, 1996, 12(6): 673-706.

[37] Melenk J M, Babuška I. The partition of unity finite element method: Basic theory and applications[J]. Computer Methods in Applied Mechanics and Engineering, 1996, 139(1-4): 289-314.

[38] Sibson R. A brief description of the natural neighbour interpolation// Barnett V. Interpolating Multivariate Data[M]. Hoboken: Wiley, 1978.

[39] Braun J, Sambridge M A. A numerical method for solving partial differential equations on highly irregular evolving grids[J]. Nature, 1995, 376: 655-660.

[40] Braun J, Sambridge M A, McQueen H. Geophysical parametrization and interpolation of irregular data using natural neighbors[J]. Geophys. J. Int., 1995, 122: 837-857.

[41] Belikov V V, Ivanov V D, Kontorovich V K, et al. The non-Sibsonian interpolation: A new method of interpolation of the values of a function on an arbitrary set of points[J]. Computational Mathematics and Mathematical Physics, 1997, 37(1): 9-15.

[42] Sukumar N, Moran B, Yu Semenov A, et al. Natural neighbour Galerkin methods[J]. International Journal for Numerical Methods in Engineering, 2001, 50(1): 1-27.

[43] Farin G. Surfaces over Dirichlet tessellations[J]. Computer Aided Geometric Design, 1990, 7(1-4): 281-292.

[44] Sukumar N, Moran B. C1 natural neighbor interpolant for partial differential equations[J]. Numerical Methods for Partial Differential Equations, 1999, 15(4): 417-447.

[45] Matheron G. The Theory of Regionalized Variables and Its Applications[M]. France: École National Supérieure Des Mines, 1971.

[46] Oliver M A, Webster R. Kriging: A method of interpolation for geographical information systems[J]. International Journal of Geographical Information System, 1990, 4(3): 313-332.

[47] Gu L. Moving kriging interpolation and element-free Galerkin method[J]. International Journal for Numerical Methods in Engineering, 2003, 56(1): 1-11.

[48] Dai K Y, Liu G R, Lim K M, et al. Comparison between the radial point interpolation and the Kriging interpolation used in meshfree methods[J]. Computational Mechanics, 2003, 32(1-2): 60-70.

[49] Tongsuk P, Kanok-Nukulchai W. Further investigation of element-free Galerkin method using moving Kriging interpolation[J]. International Journal of Computational Methods, 2004, 1(02): 345-365.

[50] Lam K Y, Wang Q X, Li H. A novel meshless approach-Local Kriging (LoKriging) method with two-dimensional structural analysis[J]. Computational Mechanics, 2004, 33(3): 235-244.

[51] Girault V. Theory of a GDM on irregular net works[J]. SIAM J. Num. Anal., 1974, 11: 260-282.

[52] Perrone N, Kao R. A general finite difference method for arbitrary meshes[J]. Computers & Structures, 1975, 5(1): 45-57.

[53] Liszka T, Orkisz J. Finite Difference Method for Arbitrary Irregular Meshes in Nonlinear Problems of Applied Mechanics[M]. San Francisco: IV SMiRt, 1977.

[54] Liszka T, Orkisz J. The finite difference method at arbitrary irregular grids and its application in applied mechanics[J]. Comput. Struct., 1980, 11: 83-95.

[55] Liszka T. An interpolation method for an irregular net of nodes[J]. International Journal for Numerical Methods in Engineering, 1984, 20(9): 1599-1612.

[56] Liu W K, Li S, Belytschko T. Moving least-square reproducing kernel methods (I) methodology and convergence[J]. Computer Methods in Applied Mechanics and Engineering, 1997, 143(1-2): 113-154.

[57] Liu W K, Chen Y. Wavelet and multiple scale reproducing kernel methods[J]. International Journal for Numerical Methods in Fluids, 1995, 21(10): 901-931.

[58] Li S, Hao W, Liu W K. Numerical simulations of large deformation of thin shell structures using meshfree methods[J]. Computational Mechanics, 2000, 25(2-3): 102-116.

[59] Liu W K, Chen Y, Uras R A, et al. Generalized multiple scale reproducing kernel particle methods[J]. Computer Methods in Applied Mechanics and Engineering, 1996, 139(1-4): 91-157.

第 4 章 MLS 稳定性及其导数近似

移动最小二乘 (moving least square, MLS) 法是无网格法中使用最广泛的形函数构造方法之一，比如著名的无单元伽辽金 (element-free Galerkin, EFG) 法 [1,2]、无网格局部彼得罗夫–伽辽金 (meshless local Petrov-Galerkin, MLPG) 法 [3] 等均将 MLS 选为首要的形函数构造法。MLS 能得到广泛应用，有其固有的诸多优势 [4]。然而，经典的 MLS 在保证其近似精确性，或提高其近似稳定性方面，仍然有诸多改进的空间，目前已经有不少新的研究成果。本章将对这方面的内容进行专门的介绍和讨论。

4.1 MLS 的构造思想

为简单起见，以一维问题为例来进行阐述，如图 4-1 所示。在求解域 Ω 上，分布的场函数 $u(x)$ 可能足够复杂，通常不能用简单函数表达，或者无法获得其函数表达形式。但在一个局部的区域 Ω_S 内，只要这个局部域足够小，由其截取的局部场函数 $u(x)$ 将会是一个足够简单的形状，这部分截取的场函数 $u(x)$ 可能被一个初等函数 $u^h(x)$ 近似地描述。因为多项式函数有易于计算机表达的优点，所以这个近似函数 $u^h(x)$ 通常选为一个完备的二次多项式函数，即

$$u(x) \approx u^h(x) = a_1 + a_2 x + a_3 x^2, \quad \forall x \in \Omega_S \tag{4-1}$$

这个式子扩展到多维的情况，并用向量形式表示，即 3.2 节的式 (3-6)，

$$u(\boldsymbol{x}) \approx u^h(\boldsymbol{x}) = \sum_{t=1}^{m} p_t(\boldsymbol{x}) a_t(\boldsymbol{x}) = \boldsymbol{p}^{\mathrm{T}}(\boldsymbol{x}) \boldsymbol{a}(\boldsymbol{x}) \tag{4-2}$$

其中，由式 (4-1) 可知，$\boldsymbol{a}(\boldsymbol{x})$ 本质上表征常数系数向量，不管其表达形式多么复杂，仍然代表的是多项式中的常系数。因为最终获得的 $\boldsymbol{a}(\boldsymbol{x})$ 是用函数形式表示的，因此被写成函数向量的形式。为了获得合理的 $\boldsymbol{a}(\boldsymbol{x})$，MLS 的关键一步是构造加权最小二乘泛函，即

$$J(\boldsymbol{a}) = \sum_{J=1}^{N} w_J(\boldsymbol{x}) \left[\boldsymbol{p}_J^{\mathrm{T}}(\boldsymbol{x}) \boldsymbol{a}(\boldsymbol{x}) - \hat{u}_J\right]^2 \tag{4-3}$$

式中，额外引入一个权函数 $w(\boldsymbol{x})$，这个权函数也不是近似函数 $u^h(\boldsymbol{x})$ 所固有的，但它对 MLS 发挥着重要作用。通过对上式寻求关于 $\boldsymbol{a}(\boldsymbol{x})$ 的驻值，即令 $\partial J/\partial \boldsymbol{a} = 0$，则可解得 $\boldsymbol{a}(\boldsymbol{x})$ 的表达式，有关内容参见 3.2 节。

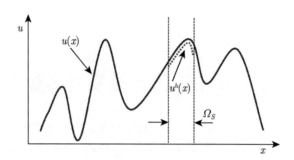

图 4-1 MLS 近似图解

Fig. 4-1 Schematics of MLS approximation

MLS 形函数具有单位分解性, 即

$$\sum_{J=1}^{N}\phi_J(\boldsymbol{x})=1 \tag{4-4}$$

此外, MLS 形函数还具有所谓的 "再生性", 即基函数中的任何函数均可由 MLS 重构, 该性质被定义为

$$\sum_{J=1}^{N}\phi_J(x)p_t(\boldsymbol{x}_I)=p_t(\boldsymbol{x}),\quad t=1,2,\cdots,m \tag{4-5}$$

显然, 具有再生性的形函数, 一定具有单位分解性。利用 MLS 的再生性, 对于一些特殊问题, 如裂纹扩展问题, 基函数可以取为

$$\boldsymbol{p}^{\mathrm{T}}(\boldsymbol{x})=\left[1,x,y,\sqrt{r}\cos\frac{\theta}{2},\sqrt{r}\sin\frac{\theta}{2},\sqrt{r}\sin\frac{\theta}{2}\sin\theta,\sqrt{r}\cos\frac{\theta}{2}\sin\theta\right] \tag{4-6}$$

Liu 等[5] 在多尺度分析中, 将基函数取为

$$\boldsymbol{p}^{\mathrm{T}}(\boldsymbol{x})=[1,\cos(kx),\sin(kx)] \tag{4-7}$$

$$\begin{aligned}\boldsymbol{p}^{\mathrm{T}}(\boldsymbol{x})=[1,&\cos(kx\cos\theta+ky\sin\theta),\sin(kx\cos\theta+ky\sin\theta)\\&\times\cos(-kx\sin\theta+ky\cos\theta),\sin(-kx\sin\theta+ky\cos\theta)]\end{aligned} \tag{4-8}$$

总而言之, MLS 能得到广泛应用, 其优势主要表现在以下几个方面。

其一, MLS 的构造形式非常简单。其构造近似函数的基本元素, 本质上仅是一个多项式函数。可以说在诸多形函数构造方法中, MLS 是最为简单的方法之一。而越是简单的方法, 在同等有效的条件下, 越容易得到推广应用。

其二, MLS 有计算稳定、精确和相容性好的特点。MLS 采用了 "最小二乘" 的最优逼近, 赋予其精确计算的自洽能力, 因此计算精度能够保证。此外, MLS 的支

撑节点数量 N 是大于多项式基向量的项数 m 的，即 $N > m$，这种最优逼近对支撑节点的数量和分布是比较宽容的，因此近似计算具有良好的稳定性和相容性。

其三，MLS 具有在全局域上近似的相容性。MLS 引入的权函数，实际上是对近似精度在局部域上的一种协调，从而提高了近似的协调性，如果强制要求近似函数能够在所有支撑点上精确近似，一旦有限近似能力的近似函数达不到要求，将会产生严重的奇异性。因此，权函数的使用，一方面保证了就近精确近似，在计算点上将得到最优的近似值；另一方面，当局部支撑域 Ω_S 伴随计算节点移动时，对应的支撑节点的进、出是光滑平顺的，故可保证 MLS 形函数在整个求解域上的相容性。

其四，MLS 具有灵活的近似能力。利用 MLS 的再生性，可以灵活采用多种不同形式的基函数，以适应特定问题求解的需要。

其五，MLS 是高阶连续的。MLS 的高阶连续性主要依赖于多项式基函数的阶数，或者是其他形式基函数的连续性。MLS 的高阶连续性，很容易适应高阶导数近似的需要。

4.2 MLS 的权函数

权函数 $w(\boldsymbol{x})$ 对 MLS 形函数发挥着重要作用，而且其光滑性，也影响着 MLS 形函数的连续性。为便于讨论，将 $w(\boldsymbol{x})$ 改写为以 r 为自变量的形式，即 $w(\boldsymbol{x}) = w(r)$，其中自变量 r 的取值区间为 $r \in [0, 1]$，权函数 $w(r)$ 需具备如下性质：

$$\begin{cases} w(r) > 0, & r < 1 \\ w(r) = 0, & r \geqslant 1 \end{cases} \tag{4-9}$$

$$\{r : 0 \to 1\} \Rightarrow \{w(r) : \searrow\} \tag{4-10}$$

上式表示 $w(r)$ 在 r 的 $0 \to 1$ 区间上是单调递减的。可供选择的权函数有多种，例如，

高斯函数 (W1-Gau)：

$$w(r) = \begin{cases} \mathrm{e}^{-(\alpha \cdot r)^{2k}}, & r \leqslant 1, k \geqslant 1, \alpha > 1 \\ 0, & r > 1 \end{cases} \tag{4-11}$$

锥形函数 (W2-Con)：

$$w(r) = \begin{cases} (1 - r^2)^{2k}, & r \leqslant 1, k \geqslant 1 \\ 0, & r > 1 \end{cases} \tag{4-12}$$

指数函数 (W3-Exp)：

$$w(r) = \begin{cases} \dfrac{e^{-(\alpha \cdot r)^{2k}} - e^{-\alpha^{2k}}}{1 - e^{-\alpha^{2k}}}, & r \leqslant 1, k \geqslant 1, \alpha > 1 \\ 0, & r > 1 \end{cases} \quad (4\text{-}13)$$

样条函数 (W4-Spl):

$$w(r) = \begin{cases} 1 - 6r^2 + 8r^3 - 3r^4, & r \leqslant 1 \\ 0, & r > 1 \end{cases} \quad (4\text{-}14)$$

此外, 还有其他类型的样条函数:

$$w(r) = \begin{cases} 1 - 10r^3 + 15r^4 - 6r^5, & r \leqslant 1 \\ 0, & r > 1 \end{cases} \quad (4\text{-}15)$$

$$w(r) = \begin{cases} 1 - 3r^2 + 2r^3, & r \leqslant 1 \\ 0, & r > 1 \end{cases} \quad (4\text{-}16)$$

$$w(r) = \begin{cases} 2/3 - 4r^2 + 4r^3, & r \leqslant 1/2 \\ 4/3 - 4r + 4r^2 + 4r^3/3, & 1/2 < r \leqslant 1 \\ 0, & r > 1 \end{cases} \quad (4\text{-}17)$$

$$w(r) = \begin{cases} 1 - 2r^2, & r \leqslant 1/2 \\ 2(1-r)^2, & 1/2 < r \leqslant 1 \\ 0, & r > 1 \end{cases} \quad (4\text{-}18)$$

在这些给出的权函数中, 建议优先选择式 (4-11) 给出的高斯型函数 (W1-Gau), 理由有两个方面, 一是其形式比较简单, 二是其具有足够的光滑性。其参数一般选为 $k=1$, 而参数 α 为对其形状的影响, 如图 4-2 所示, 其取值建议为 $\alpha = [2.0, 2.5]$。

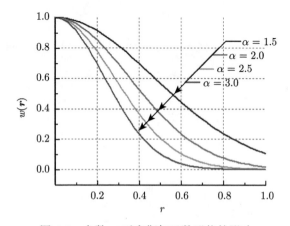

图 4-2 参数 α 对高斯权函数形状的影响

Fig. 4-2 The influence of α on Gauss weight function

4.2 MLS 的权函数

为简单起见，局部支撑域的形状较多采用圆形和矩形两种形式，如图 4-3 所示。以 r 为自变量的权函数，是使用圆形支撑域的，其定义如下：

$$r = \frac{\|\boldsymbol{x} - \boldsymbol{x}_I\|}{r_s} \tag{4-19}$$

表示任意点 $\boldsymbol{x}$ 到计算点 $\boldsymbol{x}_I$ 的"单位径向距离"，式中 r_s 表示圆形支撑域的半径。为了保证计算精度，r_s 的尺度需要合理设定。如果场节点是等距离规则离散的，节点距离用 h 表示，则 r_s 可由下式确定：

$$r_s = \alpha_s h \tag{4-20}$$

式中，$\alpha_s > 1$ 是一个关于 h 的放大系数，是数值方法中一个重要的计算参数。如果场节点是不规则随机离散的，则 r_s 可通过选取合适数量的邻近支撑点，并将距离计算点 $\boldsymbol{x}_I$ 最远的节点距离设定为 r_s，有关内容将在 4.3 节重点讨论。此外，对不规则离散的场节点，平均节点间距 $\bar{h}$ 可由局部支撑域的 Lebesgue 测度 $L(\Omega_S)$，支撑节点数 N 和空间维度数 D 确定：

$$\bar{h} = \frac{\{L(\Omega_S)\}^{\frac{1}{D}}}{N^{\frac{1}{D}} - 1} \tag{4-21}$$

对于一维问题，$L(\Omega_S)$ 表示局域的长度；二维问题，$L(\Omega_S)$ 表示局域的面积；三维问题，$L(\Omega_S)$ 表示局域的体积。当然，利用上式，用全求解域的 Lebesgue 测度和全域场节点数，也可求得全域上的平均节点间距。

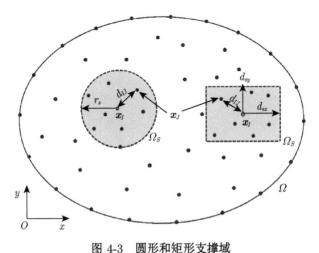

图 4-3 圆形和矩形支撑域

Fig. 4-3 Circle and rectangle support domain

对于矩形支撑域，任意点 x 到计算点 x_I 的距离需要在不同维度上进行分解，比如二维问题，式 (4-19) 对应写为

$$\begin{cases} r_x = \dfrac{|x - x_I|}{d_{sx}} \\ r_y = \dfrac{|y - y_I|}{d_{sy}} \end{cases} \quad (4\text{-}22)$$

式中，d_{sx} 和 d_{sy} 分别表示矩形域在 x 和 y 方向的半尺度，则权函数 $w(x)$ 的表达式对应写为

$$w(\boldsymbol{x}) = w_x(\boldsymbol{x}) \cdot w_y(\boldsymbol{x}) = w(r_x) \cdot w(r_y) \quad (4\text{-}23)$$

在近似计算中，有时可能用到权函数的导数形式，由复合函数求导法很容易获得，即

$$\begin{cases} w_{,i}(\boldsymbol{x}) = w_{,i}(r) = w_{,r}(r) \cdot r_{,i} \\ w_{,ij}(\boldsymbol{x}) = w_{,ij}(r) = w_{,rr}(r) \cdot r_{,i} \cdot r_{,j} + w_{,r}(r) \cdot r_{,ij} \end{cases} \quad (4\text{-}24)$$

或

$$\begin{cases} w_{,i}(\boldsymbol{x}) = \{w(r_i) \cdot w(r_j)\}_{,i} = w(r_j) \cdot w_{,r_i}(r_i) \cdot r_{i,i} \\ w_{,ii}(\boldsymbol{x}) = \{w(r_i) \cdot w(r_j)\}_{,ii} = w(r_j) \cdot w_{,r_i r_i}(r_i) \cdot r_{i,i} \cdot r_{i,i} \\ w_{,ij}(\boldsymbol{x}) = \{w(r_i) \cdot w(r_j)\}_{,ij} = \{w(r_j) \cdot w_{,r_i}(r_i) \cdot r_{i,i}\} \cdot \{w(r_i) \cdot w_{,r_j}(r_j) \cdot r_{j,j}\} \end{cases} \quad (4\text{-}25)$$

式 (4-24) 是圆域权函数的导数计算公式，而式 (4-25) 是矩形域权函数的导数计算公式。实际上，权函数的导数计算，也增加了 MLS 近似计算的复杂性，对于其必要性将在后面予以进一步讨论。由式 (4-11)～ 式 (4-14) 给出的四种权函数及其导数图像如图 4-4 所示。有关参数取值为 $\alpha=2.5$, $k=1$。

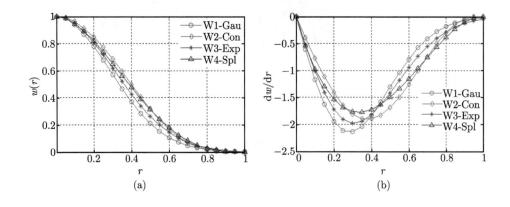

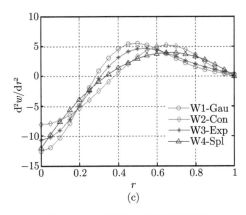

图 4-4 权函数及其导数的图像

Fig. 4-4 The functional image of weight functions and their derivatives

对局部支撑域 Ω_S 的形状选择，采用圆形和矩形各有其利弊。使用圆形支撑域的好处是计算形式比较简单，更适用于随机散乱节点分布的情况，尤其是对维度方向离散尺度无偏的情况。对于多维问题域中，如果不同维度方向上的节点离散是尺度明显不同的，尤其是规则离散的情况，则使用矩形支撑域有一定优势。对于不同维度方向上的节点离散尺度不均衡，而且是随机分布的情况，如果要使用圆形支撑域，在确定某一支撑点的径向单位距离 r 时，需要特别进行各维度离散尺度均一化处理。

4.3 MLS 稳定近似的几何条件

式 (4-20) 表示的支撑域尺度 r_s 是一个非常重要的概念，它的合理取值与 MLS 近似精确性及数值方法的计算精度密切相关，实际上也与选择合适的支撑节点数 N 有关[6]。此外，MLS 需要对近似函数中的矩阵 $\bm{A}$ 进行求逆运算，奇异的矩阵将导致计算崩塌，病态的矩阵将导致近似精度不能满足要求。然而，如何保证 MLS 近似稳定性，还缺乏足够简洁有效的判别方法。对此问题，Liu 和 Gu[7] 明确指出："为确保 $\bm{A}^{-1}$ 存在及条件良好的 $\bm{A}$，我们通常使 $N \gg m$(支撑点数远大于基向量项数)；遗憾的是 N 没有理论最佳值，只能通过数值试验确定。" 实际上，$N \gg m$ 条件不仅缺乏实际指导意义，而且其取值过大会导致支撑域尺度太大而降低近似能力，并且在数值方法中会影响计算效率。因此，MLS 的近似稳定性问题有必要进一步研究，经验性的应用条件需要更严格的理论规则来限定。

近似误差估计对域内任意支撑点及目标点均有意义。为了便于讨论 MLS 近似的影响因素，我们需要先在 $\bm{A}$ 非奇异及良态的条件下给出一个近似误差估计。MLS

需保证目标点 $\boldsymbol{x}_I$ 上得到最精确近似，因此给出一个假设条件：

$$\left|\left[u_I^h(\boldsymbol{x})\right]^{(d)} - \left[u_I(\boldsymbol{x})\right]^{(d)}\right|_{d=0}^k \to 0 \tag{4-26}$$

式中，上标符号 (d) 表示 d 阶导数。根据 Taylor 定理，满足该假设的近似函数 $u^h(\boldsymbol{x})$ 必然也是 k 项 Taylor 展开式的近似，因此近似函数在 Ω_S 上有如下误差估计：

$$R(u^h) = |u^h(\boldsymbol{x}) - u(\boldsymbol{x})| \cong \frac{|u^{(k+1)}(\xi)|}{(k+1)!} \|\boldsymbol{x} - \boldsymbol{x}_I\|^{k+1}, \quad \forall \boldsymbol{x}, \xi \in \Omega_S \tag{4-27}$$

令

$$\begin{cases} M = \max\left\{u^{(k+1)}(\xi), \forall \xi \in \Omega_S\right\} \\ d_{\max} = \max\{\|\boldsymbol{x} - \boldsymbol{x}_I\|, \forall \boldsymbol{x} \in \{\boldsymbol{x}_J\}_{J=1}^n\} \end{cases} \tag{4-28}$$

则有

$$R(u^h) \leqslant \frac{M}{(k+1)!} d_{\max}^{k+1} \leqslant \frac{M}{(k+1)!} r_s^{k+1}, \quad d_{\max} \leqslant r_s \tag{4-29}$$

故可推论为：

1) MLS 近似误差取决于标量函数 $u(\boldsymbol{x})$ 的光滑性，$u(\boldsymbol{x})$ 越光滑，$u^{(k+1)}(\xi)$ 的振荡才越小，M 才可能取到一个较小的值，近似误差 $R(u^h)$ 才越容易控制。

2) MLS 近似误差取决于局部支撑域 Ω_S 的半径尺度 r_s，其值越小，误差 $R(u^h)$ 也相应越小。在数值方法中，该要求称为近似函数的紧支性。

3) MLS 近似误差还取决于多项式基的次数 k，较高的 k 更有利于控制近似误差。通常要求 k 不低于待解微分方程的求导阶数，在数值方法中，该要求称为近似函数的 C^k 一致性。

4) 当 $r_s \leqslant 1$ 时，误差 $R(u^h)$ 是关于完备多项式基函数次数 k 的严格单减函数，即近似误差随 k 的增加而减小；当 $r_s > 1$ 时，误差 $R(u^h)$ 可能是关于 k 的增函数，即近似误差可能随 k 的增加而增大，这一点在数值实施中应特别注意，应通过一些技术手段使 $r_s \leqslant 1$ 条件成立。

以上关于近似误差的一般讨论，是以近似函数可合理近似为基础的。而实际应用中，由于近似函数是通过支撑点集 $\{\boldsymbol{x}_J\}_{J=1}^N$ 来构造的，不合理的点集选择方案可能构造出不合理的近似函数，即近似函数本身的稳定性问题。

由式 (3-16)，求解 $\boldsymbol{a}(\boldsymbol{x})$ 需要保证 $\boldsymbol{A}^{-1}$ 存在，即 $\boldsymbol{A}$ 非奇异。因此，若要保证 MLS 的稳定近似，我们首先需要从理论上确知 $\boldsymbol{A}^{-1}$ 存在的基本条件。数值方法中问题域不会超过 3 个维度，因此假定 MLS 的问题域 Ω 包含于常规欧几里得空间 S，即 $\Omega \subset S$。我们将讨论 $\boldsymbol{A}$ 非奇异时支撑点集 $\{\boldsymbol{x}_J\}_{J=1}^N$ 的几何特性。首先，将 $\boldsymbol{A}$ 写为矩阵表示形式：

$$\boldsymbol{A} = \boldsymbol{P}^\mathrm{T} \boldsymbol{W} \boldsymbol{P} \tag{4-30}$$

4.3 MLS 稳定近似的几何条件

其中，

$$\boldsymbol{W} = \mathrm{diag}\{w_1(\boldsymbol{x}), w_2(\boldsymbol{x}), \cdots, w_N(\boldsymbol{x})\}$$

$$= \begin{bmatrix} w_1(\boldsymbol{x}) & & & 0 \\ & w_2(\boldsymbol{x}) & & \\ & & \ddots & \\ 0 & & & w_N(\boldsymbol{x}) \end{bmatrix} \quad (4\text{-}31)$$

式中，$w_J(\boldsymbol{x}) = w(\boldsymbol{x}_J - \boldsymbol{x}_I)$，及

$$\begin{aligned}
\boldsymbol{P} &= \left[\boldsymbol{p}_J^\mathrm{T}(\boldsymbol{x});\right]_{J=1}^N \\
&= \left[\boldsymbol{p}_1^\mathrm{T}(\boldsymbol{x}), \boldsymbol{p}_2^\mathrm{T}(\boldsymbol{x}), \cdots, \boldsymbol{p}_N^\mathrm{T}(\boldsymbol{x})\right]^\mathrm{T} \\
&= \left[\boldsymbol{p}_{\langle 1 \rangle}(\boldsymbol{x}), \boldsymbol{p}_{\langle 2 \rangle}(\boldsymbol{x}), \cdots, \boldsymbol{p}_{\langle m \rangle}(\boldsymbol{x})\right] \\
&= \begin{bmatrix} p_1(\boldsymbol{x}_1) & p_2(\boldsymbol{x}_1) & \cdots & p_m(\boldsymbol{x}_1) \\ p_1(\boldsymbol{x}_2) & p_2(\boldsymbol{x}_2) & \cdots & p_m(\boldsymbol{x}_2) \\ \vdots & \vdots & \ddots & \vdots \\ p_1(\boldsymbol{x}_N) & p_2(\boldsymbol{x}_N) & \cdots & p_m(\boldsymbol{x}_N) \end{bmatrix}
\end{aligned} \quad (4\text{-}32)$$

式中，

$$\boldsymbol{p}_{\langle t \rangle}(\boldsymbol{x}) = \left[p_t(\boldsymbol{x}_1), p_t(\boldsymbol{x}_2), \cdots, p_t(\boldsymbol{x}_N)\right]^\mathrm{T} \quad (4\text{-}33)$$

表示基函数的第 t 项按 N 个节点组成的列向量。因此，$\boldsymbol{P} \in \boldsymbol{R}^{N \times m}$，又可写成列向量组的形式 $\boldsymbol{P} : \left(\boldsymbol{p}_{\langle 1 \rangle}, \boldsymbol{p}_{\langle 2 \rangle}, \cdots, \boldsymbol{p}_{\langle m \rangle}\right)$。我们可以给出以下 3 个 $\boldsymbol{A}^{-1}$ 存在的判定定理。

定理 1 基函数向量组 $\boldsymbol{P} : \left(\boldsymbol{p}_{\langle 1 \rangle}, \boldsymbol{p}_{\langle 2 \rangle}, \cdots, \boldsymbol{p}_{\langle m \rangle}\right)$ 线性无关是 $\boldsymbol{A}^{-1}$ 存在的充分必要条件。

证 先证必要条件。若 $\boldsymbol{A}^{-1}$ 存在，则 $\mathrm{rank}(\boldsymbol{A}) = m$，并有下列递推关系：

$$m = \mathrm{rank}(\boldsymbol{A}) = \mathrm{rank}\left(\boldsymbol{P}^\mathrm{T} \boldsymbol{W} \boldsymbol{P}\right) \leqslant \mathrm{rank}(\boldsymbol{P}) \quad (4\text{-}34)$$

当且仅当 $\mathrm{rank}(\boldsymbol{P}) = m$ 及 $N \geqslant m$ 时，以上关系式恒成立。所以，$\boldsymbol{P}$ 组线性无关。

再证充分条件。当 $\boldsymbol{P}$ 组线性无关时，有 $\mathrm{rank}(\boldsymbol{P}) = m$ 及 $N \geqslant m$。据已知权函数在局部支撑域上的定义条件 $\forall w_J \neq 0$，及 $\boldsymbol{W} = \mathrm{diag}\{w_1(\boldsymbol{x}), w_2(\boldsymbol{x}), \cdots, w_N(\boldsymbol{x})\}$，可得

$$m = \mathrm{rank}(\boldsymbol{P}) = \mathrm{rank}\left(\boldsymbol{P}^\mathrm{T} \boldsymbol{W} \boldsymbol{P}\right) = \mathrm{rank}(\boldsymbol{A}) \quad (4\text{-}35)$$

所以，$\mathrm{rank}(\boldsymbol{A}) = m$，则 $\boldsymbol{A}^{-1}$ 必存在。证毕。

推论 1.1 对于完备的 k 次多项式基向量 $\boldsymbol{p}^{\mathrm{T}}(\boldsymbol{x})$，当二维空间上的支撑点集 $\{\boldsymbol{x}_J\}_{J=1}^{N}$ 能被一条不高于 k 次的曲线严格串接时，$\boldsymbol{A}^{-1}$ 不存在[8,9]。

释 $\boldsymbol{P}$ 组线性无关，则由 $\sum\limits_{i=1}^{m}\lambda_i\boldsymbol{p}_{\langle i\rangle}=0$ 能推得 $\forall\lambda_i=0, (i=1,2,\cdots,m)$。我们可以给出一个与 $\boldsymbol{P}$ 向量组在二维空间上相容的一个函数：

$$y(x) = \beta_0 + \beta_1 x + \beta_2 x^2 + \cdots + \beta_k x^k, \quad \forall \boldsymbol{x}: (x,y) \in \Omega_S \tag{4-36}$$

当 $\beta_0,\beta_1,\cdots,\beta_k$ 不全为 0 时，$\boldsymbol{P}$ 组线性相关。而给出的这个函数 (式中 x,y 可互换) 在 (x,y) 表示的二维空间上是一条不高于 k 次的曲线。该推论的进一步扩展：

1) 在三维空间上，支撑点集能被一个不高于 k 次的曲面严格串接时，$\boldsymbol{P}$ 组也必然线性相关。

2) 二维空间上的支撑点集逼近一不高于 k 次的曲线，三维空间上的支撑点集逼近一不高于 k 次的曲面，$\boldsymbol{A}$ 将是病态的。释毕。

给出的定理 1 是 $\boldsymbol{A}^{-1}$ 存在的完备条件，遗憾的是我们很难给出其完备的几何意义。推论 1 仅是一个特殊情况而已，而且实际上也很难直观判断。因此，$\boldsymbol{A}^{-1}$ 的存在性很难利用该定理进行判别，还需要寻求更简洁的办法。

定理 2 (第一几何条件) $N \geqslant m$ 是 $\boldsymbol{A}^{-1}$ 存在的一个必要条件。

证 若 $\boldsymbol{A}^{-1}$ 存在，则 $\mathrm{rank}(\boldsymbol{A}) = m$，据如下递推关系：

$$m = \mathrm{rank}(\boldsymbol{A}) = \mathrm{rank}\left(\boldsymbol{P}^{\mathrm{T}}\boldsymbol{W}\boldsymbol{P}\right) \leqslant \mathrm{rank}(\boldsymbol{W}) = N \tag{4-37}$$

所以 $N \geqslant m$ 必成立。证毕。

该定理是一个很显然的结论，是对支撑点集 $\{\boldsymbol{x}_J\}_{J=1}^{N}$ 在数量上的几何要求。基于经验的判断，该条件很难保证 $\boldsymbol{A}^{-1}$ 存在，我们还需要更有效的几何条件。

先给出一个定义：

定义 1 设 v 是 S 的任意一个维度，如果 N 个支撑点组成的点集 $\{\boldsymbol{x}_J\}_{J=1}^{N}$ 对应于 v 的度规 (坐标轴) 上有 g 个互不相等的值，则定义 g 为此 N 个点在 v 上的几何投影数，记为 $T(N)_v = g$。

定理 3 (第二几何条件) 对于完备的 k 次多项式基向量 $\boldsymbol{p}^{\mathrm{T}}(\boldsymbol{x})$，$T(N)_v \geqslant k+1$ 是 $\boldsymbol{A}^{-1}$ 存在的必要条件[6]。

证 由完备的 k 次多项式基向量 $\boldsymbol{p}^{\mathrm{T}}(\boldsymbol{x})$ 定义知，如果 v 表示 $\boldsymbol{x}$ 的任意一个维度，则向量

$$\left[\boldsymbol{p}^{\mathrm{T}}(\boldsymbol{x})\right]^v = [1, v, v^2, \cdots, v^k] \tag{4-38}$$

是一个与 v 之外其他维度均无关的子向量。我们用 $\boldsymbol{p}_{\langle t\rangle}^{v}$ 表示 $\boldsymbol{P}$ 组中与该子向量中第 v^t 项对应的列向量，则得到一个子向量组：

$$\boldsymbol{P}^{v}:\left(\boldsymbol{p}_{\langle 0\rangle}^{v},\boldsymbol{p}_{\langle 1\rangle}^{v},\boldsymbol{p}_{\langle 2\rangle}^{v},\cdots,\boldsymbol{p}_{\langle k\rangle}^{v}\right) \subseteq \boldsymbol{P}:\left(\boldsymbol{p}_{\langle 1\rangle},\boldsymbol{p}_{\langle 2\rangle},\cdots,\boldsymbol{p}_{\langle m\rangle}\right) \tag{4-39}$$

4.3 MLS 稳定近似的几何条件

显然, $\boldsymbol{P}^\nu \in \boldsymbol{R}^{N \times (k+1)}$。据定理 1, $\boldsymbol{A}^{-1}$ 存在时, $\boldsymbol{P}$ 组线性无关, 则其子组 $\boldsymbol{P}^\nu$ 也必然线性无关。由定义 1, 下式必然成立:

$$T(N)_v \geqslant \operatorname{rank}(\boldsymbol{P}^\nu) = k+1 \tag{4-40}$$

证毕。

定理 3 给出了支撑点集 $\{\boldsymbol{x}_J\}_{J=1}^N$ 在空间分布上的几何要求, 是一个非常有效的结论。由式 (4-29) 给出一个重要的推断, 即设置尽可能小的支撑域半径尺度 r_s 是 MLS 精确近似的一个要求, 同时也是计算效率上的要求。显然需尽可能满足条件:

$$T(N)_v = k+1 \tag{4-41}$$

根据该条件, 如果场节点是规则分布的, Ω_S 半径尺度 $r_s = \alpha_s h$ 的尺度参数 α_s 可以给出理论上的建议取值范围:

$$\alpha_s \in \begin{cases} (k, k+1), & \partial \Omega_S \cap \partial \Omega \neq 0 \\ (\operatorname{ceil}(k/2), k+1), & \partial \Omega_S \cap \partial \Omega = 0 \end{cases} \tag{4-42}$$

式中, ceil($\cdot$) 表示 "不小于", 取整数运算。

如果场节点在各空间维度上是等尺度离散的, 或者均匀随机分布的, 直接按支撑点数就近选择支撑点来设置支撑域尺度是一个更有效的方法。则基于统计意义的支撑点数 N 的建议取值范围是 (以维度数 D 为幂指数)

$$(k+1)^D \leqslant N \leqslant (k+2)^D \tag{4-43}$$

由式 (3-16) 求解 $\boldsymbol{a}(x)$ 时, 如果 $\operatorname{cond}(\boldsymbol{A}) = \|\boldsymbol{A}\| \times \|\boldsymbol{A}^{-1}\|$ 是一个大数时, 极可能得到一个不合理的解答, 甚至严重影响近似精度。因此, 对 MLS 近似的精确性而言, 仅保证 $\boldsymbol{A}$ 非奇异是不够的, 还需保证其有一个合理的条件数。所以我们需要从理论上分析 $\boldsymbol{A}$ 的条件数变差的情况, 以便在实际应用中尽量规避, 从而有效保证 $\boldsymbol{A}$ 的良态性。

根据定理 3, 在 $T(N)_v = k+1$ 的前设条件下, 因为该条件是一个合理, 且需要尽可能保证的条件, 能得出如下推论。

推论 3.1 设 v_i 是 $k+1$ 个投影中的任意一个投影, 当 v_i 逼近支撑域 Ω_S 的边界时, 即 $v_i \to \partial \Omega_S$, $\boldsymbol{A}$ 必然是病态的。解释为

$$\lim_{v_i \to \partial \Omega_S} (T(N)_v = k+1) \to (T(N)_v = k) \tag{4-44}$$

推论 3.2 设 v_i 和 v_{i+1} 是 $k+1$ 个投影中任意两个相邻的投影, 当 v_i 和 v_{i+1} 相互无限逼近时, 即 $|v_i - v_{i+1}| \to 0$, $\boldsymbol{A}$ 必然是病态的。解释为

$$\lim_{\delta \to 0} (T(N)_v = k+1) \to (T(N)_v = k), \quad \delta = |v_i - v_{i+1}| \tag{4-45}$$

推论 3.3 当支撑域 Ω_S 的尺度无限缩小时，A 必然是病态的。解释为

$$\lim_{r(\Omega_S)\to 0}(T(N)_v = k+1) \to (T(N)_v = 1) \tag{4-46}$$

推论 3.4 在恒定的支撑域尺度上，当使用更高次的 k 时，A 更容易产生病态。

在恒定的支撑域尺度上，更大的 k 意味着投影密度增大，$k+1$ 个投影中任意两个相邻的投影 v_i 和 v_{i+1} 必然是趋近的，由推论 3.2 可知，A 容易产生病态。

算例 小片试验。为了验证本书给出的几何条件判定方法，使用 2×2 小片模型进行检验，如图 4-5 所示。近似计算中我们采用 2 次完备多项式核基，即 $k=2$，$m=6$。图中的计算点同时用作支撑点。

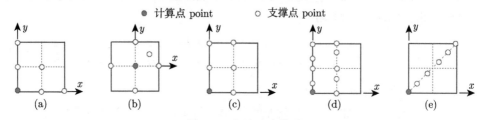

图 4-5 小片试验模型

Fig. 4-5 Patch test examples

由定理 2 给出的第一几何条件，MLS 近似至少需要 $N = m=6$ 个支撑点，当计算点同时作为支撑点时，显然，给出的 5 个小片模型全部满足该条件。由定理 3 给出的第二几何条件，点集在 x 或 y 上的独立投影数至少需要 $k+1=3$ 个，显然 (a)，(b) 和 (e) 这 3 个模型满足该条件，而 (c) 和 (d) 违反了该条件。通过近似计算，可以验证 (a) 和 (b) 能通过试验，而 (c) 和 (d) 是失败的，即便 (d) 使用了更多的支撑点。我们也可发现 (e) 模型也不能通过试验，虽然它满足给定的两个几何条件。其原因是违反了定理 1，支撑点被一条直线 (不高于 2 次的曲线) 串接。这个失败的试验旨在说明给出的两个几何条件仅是必要而非充分条件。然而，在实际应用中，支撑点通常是较为稠密且近于均匀地分布，这是数值方法的基本要求。因此，使用给出的几何条件去判定 A 的奇异性通常足够有效。

4.4 MLS 核近似法

推论 3.3 和推论 3.4 又提出了新的问题，即 MLS 近似在收敛性计算中和使用高次基的情况下，A 很容易产生病态。特别是收敛性计算 (通过加密场节点来缩小支撑域尺度) 在数值方法中是不可避免的。那么如何消除这种不稳定近似现象呢？

4.4 MLS 核近似法

一种简单有效的方法就是采用移动最小二乘核 (moving least squares core, MLSc) 近似 [6]，如图 4-6 所示。

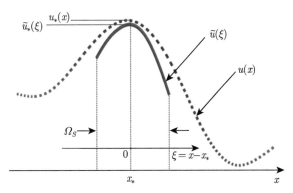

图 4-6 移动最小二乘核 (MLSc) 近似图示

Fig. 4-6 Schematics of the moving least squares core (MLSc) approximation

相比于经典的 MLS，MLSc 仅是对其中的基函数 $p^{\mathrm{T}}(x)$ 进行了改造。根据 4.3 节的定义 1 及定理 3，为了保证得到 "清晰" 的投影，我们希望投影是在计算点 x_I 的局部坐标系上进行，即将近似函数 $u^h(x)$ 定义为关于 x_I 的 "局部" 函数。而构造局部近似函数的一个简单办法就是使用局部基函数，我们称为 "核基" (core basis)。显然，由核基构造的近似函数定义在关于 x_I 的局部坐标系 $\xi = x - x_I$ 上，对应地，$u^h(x) \Rightarrow u^h(\xi)$。

对于一维问题，完备多项式的核基 $p^{\mathrm{T}}(x - x_I)$ 写为

$$p^{\mathrm{T}}(x - x_I) = \begin{cases} [1, (x - x_I)], & k = 1 \\ \left[1, (x - x_I), (x - x_I)^2\right], & k = 2 \end{cases} \tag{4-47}$$

对于二维问题，完备多项式的核基 $p^{\mathrm{T}}(x - x_I)$ 写为

$$\begin{aligned}&p^{\mathrm{T}}(x - x_I) \\ &= \begin{cases} [1, (x - x_I), (y - y_I)], & k = 1 \\ \left[1, (x - x_I), (y - y_I), (x - x_I)^2, (x - x_I) \cdot (y - y_I), (y - y_I)^2\right], & k = 2 \end{cases}\end{aligned} \tag{4-48}$$

对于三维问题，完备多项式的核基 $p^{\mathrm{T}}(x - x_I)$ 写为

$$\begin{cases} [1, (x - x_I), (y - y_I), (z - z_I)], & k = 1 \\ \big[1, (x - x_I), (y - y_I), (z - z_I), (x - x_I)^2, (x - x_I) \cdot (y - y_I), \\ (y - y_I)^2, (y - y_I) \cdot (z - z_I), (z - z_I)^2, (z - z_I) \cdot (x - x_I)\big], & k = 2 \end{cases} \tag{4-49}$$

则 MLSc 中用完备多项式表示的近似公式写为

$$u_I(\boldsymbol{x}) \approx u_I^h(\boldsymbol{x}-\boldsymbol{x}_I) = \boldsymbol{p}_I^{\mathrm{T}}(\boldsymbol{x}-\boldsymbol{x}_I)\boldsymbol{a}(\boldsymbol{x}-\boldsymbol{x}_I), \quad \forall \boldsymbol{x} \in \Omega_S(\boldsymbol{x}_*) \tag{4-50}$$

基于散点集 $\{\boldsymbol{x}_J\}_{J=1}^N$ ($\forall \boldsymbol{x}_J \in \Omega_S(\boldsymbol{x}_I)$) 构造如下加权最小二乘泛函：

$$J(\boldsymbol{a}) = \sum_{J=1}^N w_J(\boldsymbol{x}-\boldsymbol{x}_I)\left[\boldsymbol{p}_J^{\mathrm{T}}(\boldsymbol{x}-\boldsymbol{x}_I)\boldsymbol{a}(\boldsymbol{x}-\boldsymbol{x}_I) - \hat{u}_J\right]^2 \tag{4-51}$$

令 $\partial J/\partial \boldsymbol{a} = 0$，则可解得

$$\boldsymbol{a}(\boldsymbol{x}-\boldsymbol{x}_*) = \boldsymbol{A}^{-1}(\boldsymbol{x}-\boldsymbol{x}_*)\boldsymbol{B}(\boldsymbol{x}-\boldsymbol{x}_*)\hat{\boldsymbol{u}} \tag{4-52}$$

式中，

$$\left.\begin{array}{l}\boldsymbol{A}(\boldsymbol{x}-\boldsymbol{x}_I) = \displaystyle\sum_{J=1}^N \boldsymbol{A}_J(\boldsymbol{x}-\boldsymbol{x}_I) \\ \boldsymbol{A}_J(\boldsymbol{x}-\boldsymbol{x}_I) = w_J(\boldsymbol{x}-\boldsymbol{x}_I)\boldsymbol{p}_J(\boldsymbol{x}-\boldsymbol{x}_I)\boldsymbol{p}_J^{\mathrm{T}}(\boldsymbol{x}-\boldsymbol{x}_I)\end{array}\right\} \tag{4-53}$$

$$\left.\begin{array}{l}\boldsymbol{B}(\boldsymbol{x}-\boldsymbol{x}_I) = [\boldsymbol{B}_J(\boldsymbol{x}-\boldsymbol{x}_I)]_{J=1}^N \\ \boldsymbol{B}_J(\boldsymbol{x}-\boldsymbol{x}_I) = w_J(\boldsymbol{x}-\boldsymbol{x}_I)\boldsymbol{p}_J(\boldsymbol{x}-\boldsymbol{x}_I)\end{array}\right\} \tag{4-54}$$

$$\hat{\boldsymbol{u}}^{\mathrm{T}} = [\hat{u}_J]_{J=1}^N = \begin{bmatrix} \hat{u}_1 & \hat{u}_2 & \cdots & \hat{u}_N \end{bmatrix} \tag{4-55}$$

则 MLSc 在计算点 $\boldsymbol{x}_I$ 处的近似函数公式写为

$$u_I^h(\boldsymbol{x}-\boldsymbol{x}_I) = \boldsymbol{p}_I^{\mathrm{T}}(\boldsymbol{x}-\boldsymbol{x}_I)\boldsymbol{A}^{-1}(\boldsymbol{x}-\boldsymbol{x}_I)\boldsymbol{B}(\boldsymbol{x}-\boldsymbol{x}_I)\hat{\boldsymbol{u}} = \sum_{J=1}^N \phi_J(\boldsymbol{x})\hat{u}_J \tag{4-56}$$

式中，$\phi_J(\boldsymbol{x})$ 为支撑点 $\boldsymbol{x}_J$ 处的 MLSc 形函数，并定义为

$$\phi_J(\boldsymbol{x}) \equiv \phi_J(\boldsymbol{x}-\boldsymbol{x}_I) = \boldsymbol{p}_I^{\mathrm{T}}(\boldsymbol{x}-\boldsymbol{x}_I)\left[\boldsymbol{A}^{-1}(\boldsymbol{x}-\boldsymbol{x}_I)\boldsymbol{B}_J(\boldsymbol{x}-\boldsymbol{x}_I)\right] \tag{4-57}$$

其一阶导数 $\phi_{J,i}$ 和二阶导数 $\phi_{J,ij}$ 分别写为

$$\left\{\begin{array}{l}\phi_{J,i}(\boldsymbol{x}) \\ \phi_{J,ij}(\boldsymbol{x})\end{array}\right\} = \left\{\begin{array}{l}\boldsymbol{p}_{I,i}^{\mathrm{T}}(\boldsymbol{x}-\boldsymbol{x}_I) \\ \boldsymbol{p}_{I,ij}^{\mathrm{T}}(\boldsymbol{x}-\boldsymbol{x}_I)\end{array}\right\}\left[\boldsymbol{A}^{-1}(\boldsymbol{x}-\boldsymbol{x}_I)\boldsymbol{B}_J(\boldsymbol{x}-\boldsymbol{x}_I)\right] \tag{4-58}$$

更一般的形式，$\phi_J(\boldsymbol{x})$ 的 d 阶导数写为

$$\phi_J^{(d)}(\boldsymbol{x}) = \left[\boldsymbol{p}_I^{\mathrm{T}}(\boldsymbol{x}-\boldsymbol{x}_I)\right]^{(d)}\left[\boldsymbol{A}^{-1}(\boldsymbol{x}-\boldsymbol{x}_I)\boldsymbol{B}_J(\boldsymbol{x}-\boldsymbol{x}_I)\right] \tag{4-59}$$

此外，如果在基函数中使用其他非多项式函数，同样可以用局部坐标转换的方式，将其变为核基函数。

4.4 MLS 核近似法

接下来将从理论上进一步解释 MLSc 近似相比于经典 MLS 近似的优势。为了讨论方便，用 $\boldsymbol{A} = (\varLambda_{ij})_{m \times m}$ 和 $\boldsymbol{A}^* = (\varLambda_{ij}^*)_{m \times m}$ 分别表示使用普通多项式基和使用多项式核基的 "$\boldsymbol{A}$" 矩阵。在接下来的论述中，统一用 $(g)^*$ 表示使用核基的项或量。

根据 Gerschgorin 定理，矩阵 $\boldsymbol{A} = (\varLambda_{ij})_{m \times m}$ 的全体特征值都在它的 m 个盖尔圆的并集之中，任一盖尔圆由下式定义：

$$G_i = \left\{ z \left| |z - \varLambda_{ii}| \leqslant \left(R_i = \sum_{\substack{j=1 \\ j \neq i}}^{m} |\varLambda_{ij}| \right) \right. \right\} \tag{4-60}$$

以二维问题的完备二次多项式基为例，MLS 中 $\boldsymbol{A}$ 的展开形式为

$$\boldsymbol{A} = \sum_{J=1}^{N} \boldsymbol{A}_J(\boldsymbol{x}) = \boldsymbol{P}^{\mathrm{T}} \boldsymbol{W} \boldsymbol{P}$$

$$= \begin{bmatrix} \sum_{J=1}^{N} w_J & \sum_{J=1}^{N} w_J x_J & \sum_{J=1}^{N} w_J y_J & \sum_{J=1}^{N} w_J x_J^2 & \sum_{J=1}^{N} w_J x_J y_J & \sum_{J=1}^{N} w_J y_J^2 \\ \sum_{J=1}^{N} w_J x_J & \sum_{J=1}^{N} w_J x_J^2 & \sum_{J=1}^{N} w_J x_J y_J & \sum_{J=1}^{N} w_J x_J^3 & \sum_{J=1}^{N} w_J x_J^2 y_J & \sum_{J=1}^{N} w_J x_J y_J^2 \\ \sum_{J=1}^{N} w_J y_J & \sum_{J=1}^{N} w_J x_J y_J & \sum_{J=1}^{N} w_J y_J^2 & \sum_{J=1}^{N} w_J x_J^2 y_J & \sum_{J=1}^{N} w_J x_J y_J^2 & \sum_{J=1}^{N} w_J y_J^3 \\ \sum_{J=1}^{N} w_J x_J^2 & \sum_{J=1}^{N} w_J x_J^3 & \sum_{J=1}^{N} w_J x_J^2 y_J & \sum_{J=1}^{N} w_J x_J^4 & \sum_{J=1}^{N} w_J x_J^3 y_J & \sum_{J=1}^{N} w_J x_J^2 y_J^2 \\ \sum_{J=1}^{N} w_J x_J y_J & \sum_{J=1}^{N} w_J x_J^2 y_J & \sum_{J=1}^{N} w_J x_J y_J^2 & \sum_{J=1}^{N} w_J x_J^3 y_J & \sum_{J=1}^{N} w_J x_J^2 y_J^2 & \sum_{J=1}^{N} w_J x_J y_J^3 \\ \sum_{J=1}^{N} w_J y_J^2 & \sum_{J=1}^{N} w_J x_J y_J^2 & \sum_{J=1}^{N} w_J y_J^3 & \sum_{J=1}^{N} w_J x_J^2 y_J^2 & \sum_{J=1}^{N} w_J x_J y_J^3 & \sum_{J=1}^{N} w_J y_J^4 \end{bmatrix} \tag{4-61}$$

为了直观起见，将 $\boldsymbol{A}$ 的主对角线元素从 $\varLambda_{11} \to \varLambda_{mm}$ 拉直并写为

$$\{\Lambda_{ii}\}_{i=1}^m = \left(\sum_{J=1}^N w_J, \sum_{J=1}^N w_J x_J^2, \sum_{J=1}^N w_J y_J^2, \cdots, \sum_{J=1}^N w_J y_J^{2k}\right) \quad (4\text{-}62)$$

类似地，MLSc 中 $\boldsymbol{A}^*$ 的主对角线元素从 $\Lambda_{11}^* \to \Lambda_{mm}^*$ 拉直并写为

$$\{\Lambda_{ii}^*\}_{i=1}^m = \left(\sum_{J=1}^N w_J, \sum_{J=1}^N w_J(x_J-x_I)^2, \sum_{J=1}^N w_J(y_J-y_I)^2, \cdots, \sum_{J=1}^N w_J(y_J-y_I)^{2k}\right) \quad (4\text{-}63)$$

在以 $\boldsymbol{x}_I$ 为中心，半径为 r_s 的支撑域上，有下列不等式：

$$\begin{cases} |x| \leqslant \|\boldsymbol{x}\| \\ |y| \leqslant \|\boldsymbol{x}\| \\ |x-x_I| \leqslant \|\boldsymbol{x}-\boldsymbol{x}_I\| \leqslant r_s \quad (\forall \boldsymbol{x} \in \Omega_S(\boldsymbol{x}_I, r_s)) \\ |y-y_I| \leqslant \|\boldsymbol{x}-\boldsymbol{x}_I\| \leqslant r_s \\ |\|\boldsymbol{x}_I\| - r_s| \leqslant \|\boldsymbol{x}\| \leqslant (\|\boldsymbol{x}_I\| + r_s) \end{cases} \quad (4\text{-}64)$$

特别定义 Λ_{mm} (Λ_{mm}^*) 是居于极限位置 (最左或最右) 的第 m 个盖尔圆 G_m 的圆心，则有

$$1 < \Lambda_{11} = \Lambda_{11}^* = \sum_{J=1}^N w_J < N, \quad (\boldsymbol{x}_I \in \{\boldsymbol{x}_J\}_{J=1}^N, w_J \leqslant 1) \quad (4\text{-}65)$$

$$\Lambda_{mm} = \sum_{J=1}^N w_J v_J^{2k} < \sum_{J=1}^N w_J(\|\boldsymbol{x}_I\|+r_s)^{2k} < N(\|\boldsymbol{x}_I\|+r_s)^{2k}, \quad v=x\,|\,y \quad (4\text{-}66)$$

$$\Lambda_{mm}^* = \sum_{J=1}^N w_J(v_J-v_I)^{2k} < \sum_{J=1}^N w_J \|\boldsymbol{x}_J-\boldsymbol{x}_I\|^{2k} < \sum_{J=1}^N w_J r_s^{2k} < N r_s^{2k}, \quad v=x\,|\,y$$
$$(4\text{-}67)$$

式中，$v = \boldsymbol{x}\,|\,y$ 表示参考 $\boldsymbol{x}_I$ 选用绝对坐标值较大的维度。同理，可推得

$$\begin{cases} \Lambda_{mm} > (\|\boldsymbol{x}_I\| - r_s)^{2k} \\ \Lambda_{mm}^* > 0 \end{cases} \quad (4\text{-}68)$$

另有

$$R_m = \sum_{j=1}^{m-1} |\Lambda_{mj}| = \sum_{J=1}^N w_J y_J^k + \sum_{J=1}^N w_J x_J y_J^k + \sum_{J=1}^N w_J y_J^{k+1} + \cdots + \sum_{J=1}^N w_J x_J y_J^{2k-1}$$

$$< (m-1) \sum_{J=1}^N w_J(\|\boldsymbol{x}_I\|+r_s)^{2k} \quad \Leftarrow (v=\boldsymbol{x}\,|\,y)$$

$$< N(m-1)(\|\boldsymbol{x}_I\|+r_s)^{2k} \quad (4\text{-}69)$$

4.4 MLS 核近似法

同理可推得

$$R_m^* < N(m-1) r_s^{2k} \tag{4-70}$$

因 "A" 矩阵为实对称阵，m 个特征值全为实数 (任一特征值包含于对应盖尔圆确定的实数轴上)。其条件数可定义为 $\text{cond}_2(A) = |\lambda_{\max}/\lambda_{\min}|$，$\lambda_{\max}$ 和 $\lambda_{\min}$ 是按模的最大和最小特征值，所以 G_m 必然包含一个极限特征值 ($\lambda_{\max}$ 和 $\lambda_{\min}$ 其中之一)。

显然，有

$$\{\|x_I\| \gg r_s\} \Rightarrow \left\{ \begin{array}{c} \Lambda_{mm} \gg \Lambda_{mm}^* \\ R_m \gg R_m^* \end{array} \right\} \tag{4-71}$$

这意味着在第 m 个盖尔圆 G_m 中，A 比 A^* 可能取得一个大得多的特征值。因此，$\text{span}(\lambda) \supset \text{span}(\lambda^*)$，如图 4-7 所示，故 $\text{cond}_2(A)$ 比 $\text{cond}_2(A^*)$ 可能要大得多。即

$$\text{cond}_2(A) \geqslant \text{cond}_2(A^*) \tag{4-72}$$

需注意到假设条件 $\|x_I\| \gg r_s$ 反映的是问题域与局部支撑域的关系，而该假设条件是有普遍意义的。因此，通常情况下，使用核基的 "A" 矩阵更容易保证是良态的。

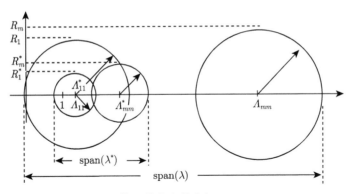

图 4-7 特征值的取值空间，$\|x_I\| \gg r_s$

Fig. 4-7 Span of the eigenvalues, $\|x_I\| \gg r_s$

为了建立一个数量上的关系，我们把第 m 个盖尔圆的圆心值域上界定义为可能的最大特征值，则有

$$\frac{\text{cond}_2(A)}{\text{cond}_2(A^*)} = \frac{|\lambda_m/\lambda_1|}{|\lambda_m^*/\lambda_1^*|} = \left|\frac{\lambda_1^*}{\lambda_1}\right| \cdot \left|\frac{\lambda_m}{\lambda_m^*}\right| \propto \left|\frac{\lambda_1^*}{\lambda_1}\right| \cdot \left|\frac{\Lambda_{mm}}{\Lambda_{mm}^*}\right| \propto \left|\frac{\lambda_1^*}{\lambda_1}\right|$$
$$\cdot \left(\frac{\|x_I\| + r_s}{r_s}\right)^{2k} \underset{r_s \ll \|x_I\|}{\propto} \left|\frac{\lambda_1^*}{\lambda_1}\right| \cdot \left(\frac{\|x_I\|}{r_s}\right)^{2k} \tag{4-73}$$

据此可推论为当 $\|x_I\| \gg 1$，且 $r_s \ll \|x_I\|$ 时，相比于 A^*，A 的条件数随 $\|x_I\|$ 的增大而增大 (恶化)，随 k 的增大而增大 (恶化)，随 r_s 的缩小而增大 (恶化)。这个结论与推论 3.3 和推论 3.4 一致。

算例 矩阵 A 的条件数计算。为了检验矩阵条件数的影响因素，及使用核基的稳定作用，首先通过一个局部域模型来构建矩阵 A，如图 4-8(a) 所示。在比较计算中，考虑了如下几个方面的因素：

1) 采用核基 $p^T(x-x_I)$ 和采用普通基 $p^T(x)$ 的比较，考察 MLSc 和 MLS 的 A 矩阵的条件数，并检验 MLSc 的优势。

2) 考察 $\|x_I\|/r_s$ 的值对 A 条件数的影响。对散点采用了两种坐标情况 (Case 1 和 Case 2)，如图 4-8(b) 的列表所示。对 Case 1 坐标系，r_s 取为 0.65；对 Case 2 坐标系，r_s 取为 1.3。

3) 考察基函数次数 k 对 A 的条件数的影响，计算中分别考虑了 $k=1,2,3$ 三种情况。

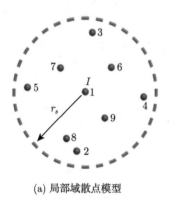

No. J	$\{x_J\}$,Case 1	$\{x_J\}$,Case 2
1,(I)	(0.50, 0.10)	(100.0, 5.0)
2	(0.40, −0.35)	(99.8, 4.1)
3	(0.55, 0.60)	(100.1, 6.0)
4	(1.00, 0.05)	(101.0, 4.9)
5	(0.05, 0.15)	(99.1, 5.1)
6	(0.70, 0.35)	(100.4, 5.5)
7	(0.25, 0.30)	(99.5, 5.4)
8	(0.35, −0.30)	(99.7, 4.2)
9	(0.75, −0.20)	(100.5, 4.4)

(a) 局部域散点模型　　(b) 散点坐标

图 4-8　构建矩阵 A 的局部域

Fig. 4-8　Local domain to construct the matrix A

计算结果如表 4-1 所示。结果表明：① 当 $k>1$ 时，如果 $\|x_I\|/r_s$ 是一个较大的值，则 A 的条件数远大于 A^*，表明 MLSc 比 MLS 更容易保证 A 矩阵是良态的；② $\|x_I\|/r_s$ 的值对 A 的条件数有显著影响，其值较大时，A 将是一个大的条件

表 4-1　矩阵 A 的条件数

Tab. 4-1　The condition numbers of the matrix A

cond$_2$	Case 1 (r_s=0.65, $\|x_I\|/r_s$=0.7845)			Case 2 (r_s=1.3, $\|x_I\|/r_s$=77.02)		
	$k=1$	$k=2$	$k=3$	$k=1$	$k=2$	$k=3$
cond$_2(A)$: $p^T(x)$	4.2834×10	3.2125×10^3	1.3279×10^{17}	1.0726×10	4.3716×10^{17}	1.1977×10^{26}
cond$_2(A^*)$: $p^T(x-x_I)$	4.2834×10	1.5487×10^3	4.4825×10^{17}	1.0726×10	1.0019×10^2	5.0830×10^{16}
cond$_2(A)$/cond$_2(A^*)$	1	2.07	2.96×10^{-1}	1	4.3633×10^{15}	2.36×10^9

数；③基函数次数 k 对 A 的条件数有显著影响，其值较大时，A 将是一个大的条件数。显然，这些结论符合 4.3 节和 4.4 节中的理论预期。

4.5 改进的 MLS 近似法

4.5.1 改进的 MLS 法

为了避免对 A 进行直接求逆运算，一种应对的方法就是采用带权正交函数作为基函数，陈美娟等 [10-12] 将其称为改进的移动最小二乘 (improved least-square, IMLS) 法。

对于点集 $\{x_J\}_{J=1}^N$ 及其对应的权函数 $w_J(x)$，首先引入一个带权函数积的记号：

$$(\varphi_i, \varphi_j) = \sum_{J=1}^N w_J(x) \cdot \varphi_i(x_J) \cdot \varphi_j(x_J) \tag{4-74}$$

如果函数 $\varphi(x)$ 满足：

$$(\varphi_i, \varphi_j) = \begin{cases} 0, & i \neq j \\ A_i, & i = j \end{cases} \quad (i, j = 1, 2, \cdots, m) \tag{4-75}$$

式中，A_i 不恒为 0。则称 $\varphi(x)$ 为关于点集 $\{x_J\}_{J=1}^N$ 带权 $\{w_J\}_{J=1}^N$ 的正交函数族。利用 Schmidt 正交化方法，带权的正交基函数可构造如下：

$$\left.\begin{array}{l} p_1 = 1, \\ p_t = r^{t-1} - \sum_{k=1}^{t-1} \dfrac{(r^{t-1}, p_k)}{(p_k, p_k)} p_k, \quad t = 2, 3, \cdots \end{array}\right\} \tag{4-76}$$

式中，

$$\left.\begin{array}{ll} r = x, & \text{1D} \\ r = \sqrt{x^2 + y^2}, & \text{2D} \\ r = \sqrt{x^2 + y^2 + z^2}, & \text{3D} \end{array}\right\} \tag{4-77}$$

或

$$\left.\begin{array}{ll} r = x, & \text{1D} \\ r = x + y, & \text{2D} \\ r = x + y + z, & \text{3D} \end{array}\right\} \tag{4-78}$$

另外，利用 Schmidt 正交化方法，也可对常用的完备的多项式基函数进行正交化改造，其第 t 项 p_t 对应的正交化基函数写为

$$\langle p_t \rangle = p_t - \sum_{k=1}^{t-1} \dfrac{(p_t, p_k)}{(p_k, p_k)} p_k, \quad t = 1, 2, 3, \cdots \tag{4-79}$$

由式 (4-3) 得到的 $\boldsymbol{A}$ 非逆运算的线性关系为

$$\boldsymbol{A}\boldsymbol{a} = \boldsymbol{B}\hat{\boldsymbol{u}} \tag{4-80}$$

由式 (4-74) 定义，上式可改写为

$$\begin{bmatrix} (p_1,p_1) & (p_1,p_2) & \cdots & (p_1,p_m) \\ (p_2,p_1) & (p_2,p_2) & \cdots & (p_2,p_m) \\ \vdots & \vdots & \ddots & \vdots \\ (p_m,p_1) & (p_m,p_2) & \cdots & (p_m,p_m) \end{bmatrix} \begin{bmatrix} a_1(\boldsymbol{x}) \\ a_2(\boldsymbol{x}) \\ \vdots \\ a_m(\boldsymbol{x}) \end{bmatrix} = \begin{bmatrix} (p_1,\hat{u}_J) \\ (p_2,\hat{u}_J) \\ \vdots \\ (p_m,\hat{u}_J) \end{bmatrix} \tag{4-81}$$

若基函数 $p_t(\boldsymbol{x})$ 为带权正交基函数，则有：若 $i \neq j$，则 $(p_i,p_j) = 0$。则式 (4-81) 可改写为

$$\begin{bmatrix} (p_1,p_1) & 0 & \cdots & 0 \\ 0 & (p_2,p_2) & \cdots & 0 \\ \vdots & \vdots & \ddots & \vdots \\ 0 & 0 & \cdots & (p_m,p_m) \end{bmatrix} \begin{bmatrix} a_1(\boldsymbol{x}) \\ a_2(\boldsymbol{x}) \\ \vdots \\ a_m(\boldsymbol{x}) \end{bmatrix} = \begin{bmatrix} (p_1,\hat{u}_J) \\ (p_2,\hat{u}_J) \\ \vdots \\ (p_m,\hat{u}_J) \end{bmatrix} \tag{4-82}$$

这样，不必对矩阵 $\boldsymbol{A}$ 求逆，系数 $a_t(\boldsymbol{x})$ 就可直接得到，即

$$a_t(\boldsymbol{x}) = \frac{(p_t,\hat{u}_J)}{(p_t,p_t)} \tag{4-83}$$

写成矩阵形式：

$$\boldsymbol{a}(\boldsymbol{x}) = \boldsymbol{A}^*(\boldsymbol{x})\boldsymbol{B}(\boldsymbol{x})\hat{\boldsymbol{u}} \tag{4-84}$$

式中，

$$\boldsymbol{A}^*(\boldsymbol{x}) = \begin{bmatrix} \dfrac{1}{(p_1,p_1)} & 0 & \cdots & 0 \\ 0 & \dfrac{1}{(p_2,p_2)} & \cdots & 0 \\ \vdots & \vdots & \ddots & \vdots \\ 0 & 0 & \cdots & \dfrac{1}{(p_m,p_m)} \end{bmatrix} \tag{4-85}$$

即 $\boldsymbol{A}^*$ 可直接组装得到，而非求逆运算得到。将式 (4-84) 代入式 (3-13)，需注意，式 (3-13) 中的 $\boldsymbol{p}_I^T(\boldsymbol{x})$ 将是对应的带权正交基函数，便可得到类似 $u_I^h(\boldsymbol{x}) = \sum_{J=1}^{N} \phi_J(\boldsymbol{x})\hat{u}_J$ 近似函数的 IMLS 形函数表示形式。

4.5 改进的 MLS 近似法

IMLS 法的优势主要体现在两个方面。其一, 避免了对矩阵 A 的求逆运算, A^{-1} 是否存在将不再是一个技术问题, 即便 A 是弱病态的, 也不会因求逆运算放大计算误差, 使得 MLS 计算具有更好的稳定性。其二, 可以显著提高计算效率, 如果选取式 (4-76) 作为基函数, 则在试函数阶次相同的情况下, 试函数中的待定系数的个数比原来要少; 对线性基, 原来的待定系数是 3 个, 现在是 2 个; 对二次基, 原来的待定系数是 6 个, 现在是 3 个; 这样对任意一个计算点而言, 其支撑域中所要求的支撑点数就明显减少了, 进而在整个求解域上所需选取的节点数将显著减少, 所以, 计算效率会大幅度提高。

4.5.2 复变量 MLS 法

为了提高 MLS 的计算效率, 程玉民等[13,14] 提出了复变量移动最小二乘 (complex variable moving least-square, CVMLS) 法, 对二维问题, 其试函数取为

$$u^h(z) = u_1^h(z) + \mathrm{i} u_2^h(z) = \sum_{t=1}^{m} p_t(z) \cdot a_t(z) = \boldsymbol{p}^{\mathrm{T}}(z) \cdot \boldsymbol{a}(z) \tag{4-86}$$

式中,

$$z = x + \mathrm{i} y \tag{4-87}$$

基于散点集 $\{z_J\}_{J=1}^{N} (\forall z_J \in \Omega_S(z_I))$ 定义泛函:

$$\begin{aligned} J(\boldsymbol{a}) &= \sum_{J=1}^{N} w_J(z - z_I) \left[u_J^h(z) - \hat{u}_J(z) \right]^2 \\ &= \sum_{J=1}^{N} w_J(z - z_I) \left[\sum_{t=1}^{m} p_t(z_J) \cdot a_t(z) - \hat{u}_J(z) \right]^2 \\ &= (\boldsymbol{P}\boldsymbol{a} - \boldsymbol{u}^*)^{\mathrm{T}} \boldsymbol{W} (\boldsymbol{P}\boldsymbol{a} - \boldsymbol{u}^*) \end{aligned} \tag{4-88}$$

式中,

$$\hat{u}_J(z) = \hat{u}_1(z_J) + \mathrm{i} \hat{u}_2(z_J) \tag{4-89}$$

$$\boldsymbol{u}^* = [\hat{u}_{J=1}(z), \hat{u}_{J=2}(z), \cdots, \hat{u}_{J=N}(z)]^{\mathrm{T}} = \boldsymbol{Q}\hat{\boldsymbol{u}} \tag{4-90}$$

$$\hat{\boldsymbol{u}} = [\hat{u}_1(z_1), \hat{u}_2(z_1), \hat{u}_1(z_2), \hat{u}_2(z_2), \cdots, \hat{u}_1(z_N), \hat{u}_2(z_N)]^{\mathrm{T}} \tag{4-91}$$

$$\boldsymbol{Q} = \begin{bmatrix} 1 & \mathrm{i} & 0 & 0 & 0 & 0 & \cdots & 0 & 0 \\ 0 & 0 & 1 & \mathrm{i} & 0 & 0 & \cdots & 0 & 0 \\ 0 & 0 & 0 & 0 & 1 & \mathrm{i} & \cdots & 0 & 0 \\ \vdots & \vdots & \vdots & \vdots & \vdots & \vdots & \ddots & \vdots & \vdots \\ 0 & 0 & 0 & 0 & 0 & 0 & \cdots & 1 & \mathrm{i} \end{bmatrix}_{N \times 2N} \tag{4-92}$$

$$\boldsymbol{P} = \begin{bmatrix} p_1(z_1) & p_2(z_1) & \cdots & p_m(z_1) \\ p_1(z_2) & p_2(z_2) & \cdots & p_m(z_2) \\ \vdots & \vdots & \ddots & \vdots \\ p_1(z_N) & p_2(z_N) & \cdots & p_m(z_N) \end{bmatrix} \tag{4-93}$$

$$\boldsymbol{W} = \begin{bmatrix} w_1(z) & & & 0 \\ & w_2(z) & & \\ & & \ddots & \\ 0 & & & w_N(z) \end{bmatrix} \tag{4-94}$$

令 $\partial J/\partial \boldsymbol{a} = 0$, 则可解得

$$\boldsymbol{a}(z) = \boldsymbol{A}^{-1}(z)\boldsymbol{B}(z)\boldsymbol{u}^* \tag{4-95}$$

上式与式 (3-16) 类似, 不予赘述。则 CVMLS 的近似函数表达式写为

$$u^h(z) = \boldsymbol{\Phi}(z)\boldsymbol{u}^* = \sum_{J=1}^{N} \phi_J(z)\hat{u}_J(z) \tag{4-96}$$

则有

$$\left.\begin{aligned} u_1^h(z) &= \text{Re}\left[\boldsymbol{\Phi}(z)\boldsymbol{u}^*\right] = \text{Re}\left[\sum_{J=1}^{N} \phi_J(z)\hat{u}_J(z)\right] \\ u_2^h(z) &= \text{Im}\left[\boldsymbol{\Phi}(z)\boldsymbol{u}^*\right] = \text{Im}\left[\sum_{J=1}^{N} \phi_J(z)\hat{u}_J(z)\right] \end{aligned}\right\} \tag{4-97}$$

可见, CVMLS 在形式上较经典的 MLS 复杂, 但其具有计算效率高的优势。在试函数阶次相同的情况下, 试函数中的待定系数的个数比原来要少。对线性基, 原来的基函数是 $\boldsymbol{p}^{\text{T}} = [1, x, y]$, 待定系数是 3 个, 现在的基函数是 $\boldsymbol{p}^{\text{T}} = [1, z]$, 待定系数是两个。对二次基, 原来的基函数是 $\boldsymbol{p}^{\text{T}} = [1, x, y, x^2, xy, y^2]$, 待定系数是 6 个, 现在的基函数是 $\boldsymbol{p}^{\text{T}} = [1, z, z^2]$, 待定系数是 3 个。这样, 对任意一个计算点而言, 其支撑域中所要求的支撑点数就明显减少了, 进而大幅度提高了计算效率。此外, CVMLS 通常采用带权正交基函数, 因此, 也会将 IMLS 无须对 $\boldsymbol{A}$ 矩阵求逆运算的优势引入。

4.6 MLS 导数近似的讨论

数值方法中, 在构造离散系统方程时, MLS 通常需要执行导数近似[15]。如果

4.6 MLS 导数近似的讨论

要执行场函数导数的近似，则通过形函数 ϕ_J 的求导来实现，即

$$\begin{cases} u_{,i}^h(\boldsymbol{x}) = \sum_{J=1}^{N} \phi_{J,i}(\boldsymbol{x}) \hat{u}_J \\ u_{,ij}^h(\boldsymbol{x}) = \sum_{J=1}^{N} \phi_{J,ij}(\boldsymbol{x}) \hat{u}_J \end{cases} \quad (4\text{-}98)$$

更一般的形式写为

$$[u^h(\boldsymbol{x})]^{(d)} = \sum_{J=1}^{N} \phi_J^{(d)}(\boldsymbol{x}) \hat{u}_J \quad (4\text{-}99)$$

式中，上标 (d) 表示求 d 阶导数 (偏导数)。

对 ϕ_J 的求导，原先 Nayroles 等[1] 建议为

$$\begin{cases} \phi_{J,i} = \boldsymbol{p}_{I,i}^{\mathrm{T}} \boldsymbol{A}^{-1} \boldsymbol{B}_J \\ \phi_{J,ij} = \boldsymbol{p}_{I,ij}^{\mathrm{T}} \boldsymbol{A}^{-1} \boldsymbol{B}_J \end{cases} \quad (4\text{-}100)$$

该求导运算中，系数向量 $\boldsymbol{a}(\boldsymbol{x})$ 被当作常数项处理，不会参与导数运算。图 4-9 给

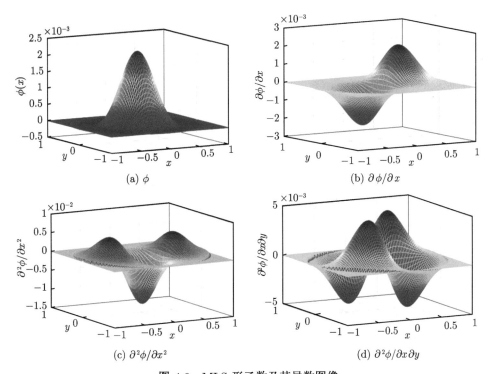

图 4-9 MLS 形函数及其导数图像

Fig. 4-9 The functional image of MLS shape function and their derivatives

出的是使用高斯型权函数和圆形支撑域，且 $a(x)$ 不参与导数运算的 MLS 形函数及其导数图像。

而 Belytschko 等[2] 在提出 EFG 法时，指出 MLS 导数近似公式中应将 $a(x)$ 作为函数向量考虑，并参与导数运算，即有

$$\begin{cases} \phi_{J,i} = \boldsymbol{p}_{I,i}^{\mathrm{T}} \boldsymbol{A}^{-1} \boldsymbol{B}_J + \boldsymbol{p}_I^{\mathrm{T}} \left(\boldsymbol{A}_{,i}^{-1} \boldsymbol{B}_J + \boldsymbol{A}^{-1} \boldsymbol{B}_{J,i} \right) \\ \phi_{J,ij} = \boldsymbol{p}_{I,ij}^{\mathrm{T}} \boldsymbol{A}^{-1} \boldsymbol{B}_J + \boldsymbol{p}_{I,i}^{\mathrm{T}} \left(\boldsymbol{A}_{,j}^{-1} \boldsymbol{B}_J + \boldsymbol{A}^{-1} \boldsymbol{B}_{J,j} \right) + \boldsymbol{p}_{I,j}^{\mathrm{T}} \left(\boldsymbol{A}_{,i}^{-1} \boldsymbol{B}_J + \boldsymbol{A}^{-1} \boldsymbol{B}_{J,i} \right) \\ \quad + \boldsymbol{p}_I^{\mathrm{T}} \left(\boldsymbol{A}_{,ij}^{-1} \boldsymbol{B}_J + \boldsymbol{A}_{,i}^{-1} \boldsymbol{B}_{J,j} + \boldsymbol{A}_{,j}^{-1} \boldsymbol{B}_{J,i} + \boldsymbol{A}^{-1} \boldsymbol{B}_{J,ij} \right) \end{cases}$$

(4-101)

显然，式 (4-101) 要比式 (4-100) 复杂得多。

上述两种导数近似方法，其本质的区别就是系数向量 $a(x)$ 作常数考虑还是作非常数考虑，对该分歧讨论如下：

1) 多项式表示的 MLS 试函数 $u^h(x)$ 中，$a(x)$ 原初就是表示常数系数的。将其作常数考虑是合理的。

2) MLS 在构造式 (3-14) 表示的离散加权 L_2 泛函时，引入的权函数使得试函数有就近精确近似的优先意义，提高了近似计算的相容性。这个权函数本质上与系数向量 $a(x)$ 无关，不应当改变系数向量的原初假定和本质属性。

3) $a(x)$ 不作常数考虑，则除了基函数需求导外，还需要对权函数求导，这无疑会增加对权函数连续性的要求。通常，权函数的连续性要高于基函数的连续性，权函数求导并不会提高近似函数的连续性，因为近似函数的连续性由基函数的连续性决定。

4) $a(x)$ 不作常数考虑时，形函数的导数中将会出现 A 矩阵及其导数求逆的累加。这会造成误差累计，对计算精度应当是不利的。

5) MLS 本质上是非常简单的一种方法，但 $a(x)$ 参与导数运算后，使算法变得非常复杂，这样会削弱 MLS 的竞争性。

6) 如果 $a(x)$ 不作常数考虑确定是对 EFG 法有利，但对更一般的方法，这种优势是存疑的。

接下来，我们将通过进一步的数值试验来进行审视。在后续给出的图例中约定为："MLS, $a(x)$" 表示使用经典 MLS 法，而向量 $a(x)$ 视为非常数参与导数运算；"MLS, a" 表示使用经典 MLS 法，而向量 $a(x)$ 不参与导数运算；"MLSc, a" 表示使用改进后的核近似法，即 MLSc 法，且向量 $a(x)$ 不参与导数运算。

算例 1　曲面拟合问题对近似方法的检验更为直接，因此通常被用于近似算法检验。首先给出第一个简单的曲面拟合问题，设在 $x \in [0,1]$，$y \in [0,1]$ 的问题域上曲面函数为

$$u(x,y) = \sin(x) \sin(y) \tag{4-102}$$

4.6 MLS 导数近似的讨论

其函数图像如图 4-10 所示。该函数的偏导数很容易得到,此处从略。在近似计算中,为了避免散乱节点的扰动,所以采用规则节点离散方案。节点间距 $h = 0.2$ 的场节点离散方案如图 4-11 所示。

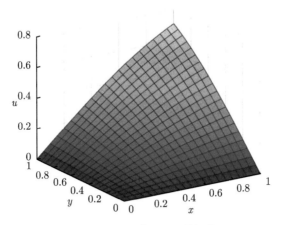

图 4-10　场函数 $u(\boldsymbol{x})$ 的图像

Fig. 4-10　Image of the field function $u(\boldsymbol{x})$

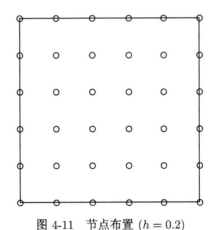

图 4-11　节点布置 $(h = 0.2)$

Fig. 4-11　Nodal arrangement $(h = 0.2)$

近似计算中,任一目标节点 $\boldsymbol{x}_I$ 的场函数 u、一阶导数 $u_{,x}$、二阶导数 $u_{,xx}$,由其他节点的场函数值通过近似计算得到。为了评价近似计算的精度和收敛性,我们定义如下的平均 L2 误差范数 (error norm):

$$e(\xi) = \frac{1}{N} \sqrt{\sum_{I=1}^{N} (\xi_I^{\text{num}} - \xi_I^{\text{exa}})^2 \Big/ \sum_{I=1}^{N} (\xi_I^{\text{exa}})^2} \qquad (4\text{-}103)$$

式中，ξ_I^{num} 表示数值计算得到的值；ξ_I^{exa} 表示解析得到的精确值。N 表示考察的计算点数量，对曲面拟合问题，N 为所有的场节点；对无网格法求解问题，N 为结果输出路径上的取值点。

设置一组疏密不同的节点离散方案 (节点间距用 h 来表示)，然后将原函数，及一阶和二阶导数的近似误差汇总到图 4-12 中一起表示。忽略一些细节问题，由图可以解读为两个主要结论：①$a(x)$ 参与导数运算并没有表现出预期的优势；②MLS 法在节点稠密时，其收敛性将变差，而 MLSc 总是表现为严格的收敛性。

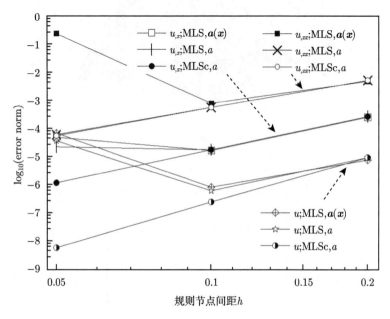

图 4-12 算例 1 的近似误差

Fig. 4-12 Approximation errors of Example 1

算例 2 接下来考虑一个形状上较为复杂的，或者说场函数光滑性较弱的曲面拟合问题。设在 $x \in [-8, 8]$，$y \in [-8, 8]$ 的问题域上曲面函数为

$$u(R) = \frac{\sin R}{R}, \quad R = \sqrt{x^2 + y^2} + 10^{-5} \tag{4-104}$$

其偏导数写为

$$\frac{\partial u}{\partial x} = x \left(\frac{\cos R}{R^2} - \frac{\sin R}{R^3} \right) \tag{4-105}$$

$$\frac{\partial^2 u}{\partial x^2} = \frac{x^2}{R^2} \left(\frac{2 \sin R}{R^3} - \frac{2 \cos R}{R^2} - \frac{\sin R}{R} \right) + \left(\frac{1}{R} - \frac{x^2}{R^3} \right) \left(\frac{\cos R}{R} - \frac{\sin R}{R^2} \right) \tag{4-106}$$

4.6 MLS 导数近似的讨论

其函数图像如图 4-13 所示。对一组疏密不同的规则节点离散方案，场函数及其导数的近似计算误差绘制于图 4-14 中。由其得出的结论与算例 1 保持一致：① $a(x)$ 参与导数运算在近似精度和收敛性上，比其不参与导数运算更差；②MLS 法随节点加密而丧失收敛性，MLSc 总是表现为更高的计算精度和严格的收敛性。

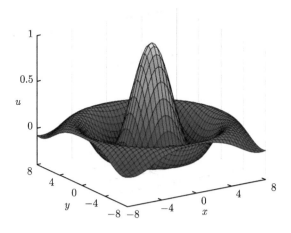

图 4-13 场函数 $u(x)$ 的图像

Fig. 4-13 Image of the field function $u(x)$

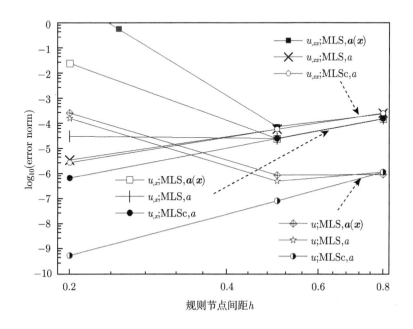

图 4-14 算例 2 的近似误差

Fig. 4-14 Approximation errors of Example 2

算例 3　接下来结合一种具体的无网格方法，来进一步检验，所使用的方法为有限点法 (FPM)[16,17]。选择 FPM 主要考虑到：该法是直接的配点法，属于最简单的无网格法之一，对近似计算的验证受数值离散算法的影响较小。更为重要的是，FPM 属于强式法，会使用到二阶导数近似，对问题的验证会更为完整。

考虑一个常用的悬臂梁问题算例，其结构模型如图 4-15 所示。计算中采用无量纲化处理 (或者默认为均采用标准国际单位)，计算参数取为：梁宽 $D=2$，梁长 $L=12$，梁端荷载总量 (表面抛物线型剪切荷载的积分)$P=6$，弹性模量 $E=10^4$，泊松比 $\nu=1/3$。该问题存在如下精确解 [18]：

$$\begin{cases} u_x = -\dfrac{Py}{6EI}\left[(6L-3x)x + (2+\nu)\left(y^2 - \dfrac{D^2}{4}\right)\right] \\ u_y = \dfrac{P}{6EI}\left[3\nu y^2(L-x) + (4+5\nu)\dfrac{D^2 x}{4} + (3L-x)x^2\right] \end{cases} \quad (4\text{-}107)$$

$$\begin{cases} \sigma_x = -\dfrac{P(L-x)y}{I} \\ \sigma_y = 0 \\ \sigma_{xy} = \dfrac{P}{2I}\left(\dfrac{D^2}{4} - y^2\right) \end{cases} \quad (4\text{-}108)$$

式中，I 表示惯性矩，对单位厚度的矩形截面梁有

$$I = \dfrac{D^3}{12} \quad (4\text{-}109)$$

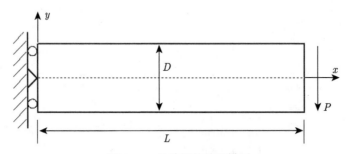

图 4-15　悬臂梁的结构模型

Fig. 4-15　Structural model of the cantilever beam

数值计算中为避免散乱节点扰动，采用规则节点离散方案，但各坐标方向的标准节点间距不同，并有 $h_x=2h_y$。$h_y=0.4$ 的场节点离散方案如图 4-16 所示。在数值验算中，将 $y=0$ 路径上的纵向位移 u_y 和 $x=L/2$ 路径上的剪切应力 σ_{xy} 作为计算指标。为了便于读者对 FPM 求解该问题的效果有一个直观的了解，将 $h_y=0.2$ 离散方案的计算结果给出，如图 4-17 所示。

4.6 MLS 导数近似的讨论

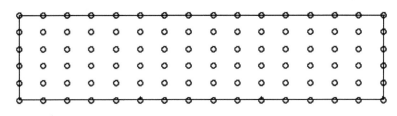

图 4-16　悬臂梁的节点布置

Fig. 4-16　Nodal arrangement for the cantilever beam

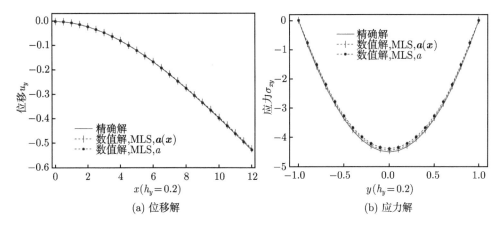

(a) 位移解　　　　　　　　　　(b) 应力解

图 4-17　数值解与精确解的比较

Fig. 4-17　Comparison between exact and numerical solutions

数值计算收敛性结果 (最密节点间距取为 $h_y=0.05$) 如图 4-18 所示。可以看出，$a(x)$ 不作常数考虑时，无论是位移解，还是应力解，其计算收敛性要比另外的方案明显变差。只是在节点较为稀疏时，似乎 $a(x)$ 不作常数考虑的解更精确。但需注意到，在节点非常稀疏的情况下，求解精度本身是非常粗糙的。本算例的结论依然是：$a(x)$ 参与导数运算时，求解收敛性并不具有优势。

无网格方法对局部支撑域尺度参数的敏感性，不仅直接反映数值方法计算的稳定性，也可以反映出近似计算的稳定性。因此，对支撑域尺度参数的数值计算影响效果进行了分析计算，其结果如图 4-19 所示。显然，无论从位移解，还是从应力解来看，$a(x)$ 参与导数计算的数值结果明显要差得多。换句话说，$a(x)$ 参与导数计算的情况下，数值方法的计算稳定性会变差。需注意的是，支撑域尺度参数 $\alpha_s=3.4$ 时，$a(x)$ 参与导数计算的效果出现偶发性的好转，而图 4-17 的结果正是采用这一参数。此处的结论依然类似，$a(x)$ 参与导数计算时，在数值计算的稳定性上不具有优势。

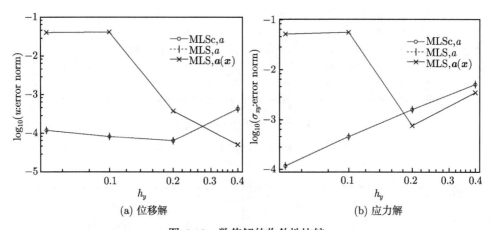

图 4-18　数值解的收敛性比较

Fig. 4-18　Comparison of the numerical convergence

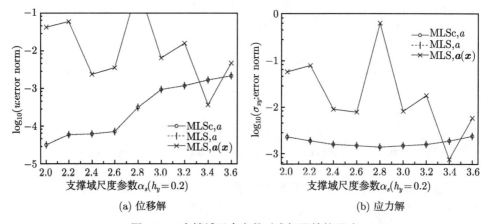

图 4-19　支撑域尺度参数对求解误差的影响

Fig. 4-19　Effect of support-domain parameter on numerical-solution errors

算例 4　在矩形域 $\Omega: (0 \leqslant x \leqslant a, -b/2 \leqslant y \leqslant b/2)$ 上，此处取 $a = b = 1$，一个泊松方程定义为

$$\begin{cases} \dfrac{\partial^2 u(\boldsymbol{x})}{\partial x^2} + \dfrac{\partial^2 u(\boldsymbol{x})}{\partial y^2} = -x^2 y, & \boldsymbol{x} \in \Omega \\ u(\boldsymbol{x}) = 0, & \boldsymbol{x} \in \Gamma \end{cases} \quad (4\text{-}110)$$

该方程存在如下解析解[19]：

$$u(\boldsymbol{x}) = \frac{xy}{12}(a^3 - x^3) + \sum_{n=1}^{\infty} \frac{a^4 b \left[(-1)^n (n^2 \pi^2 - 2) + 2 \right]}{n^5 \pi^5 \sinh(n\pi b / 2a)} \sinh\left(\frac{n\pi y}{a}\right) \sin\left(\frac{n\pi x}{a}\right) \quad (4\text{-}111)$$

4.6 MLS 导数近似的讨论

该泊松方程的函数图像如图 4-20 所示。同样采用 FPM 来求解该问题。问题域采用不规则节点离散，如图 4-21 所示。

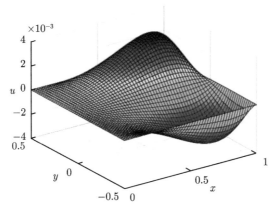

图 4-20　场函数 $u(x)$ 的图像

Fig. 4-20　Image of the field function $u(x)$

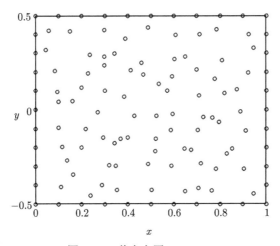

图 4-21　节点布置 (h_a=0.2)

Fig. 4-21　Nodal arrangement (h_a=0.2)

对不规则节点的平均间距给出如下定义：

$$h_a = \sqrt{S_\Omega}\big/\left(\sqrt{N}-1\right) \tag{4-112}$$

式中，S_Ω 表示问题域面积；N 表示场节点总数量。

取 x=0.8 的输出路径，并选择 21 个等间距分布的取值点。将此输出路径上的数值计算收敛性给出，如图 4-22 所示。同样可以得到与前几个算例类似的结论：

①总体而言，$a(x)$ 参与导数计算，相比于其不参与导数运算，求解结果会变得更差。虽然在随机节点高度稠密区间，即 $h_a < 0.02$ 时，$a(x)$ 参与导数计算的结果似乎更好些，但这个区间对经典 MLS 而言，数值解已经丧失精确性，其优劣的评价意义不大。②本算例再次验证了改进的 MLSc 的数值解是严格收敛的，而 MLS 会随着节点稠密而变得非常不稳定。

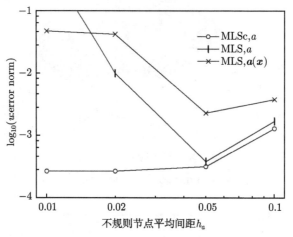

图 4-22　数值解的收敛性比较

Fig. 4-22　Comparison of the numerical convergence

综上，通过 4 个数值算例验证，均得出 $a(x)$ 参与导数运算的负面结论，这显然与 EFG 方法的建议不同。其中，两个算例是通过曲面拟合的方式，来直接检验 MLS 的导数近似效果，对问题的评价是有普遍意义的。此外，还采用一种具体的无网格方法 (FPM)，通过两个数值求解算例来评价 $a(x)$ 参与导数运算的优劣性。用一种具体的方法评价，显然有失全面性和公允性。但至少可以说明一个问题：EFG 方法中得出的 $a(x)$ 参与导数运算使数值求解更精确，这个结论不是对任何方法都适用，不具有普遍性。

在文献 [2] 中对 $a(x)$ 参与导数运算的效果评价，是通过一个简单小片试验算例进行验证的。场节点非常稀疏，精确解又是一个简单的二次多项式，加之 EFG 方法只会使用一阶导数近似。因此，其论据支持显然偏弱，说服力不够。将其结论不加鉴别地普遍推广，是不恰当的。

从原理性分析的角度来看，系数向量 $a(x)$ 不参与导数运算是完全合理的。简单地说，多项式函数的求导，与系数无关，这应当是没有问题的。将 $a(x)$ 作为函数处理，并参与导数运算，增加了问题的复杂性。更重要的是，经过以上的初步数值验证，这种处理会使情况变差，而非向好。

此外，数值算例也对建议的核近似方法 (MLSc) 和经典的 MLS 法进行了比较，进一步证明了 MLSc 法是一种严格收敛和计算稳定的有效方法。此数值检验结果完全符合文献 [6] 的理论解释。MLSc 法的改进措施非常简单，但对提升 MLS 方法的计算效果却非常显著。

4.7 本章小结

对 MLS 的稳定精确计算而言，首先要保证 A 矩阵非奇异，而且是条件良好的。在 4.3 节中给出的两个几何条件，尤其是第二几何条件，实际上给出了规范的支撑点集 $\{x_J\}_{J=1}^N$ 在局部域上的分布要求。反言之，如果场节点分布是预先确定的，由给出的两个几何条件，可以合理选择支撑点，并确定合理的支撑域尺度。给出这两个几何必要条件，其意义在于将讨论 A^{-1} 是否存在的纯代数问题，转换为考察支撑域上支撑点集空间分布的几何问题，在实际应用中便于直观判断，并能做到预先估计。根据其推论，甚至可以通过读图直观判断 A 的条件及其病态性。将其应用于数值方法，有助于合理规划场节点，并合理选择计算参数。

MLSc 是对经典 MLS 法的一种改进。MLSc 采用 $p^T(x-x_I)$ 构造试函数，能构造出随计算节点 x_I 移动的局部近似函数，则 MLS 近似法具有了更为纯粹的"移动"性。这种改进方法非常简单，但对于提高 MLS 近似稳定性而言，效果却非常显著。MLSc 本质上与 RKPM 是等价的，但其构造方法要简单得多。

改进的移动最小二乘 (IMLS) 法使用带权的正交基函数构造试函数，可以避免对 A 求逆运算，也是一种稳定化的改进算法。如果采用适当的带权正交基函数，在基函数阶次相同的情况下，可以减少基函数项数，其未知数也对应减少，所需的支撑节点数也对应减少，有助于显著提高计算效率。IMLS 是对经典 MLS 的一种更为全面的改进，表现出更具竞争性的应用潜力。

随着 EFG 法取得极大成功，系数向量 $a(x)$ 参与 MLS 形函数导数计算成为一种广泛使用的标准形式。鉴于这种导数计算公式较为复杂，为了探讨其简化计算的可能性，所以有必要对其重新审视。作者认为 $a(x)$ 参与导数计算并不是完全必要的，仅是一种经验性的用法。我们总的结论就是：MLS 导数近似中，系数向量 $a(x)$ 不应当参与导数运算，并建议采用核近似方法 (MLSc)，这样的 MLS 算法会更简单，计算会更加精确和稳定，在无网格近似方法的选择比较中将更具竞争力。

参 考 文 献

[1] Nayroles B, Touzot G, Villon P. Generalizing the finite element method: Diffuse approximation and diffuse elements[J]. Computational Mechanics, 1992, 10(5): 307-318.

[2] Belytschko T, Lu Y Y, Gu L. Element-free Galerkin methods[J]. International Journal for Numerical Methods in Engineering, 1994, 37(2): 229-256.

[3] Atluri S N, Zhu T. A new meshless local Petrov-Galerkin (MLPG) approach in computational mechanics[J]. Computational Mechanics, 1998, 22(2): 117-127.

[4] 程玉民. 移动最小二乘法研究进展与述评 [J]. 计算机辅助工程, 2009, 18(2): 5-11.

[5] Liu W K, Chen Y, Uras R A, et al. Generalized multiple scale reproducing kernel particle methods[J]. Computer Methods in Applied Mechanics and Engineering, 1996, 139(1-4): 91-157.

[6] 杨建军, 郑健龙. 移动最小二乘法的近似稳定性 [J]. 应用数学学报, 2012, 35(4): 637-648.

[7] Liu G R, Gu Y T. An introduction to meshfree methods and their programming[M]. Netherlands: Springer Science & Business Media, 2005.

[8] 左传伟, 聂玉峰, 赵美玲. 移动最小二乘方法中影响半径的选取 [J]. 工程数学学报. 2005, 22(5):833-838.

[9] 袁占斌, 聂玉峰, 欧阳洁. 基于泰勒基函数的移动最小二乘法及误差分析 [J]. 数值计算与计算机应用, 2012, 33(1): 25-31.

[10] 陈美娟, 程玉民. 改进的移动最小二乘法 [J]. 力学季刊, 2003, 24(2): 266-272.

[11] 程玉民. 移动最小二乘法研究进展与述评 [J]. 计算机辅助工程, 2009, 18(2): 5-11.

[12] 任红萍, 程玉民, 张武. 改进的移动最小二乘插值法研究 [J]. 工程数学学报, 2010, 27(6): 1021-1029.

[13] 程玉民, 彭妙娟, 李九红. 复变量移动最小二乘法及其应用 [J]. 力学学报. 2005, 37(6): 719-723.

[14] Gao H, Cheng Y. A complex variable meshless manifold method for fracture problems[J]. International Journal of Computational Methods, 2010, 7(01): 55-81.

[15] 杨建军, 杨子乐, 黄旺, 等. 移动最小二乘法导数近似讨论 [J]. 计算机辅助工程,2018,27(1): 28-34.

[16] Onate E, Idelsohn S, Zienkiewicz O C, et al. A finite point method in computational mechanics. Applications to convective transport and fluid flow[J]. International Journal for Numerical Methods in Engineering, 1996, 39(22): 3839-3866.

[17] Onate E, Idelsohn S, Zienkiewicz O C, et al. A stabilized finite point method for analysis of fluid mechanics problems[J]. Computer Methods in Applied Mechanics and Engineering, 1996, 139(1-4): 315-346.

[18] Timoshenko S P, Goodier J N. Theory of Elasticity[M]. Beijing: Tsinghua University Press, 2004.

[19] 梁昆淼. 数学物理方法 [M]. 北京: 高等教育出版社, 1998.

第 5 章　无网格全局弱式法

无网格全局弱式法采用加权残值法的伽辽金 (Galerkin) 法进行数值离散,其收敛性解释符合介点原理的推论 1。最初的无网格全局弱式法是由 Nayroles 等[1]提出的所谓扩散单元法 (diffuse element method, DEM),此后,Belytschko 等[2] 在 DEM 基础上作了部分修整,并称之为无单元伽辽金 (element-free Galerkin, EFG) 法,这两种方法均使用 MLS 构造形函数。如果采用第 2 章中介绍的其他形函数构造方法,则可以发展出多种全局弱式法。重构核粒子法 (reproducing kernel particle method, RKPM)[3]、自然单元法 (natural element method, NEM)[4]、自然邻接伽辽金法 (natural neighbour Galerkin method, NNGM)[5]、点插值法 (point interpolation method, PIM)[6]、无网格伽辽金法 (meshless Galerkin method, MGM)[7]、径向基点插值法 (radial point interpolation method, RPIM)[8]、Hp 云法[9,10]、扩展有限元法 (extended finite element method, EFEM)[11]、广义有限元法 (generalized finite element method, GFEM)[12]、误差重构无网格 (error reproducing mesh-free, ERMF) 法[13] 等都属于无网格全局弱式法。

5.1　强式和弱式

本书中经常用"强式"和"弱式"来划分无网格法的类型,因此有必要对其概念先作一个简单的说明。简而言之,所谓弱式,就是所构建的离散系统方程使用积分运算的形式;对应地,所谓强式就是所构建的离散系统方程不使用积分运算的形式。

加权残值法的配点法是一种典型的基于强式的方法,其最终得到的离散方程是由式 (2-29) 和式 (2-30) 所表示的直接令残值函数为 0 的形式,是一个单纯的微分方程形式,而非一个微分方程的积分形式。其优点主要体现在:其一,离散方程的构造非常简单,由控制方程和边界条件可以直接写出,不需要任何数学上的变换;其二,所需要的计算元素最少,求解效率很高;其三,数值实施简单,很容易适应具有复杂几何结构的求解问题。然而,在强式方法的算式中,试函数的导数阶数和控制方程的微分形式是同阶的,要求试函数有较高的连续性,而且对试函数及其导数的逼近要求是强条件的。

加权残值法的 Galerkin 法和子域法是典型的基于全局弱式和局部弱式的方法,由其导出的离散系统方程是微分方程的积分形式。弱式方法的优点有:能够将数学或物理原理通过积分的形式引入系统方程,形成富于变化的离散方程,可获得稳定而高精度的计算结果。其不利的一方面是,形成的离散系统方程比较复杂,通常与控制方程的初始形式有较大差别,数值实施较为复杂,计算效率也有不同程度的降低。

以弱式形式表示的场变量的近似函数,对连续性的要求较强式形式有显著差别。对于一些数学或物理问题中经典的二阶微分控制方程,强式要求近似函数也为二阶连续,而其对应的弱式通常仅要求具有一阶连续。因此,弱式通过积分运算,降低了近似函数连续性阶数的要求,对试函数及其导数的逼近要求是弱条件的,这就是所谓 "弱式" 这一术语的由来 [14]。

5.2 Galerkin 弱式

加权残值法的 Galerkin 法中,将权函数取为位移函数的变分,则弹性静力学的 Galerkin 弱式写为

$$\delta \varPi^{\mathrm{PE}} = -\int_{\varOmega} \delta u_i \left(\sigma_{ij,j} + b_i\right) \mathrm{d}\varOmega + \int_{\varGamma_t} \delta u_i \left(n_j \sigma_{ij} - \bar{t}_i\right) \mathrm{d}\varGamma = 0 \tag{5-1}$$

对比式 (2-57),可见上式即最小势能原理的变分形式。对上式第一项进行分部积分,并应用 Green 公式,则有

$$\begin{aligned}
&\int_{\varOmega} \delta u_i \left(\sigma_{ij,j} + b_i\right) \mathrm{d}\varOmega \\
&= \int_{\varOmega} \delta u_i \sigma_{ij,j} \mathrm{d}\varOmega + \int_{\varOmega} \delta u_i b_i \mathrm{d}\varOmega \\
&= \int_{\varGamma_t} \delta u_i n_j \sigma_{ij} \mathrm{d}\varGamma - \int_{\varOmega} \delta u_{i,j} \sigma_{ij} \mathrm{d}\varOmega + \int_{\varOmega} \delta u_i b_i \mathrm{d}\varOmega
\end{aligned} \tag{5-2}$$

将上式代入式 (5-1),得到

$$\begin{aligned}
\delta \varPi^{\mathrm{PE}} &= \int_{\varOmega} \delta u_{i,j} \sigma_{ij} \mathrm{d}\varOmega - \int_{\varOmega} \delta u_i b_i \mathrm{d}\varOmega - \int_{\varGamma_t} \delta u_i \bar{t}_i \mathrm{d}\varGamma \\
&= \int_{\varOmega} \delta \varepsilon_{ij} E_{ijkl} \varepsilon_{kl} \mathrm{d}\varOmega - \int_{\varOmega} \delta u_i b_i \mathrm{d}\varOmega - \int_{\varGamma_t} \delta u_i \bar{t}_i \mathrm{d}\varGamma = 0
\end{aligned} \tag{5-3}$$

上式即 Galerkin 弱式的一般公式。将上式写成矩阵表示的形式:

$$\begin{aligned}
&\int_{\varOmega} \delta \boldsymbol{\varepsilon}^{\mathrm{T}} \boldsymbol{D} \boldsymbol{\varepsilon} \mathrm{d}\varOmega - \int_{\varOmega} \delta \boldsymbol{u}^{\mathrm{T}} \boldsymbol{b} \mathrm{d}\varOmega - \int_{\varGamma_t} \delta \boldsymbol{u}^{\mathrm{T}} \bar{\boldsymbol{t}} \mathrm{d}\varGamma \\
&= \int_{\varOmega} (\boldsymbol{L} \delta \boldsymbol{u})^{\mathrm{T}} \boldsymbol{D} (\boldsymbol{L} \boldsymbol{u}) \mathrm{d}\varOmega - \int_{\varOmega} \delta \boldsymbol{u}^{\mathrm{T}} \boldsymbol{b} \mathrm{d}\varOmega - \int_{\varGamma_t} \delta \boldsymbol{u}^{\mathrm{T}} \bar{\boldsymbol{t}} \mathrm{d}\varGamma = 0
\end{aligned} \tag{5-4}$$

5.2 Galerkin 弱式

式中的符号与 2.1 节的表达一致。将式 (3-1) 表示的无网格近似函数公式代入,并考虑到变分符号 δ 的任意性,将得到 Galerkin 弱式的离散系统方程:

$$K\hat{u} = F \tag{5-5}$$

式中,K 为整体刚度矩阵;F 为总体力矩阵;而

$$\hat{u} = [\hat{u}_1, \hat{u}_2, \cdots, \hat{u}_n]^{\mathrm{T}} \tag{5-6}$$

为总体位移矩阵,是待求解的位移未知数向量,$\hat{u}_I$ 就是第 I 个节点的待定位移向量。

需特别注意,由式 (5-5) 得到的系统方程还不能直接求解,因为在这个方程组中,尚不包含位移边界条件,在数值求解时还需要特别的考虑。这部分内容将在 5.3 节专门介绍。

为了便于对无网格全局弱式法的数值实施和式 (5-5) 的系统方程组装有一个更好的理解,接下来以二维问题为例,给出进一步的说明。其数值离散如图 5-1 所示,图中 x_p 为第 i 个编号的积分网格 c_i 中的一个积分点,也可称之为弱介点,则 Ω_S 表示该积分点构造试函数的局部支撑域。而 x_I 和 x_J 是求解域内给出特别标识的两个场节点,其中 x_I 位于 Ω_S 内。

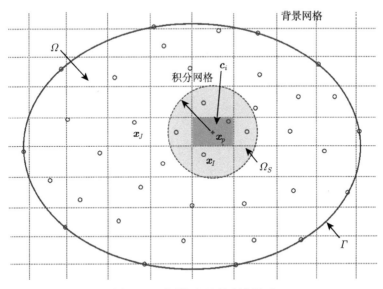

图 5-1 无网格全局弱式法图示

Fig. 5-1 Schematics of the meshless global weak-form methods

式 (5-5) 是以积分点 x_p 为基础进行组装的,因此式 (5-4) 的离散形式要与 x_p 对应,并须用场变量近似函数来代替场变量,即 $u_p(x) \approx u_p^h(x)$。如果问题域及其

边界共使用 n 个场节点离散，并对节点 $\{x_I\}_{I=1}^n$ 严格序列化，即每个节点都有其固定编号。则对积分点 x_p 的位移近似函数写为

$$u_p^h(x)=\left\{\begin{array}{c}u_{px}^h\\u_{py}^h\end{array}\right\}=\left[\begin{array}{ccccccc}\phi_1&0&\phi_2&0&\cdots&\phi_n&0\\0&\phi_1&0&\phi_2&\cdots&0&\phi_n\end{array}\right]\left\{\begin{array}{c}\hat{u}_{1x}\\\hat{u}_{1y}\\\hat{u}_{2x}\\\hat{u}_{2y}\\\vdots\\\hat{u}_{nx}\\\hat{u}_{ny}\end{array}\right\}=\boldsymbol{\Phi}\hat{\boldsymbol{u}} \quad (5\text{-}7)$$

上式也可表示为对节点求和的形式：

$$u_p^h(x)=\sum_{I=1}^n\left[\begin{array}{cc}\phi_I&0\\0&\phi_I\end{array}\right]\left\{\begin{array}{c}\hat{u}_{Ix}\\\hat{u}_{Iy}\end{array}\right\}=\sum_{I=1}^n\boldsymbol{\phi}_I\hat{\boldsymbol{u}}_I \quad (5\text{-}8)$$

需注意到，上式是全局表示的二维近似函数。在执行近似时，仍然采用局部近似的方法，将局部域 Ω_S 内任一节点 x_I 的形函数值 ϕ_I 写到相应位置上，对于局部域外节点，其形函数在上式对应位置上赋 0 值。

对积分点 x_p 的应变近似函数写为

$$\boldsymbol{\varepsilon}_p^h = \boldsymbol{L}\boldsymbol{u}_p^h = \boldsymbol{L}\boldsymbol{\Phi}\hat{\boldsymbol{u}}$$

$$=\left[\begin{array}{cc}\dfrac{\partial}{\partial x}&0\\0&\dfrac{\partial}{\partial y}\\\dfrac{\partial}{\partial y}&\dfrac{\partial}{\partial x}\end{array}\right]\left[\begin{array}{ccccccc}\phi_1&0&\phi_2&0&\cdots&\phi_n&0\\0&\phi_1&0&\phi_2&\cdots&0&\phi_n\end{array}\right]\left\{\begin{array}{c}\hat{u}_{1x}\\\hat{u}_{1y}\\\hat{u}_{2x}\\\hat{u}_{2y}\\\vdots\\\hat{u}_{nx}\\\hat{u}_{ny}\end{array}\right\}$$

$$=\left[\begin{array}{cccccccc}\dfrac{\partial\phi_1}{\partial x}&0&\dfrac{\partial\phi_2}{\partial x}&0&\cdots&\dfrac{\partial\phi_n}{\partial x}&0\\0&\dfrac{\partial\phi_1}{\partial y}&0&\dfrac{\partial\phi_2}{\partial y}&\cdots&0&\dfrac{\partial\phi_n}{\partial y}\\\dfrac{\partial\phi_1}{\partial y}&\dfrac{\partial\phi_1}{\partial x}&\dfrac{\partial\phi_2}{\partial y}&\dfrac{\partial\phi_2}{\partial x}&\cdots&\dfrac{\partial\phi_n}{\partial y}&\dfrac{\partial\phi_n}{\partial x}\end{array}\right]\left\{\begin{array}{c}\hat{u}_{1x}\\\hat{u}_{1y}\\\hat{u}_{2x}\\\hat{u}_{2y}\\\vdots\\\hat{u}_{nx}\\\hat{u}_{ny}\end{array}\right\}$$

5.2 Galerkin 弱式

$$=\boldsymbol{B}_{(3\times 2n)}\hat{\boldsymbol{u}}_{(2n\times 1)} = \sum_{I=1}^{n} \boldsymbol{B}_I \hat{\boldsymbol{u}}_I \tag{5-9}$$

式中，

$$\boldsymbol{B}_I = \boldsymbol{L}\phi_I = \begin{bmatrix} \dfrac{\partial \phi_I}{\partial x} & 0 \\ 0 & \dfrac{\partial \phi_I}{\partial y} \\ \dfrac{\partial \phi_I}{\partial y} & \dfrac{\partial \phi_I}{\partial x} \end{bmatrix} \tag{5-10}$$

则对积分点 $\boldsymbol{x}_p$ 的应力近似函数写为

$$\boldsymbol{\sigma}_p^h = \boldsymbol{D}\boldsymbol{\varepsilon}_p^h = \boldsymbol{D}_{(3\times 3)}\boldsymbol{B}_{(3\times 2n)}\hat{\boldsymbol{u}}_{(2n\times 1)} = \sum_{I=1}^{n} \boldsymbol{D}\boldsymbol{B}_I \hat{\boldsymbol{u}}_I \tag{5-11}$$

因此，式 (5-4) 第一项写成积分点 $\boldsymbol{x}_p$ 对应的形式，并将近似函数代入，则有

$$\begin{aligned}
\int_\Omega \delta\left(\boldsymbol{\varepsilon}_p^h\right)^{\mathrm{T}} \boldsymbol{D}\boldsymbol{\varepsilon}_p^h \mathrm{d}\Omega &= \int_\Omega \left(\sum_{I=1}^{n} \boldsymbol{B}_I \delta\hat{\boldsymbol{u}}_I\right)^{\mathrm{T}} \boldsymbol{D} \left(\sum_{J=1}^{n} \boldsymbol{B}_J \hat{\boldsymbol{u}}_J\right) \mathrm{d}\Omega \\
&= \int_\Omega \left\{\sum_{I=1}^{n}\sum_{J=1}^{n} \delta\hat{\boldsymbol{u}}_I^{\mathrm{T}} \left[\boldsymbol{B}_I^{\mathrm{T}} \boldsymbol{D}\boldsymbol{B}_J\right] \hat{\boldsymbol{u}}_J\right\} \mathrm{d}\Omega \\
&= \sum_{I=1}^{n}\sum_{J=1}^{n} \delta\hat{\boldsymbol{u}}_I^{\mathrm{T}} \underbrace{\left[\int_\Omega \boldsymbol{B}_I^{\mathrm{T}} \boldsymbol{D}\boldsymbol{B}_J \mathrm{d}\Omega\right]}_{\boldsymbol{K}_{IJ}} \hat{\boldsymbol{u}}_J \\
&= \sum_{I=1}^{n}\sum_{J=1}^{n} \delta\hat{\boldsymbol{u}}_I^{\mathrm{T}} \boldsymbol{K}_{IJ} \hat{\boldsymbol{u}}_J \\
&= \delta\hat{\boldsymbol{u}}^{\mathrm{T}} (\boldsymbol{K}) \hat{\boldsymbol{u}}
\end{aligned} \tag{5-12}$$

其中

$$(\boldsymbol{K}_{IJ})_{(2\times 2)} = \int_\Omega \left(\boldsymbol{B}_I^T\right)_{(2\times 3)} \boldsymbol{D}_{(3\times 3)} (\boldsymbol{B}_J)_{(3\times 2)} \mathrm{d}\Omega \tag{5-13}$$

由 $\boldsymbol{D}$ 的对称性可得到

$$\left[\boldsymbol{B}_I^{\mathrm{T}} \boldsymbol{D}\boldsymbol{B}_J\right]^{\mathrm{T}} = \left[\boldsymbol{B}_J^{\mathrm{T}} \boldsymbol{D}\boldsymbol{B}_I\right] \tag{5-14}$$

即有

$$[\boldsymbol{K}_{IJ}]^{\mathrm{T}} = \boldsymbol{K}_{JI} \tag{5-15}$$

这意味着总体刚度矩阵 $\boldsymbol{K}$ 将是对称的。将式 (5-13) 定义的子矩阵 $\boldsymbol{K}_{IJ}$ 组装到总

体矩阵的对应位置，则可得到总体刚度矩阵 K，其形式为

$$K_{(2n \times 2n)} = \begin{bmatrix} K_{11} & K_{12} & \cdots & K_{1n} \\ K_{21} & K_{22} & \cdots & K_{2n} \\ \vdots & \vdots & \ddots & \vdots \\ K_{n1} & K_{n2} & \cdots & K_{nn} \end{bmatrix} \tag{5-16}$$

式 (5-4) 的第二项对应写为

$$\int_\Omega \delta \left(u_p^h\right)^{\mathrm{T}} b_p \mathrm{d}\Omega = \int_\Omega \delta \left(\sum_{I=1}^n \phi_I \hat{u}_I\right)^{\mathrm{T}} b_p \mathrm{d}\Omega = \sum_{I=1}^n \left\{ \delta \hat{u}_I^{\mathrm{T}} \underbrace{\left(\int_\Omega \phi_I^{\mathrm{T}} b_p \mathrm{d}\Omega\right)}_{F_I^b} \right\} = \delta \hat{u}^{\mathrm{T}} F^b \tag{5-17}$$

其中，

$$b_p = [b_{px}, b_{py}]^{\mathrm{T}} \tag{5-18}$$

为积分点 x_p 的体力向量，而

$$F^b = \left[F_1^b, F_2^b, \cdots, F_n^b\right]^{\mathrm{T}} \tag{5-19}$$

是总体体力向量，其中

$$\left(F_I^b\right)_{(2\times 1)} = \int_\Omega \phi_I^{\mathrm{T}} b_p \mathrm{d}\Omega \tag{5-20}$$

为场节点 x_I 对应的节点体力向量。

式 (5-4) 的第三项对应写为

$$\int_{\Gamma_t} \delta \left(u_p^h\right)^{\mathrm{T}} \bar{t}_p \mathrm{d}\Gamma = \int_{\Gamma_t} \delta \left(\sum_{I=1}^n \phi_I \hat{u}_I\right)^{\mathrm{T}} \bar{t}_p \mathrm{d}\Gamma = \sum_{I=1}^n \left\{ \delta \hat{u}_I^{\mathrm{T}} \underbrace{\left(\int_{\Gamma_t} \phi_I^{\mathrm{T}} \bar{t}_p \mathrm{d}\Gamma\right)}_{F_I^t} \right\} = \delta \hat{u}^{\mathrm{T}} F^t \tag{5-21}$$

其中，

$$\bar{t}_p = [\bar{t}_{px}, \bar{t}_{py}]^{\mathrm{T}} \tag{5-22}$$

为边界积分点 x_p 的表面力向量，而

$$F^t = \left[F_1^t, F_2^t, \cdots, F_n^t\right]^{\mathrm{T}} \tag{5-23}$$

是总体表面力向量，其中

$$\left(F_I^t\right)_{(2\times 1)} = \int_{\Gamma_t} \phi_I^{\mathrm{T}} \bar{t}_p \mathrm{d}\Gamma \tag{5-24}$$

为场节点 x_I 对应的节点表面力向量。

将式 (5-12)、式 (5-17)、式 (5-21) 代入式 (5-4) 得到

$$\delta\hat{\boldsymbol{u}}^{\mathrm{T}}\boldsymbol{K}\hat{\boldsymbol{u}} - \delta\hat{\boldsymbol{u}}^{\mathrm{T}}\boldsymbol{F}^b - \delta\hat{\boldsymbol{u}}^{\mathrm{T}}\boldsymbol{F}^t = 0 \tag{5-25}$$

由变分的任意性，可在上式两边同乘以 $1/\delta\hat{\boldsymbol{u}}^{\mathrm{T}}$，则可消除 $\delta\hat{\boldsymbol{u}}^{\mathrm{T}}$，并令

$$\boldsymbol{F} = \boldsymbol{F}^b + \boldsymbol{F}^t \tag{5-26}$$

则有

$$\begin{cases} \boldsymbol{F} = [\boldsymbol{F}_1, \boldsymbol{F}_2, \cdots, \boldsymbol{F}_n]^{\mathrm{T}} \\ \boldsymbol{F}_I = \boldsymbol{F}_I^b + \boldsymbol{F}_I^t = \int_{\Omega} \boldsymbol{\phi}_I^{\mathrm{T}} \boldsymbol{b}_p \mathrm{d}\Omega + \int_{\Gamma_t} \boldsymbol{\phi}_I^{\mathrm{T}} \bar{\boldsymbol{t}}_p \mathrm{d}\Gamma \end{cases} \tag{5-27}$$

最终将得到式 (5-5) 所示的总体离散系统方程形式。综上，可以看出 Galerkin 弱式无网格法的计算公式与有限元法的计算格式非常相似。

5.3 位移边界条件的施加

由式 (5-5) 得到的系统方程还不能直接求解，还需要进一步考虑位移边界条件施加的问题。如果形函数 $\phi(\boldsymbol{x})$ 具有 Kronecker δ 函数性质，则位移边界条件可以直接施加，而如果形函数不具有此性质，则需要特别的处理。接下来，分两种情况予以介绍。

5.3.1 形函数具有插值特性

如果形函数 $\phi(\boldsymbol{x})$ 具有插值特性，即具有 Kronecker δ 函数性质，比如 PIM, RPIM, NNI, KIM 等类型形函数均具有该函数性质，则无须事先考虑位移边界条件的施加，而是在离散方程组装完成后，即得到式 (5-5) 的系统矩阵后，再进行处理，较为常用的有两种施加方式。

1. 直接法

设节点 x_I 给定位移，即

$$\boldsymbol{u}_I = \bar{\boldsymbol{u}}_I = \left\{ \begin{array}{c} \bar{u}_{Ix} \\ \bar{u}_{Iy} \end{array} \right\} \tag{5-28}$$

则式 (5-16) 给出的总体刚度矩阵 $\boldsymbol{K}$ 需要改写，改写的具体步骤是将 $\boldsymbol{K}_{II}$ 子阵改写为单位矩阵，即

$$\boldsymbol{K}_{II} = [\boldsymbol{I}] = \left[\begin{array}{cc} 1 & 0 \\ 0 & 1 \end{array} \right] \tag{5-29}$$

在 K_{II} 子阵对应的行和列子阵都改写为 0 矩阵，即

$$K_{IJ} = K_{JI} = [0] = \begin{bmatrix} 0 & 0 \\ 0 & 0 \end{bmatrix}, \quad J \neq I \tag{5-30}$$

则改写后的总体矩阵形式为

$$K^R = \begin{bmatrix} K_{11} & K_{12} & \cdots & K_{1I} = [0] & \cdots & K_{1n} \\ K_{21} & K_{22} & \cdots & K_{2I} = [0] & \cdots & K_{2n} \\ \vdots & \vdots & \ddots & \vdots & \ddots & \vdots \\ K_{I1} = [0] & K_{I2} = [0] & \cdots & K_{II} = [I] & \cdots & K_{In} = [0] \\ \vdots & \vdots & \ddots & \vdots & \ddots & \vdots \\ K_{n1} & K_{n2} & \cdots & K_{nI} = [0] & \cdots & K_{nn} \end{bmatrix} \tag{5-31}$$

总体力向量的相应分量也需要改写为

$$\begin{cases} F_I^R = \bar{u}_I \\ F_J^R = F_J - K_{JI}\bar{u}_I, \quad J \neq I \end{cases} \tag{5-32}$$

如果节点 x_I 上只有某一个坐标分量上的位移给定，则刚度子阵和力向量子阵只改写对应的一行即可。

采用直接法可精确施加本质边界条件，但修改刚度矩阵和力矩阵操作比较复杂，而且需要增加额外计算量。

2. 罚函数法

罚函数法是另外一种施加本质边界条件的简便方法，而且只需对 K_{II} 和 F_I 改写即可，不涉及其他与节点 x_I 相关的刚度子阵和力向量子阵，其改写规则为

$$\begin{cases} K_{II} = \alpha \cdot K_{II} \\ F_I = \alpha \cdot K_{II}\bar{u}_I \end{cases} \tag{5-33}$$

则改写后的总体矩阵形式为

$$K^R = \begin{bmatrix} K_{11} & K_{12} & \cdots & K_{1I} & \cdots & K_{1n} \\ K_{21} & K_{22} & \cdots & K_{2I} & \cdots & K_{2n} \\ \vdots & \vdots & \ddots & \vdots & \ddots & \vdots \\ K_{I1} & K_{I2} & \cdots & \alpha \cdot K_{II} & \cdots & K_{In} \\ \vdots & \vdots & \ddots & \vdots & \ddots & \vdots \\ K_{n1} & K_{n2} & \cdots & K_{nI} & \cdots & K_{nn} \end{bmatrix} \tag{5-34}$$

5.3 位移边界条件的施加

式中，α 为罚参数，其数量级需要合理选择，根据经验，其值可取为

$$\alpha = \left(10^4 \sim 10^8\right) \times \{K_{II}\}_{\max} \tag{5-35}$$

式中，$\{K_{II}\}_{\max}$ 为总体刚度矩阵中最大的对角线元素。

罚函数法施加本质边界条件更为简单，所需要的额外计算量很少。然而它仅能近似满足边界条件，其精确性易受罚参数取值的影响。

5.3.2 形函数不具有插值特性

如果形函数 $\phi(x)$ 不具有插值特性，即不具有 Kronecker δ 函数性质，比如 MLS、RKPM 等，则位移边界条件的施加需要最初就对泛函式 (5-1) 进行改写，将位移边界条件引入。有几种可行的方法，接下来予以简要介绍。

1. Lagrange 乘子法

拉格朗日 (Lagrange) 乘子法由 Belytschko 等[2] 提出，并成为 EFG 法的一般计算格式。泛函式 (5-1) 采用 Lagrange 乘子法引入位移边界条件，将变成

$$\begin{aligned}\delta\Pi = & -\int_{\Omega} \delta u_i \left(\sigma_{ij,j} + b_i\right) \mathrm{d}\Omega + \int_{\Gamma_t} \delta u_i \left(n_j \sigma_{ij} - \bar{t}_i\right) \mathrm{d}\Gamma \\ & -\int_{\Gamma_u} \delta \lambda_i \left(u_i - \bar{u}_i\right) \mathrm{d}\Gamma - \int_{\Gamma_u} \delta u_i \lambda_i \mathrm{d}\Gamma = 0\end{aligned} \tag{5-36}$$

由此推导得出矩阵形式的方程：

$$\begin{aligned}& \int_{\Omega} (\boldsymbol{L}\delta\boldsymbol{u})^{\mathrm{T}} \boldsymbol{D} (\boldsymbol{L}\boldsymbol{u}) \mathrm{d}\Omega - \int_{\Omega} \delta\boldsymbol{u}^{\mathrm{T}} \boldsymbol{b} \mathrm{d}\Omega - \int_{\Gamma_t} \delta\boldsymbol{u}^{\mathrm{T}} \bar{\boldsymbol{t}} \mathrm{d}\Gamma \\ & -\int_{\Gamma_u} \delta\boldsymbol{\lambda}^{\mathrm{T}} (\boldsymbol{u} - \bar{\boldsymbol{u}}) \mathrm{d}\Gamma - \int_{\Gamma_u} \delta\boldsymbol{u}^{\mathrm{T}} \boldsymbol{\lambda} \mathrm{d}\Gamma = 0\end{aligned} \tag{5-37}$$

式中，Lagrange 乘子 λ 实际上是一种与边界节点相关的插值形函数，其物理意义是表示位移自变函数，对于本质边界 Γ_u 上的任意一个积分点 x_p，有

$$\tilde{\boldsymbol{\lambda}}_p = \left\{\begin{array}{c} \tilde{\lambda}_{px} \\ \tilde{\lambda}_{py} \end{array}\right\} = \left[\begin{array}{ccccccc} N_1 & 0 & N_2 & 0 & \cdots & N_{n_u} & 0 \\ 0 & N_1 & 0 & N_2 & \cdots & 0 & N_{n_u} \end{array}\right] \left\{\begin{array}{c} \hat{\lambda}_{1x} \\ \hat{\lambda}_{1y} \\ \hat{\lambda}_{2x} \\ \hat{\lambda}_{2y} \\ \vdots \\ \hat{\lambda}_{n_u x} \\ \hat{\lambda}_{n_u y} \end{array}\right\}$$

$$= \boldsymbol{N}_{(2 \times 2n_u)} \hat{\boldsymbol{\lambda}}_{(2n_u \times 1)} \tag{5-38}$$

式中，n_u 表示本质边界上的节点数量。上式可改写为节点形函数的求和形式：

$$\tilde{\boldsymbol{\lambda}}_p = \sum_{I=1}^{n_u} \begin{bmatrix} N_I & 0 \\ 0 & N_I \end{bmatrix} \left\{ \begin{array}{c} \hat{\lambda}_{Ix} \\ \hat{\lambda}_{Iy} \end{array} \right\} = \sum_{I=1}^{n_u} \boldsymbol{N}_I \hat{\boldsymbol{\lambda}}_I \tag{5-39}$$

式中，$\boldsymbol{N}_I$ 表示边界节点对应的 $\boldsymbol{x}_I$ 插值形函数；$\hat{\boldsymbol{\lambda}}_I$ 为其对应的 Lagrange 乘子向量。如果将 $\boldsymbol{x}_I$ 特别定义为距离计算积分点 $\boldsymbol{x}_p$ 最近的本质边界节点，而 $\boldsymbol{x}_{I\pm1}$ 定义为次近的边界节点，则一种简单的插值形函数计算公式可定义为

$$N_I = \frac{\|\boldsymbol{x}_p - \boldsymbol{x}_I\|}{\|\boldsymbol{x}_I - \boldsymbol{x}_{I\pm1}\|} \tag{5-40}$$

由此定义，式 (5-38) 的整体形函数 $\boldsymbol{N}$ 中，只有最近的节点 $\boldsymbol{x}_I$ 的形函数 N_I 非 0，其他节点形函数均为 0。

则式 (5-37) 的第四项对应写为

$$\int_{\Gamma_u} \delta \tilde{\boldsymbol{\lambda}}_p^{\mathrm{T}} (\tilde{\boldsymbol{u}}_p - \bar{\boldsymbol{u}}_p) \mathrm{d}\Gamma$$

$$= \int_{\Gamma_u} \delta \left(\sum_{I=1}^{n_u} \boldsymbol{N}_I \hat{\boldsymbol{\lambda}}_I \right)^{\mathrm{T}} \left(\sum_{J=1}^{n} \boldsymbol{\phi}_J \hat{\boldsymbol{u}}_J \right) \mathrm{d}\Gamma - \int_{\Gamma_u} \delta \left(\sum_{I=1}^{n_u} \boldsymbol{N}_I \hat{\boldsymbol{\lambda}}_I \right)^{\mathrm{T}} \bar{\boldsymbol{u}}_p \mathrm{d}\Gamma$$

$$= \sum_{I=1}^{n_u} \sum_{J=1}^{n} \left\{ \delta \hat{\boldsymbol{\lambda}}_I^{\mathrm{T}} \underbrace{\left(\int_{\Gamma_u} \boldsymbol{N}_I^{\mathrm{T}} \boldsymbol{\phi}_J \mathrm{d}\Gamma \right)}_{-\boldsymbol{G}_{IJ}^{\mathrm{T}}} \hat{\boldsymbol{u}}_J \right\} - \sum_{I=1}^{n_u} \left\{ \delta \hat{\boldsymbol{\lambda}}_I^{\mathrm{T}} \underbrace{\left(\int_{\Gamma_u} \boldsymbol{N}_I^{\mathrm{T}} \bar{\boldsymbol{u}}_p \mathrm{d}\Gamma \right)}_{-\boldsymbol{Q}_I} \right\}$$

$$= -\sum_{I=1}^{n_u} \sum_{J=1}^{n} \left\{ \delta \hat{\boldsymbol{\lambda}}_I^{\mathrm{T}} \boldsymbol{G}_{IJ}^{\mathrm{T}} \hat{\boldsymbol{u}}_J \right\} + \sum_{I=1}^{n_u} \left\{ \delta \hat{\boldsymbol{\lambda}}_I^{\mathrm{T}} \boldsymbol{Q}_I \right\}$$

$$= \delta \hat{\boldsymbol{\lambda}}^{\mathrm{T}} \left(-\boldsymbol{G}^{\mathrm{T}} \hat{\boldsymbol{u}} + \boldsymbol{Q} \right) \tag{5-41}$$

式中，

$$\boldsymbol{G}_{IJ}^{\mathrm{T}} = -\int_{\Gamma_u} \boldsymbol{N}_I^{\mathrm{T}} \boldsymbol{\phi}_J \mathrm{d}\Gamma \tag{5-42}$$

$$\boldsymbol{Q}_I = -\int_{\Gamma_u} \boldsymbol{N}_I^{\mathrm{T}} \bar{\boldsymbol{u}}_p \mathrm{d}\Gamma \tag{5-43}$$

同理，式 (5-37) 的第五项对应写为

$$\int_{\Gamma_u} \delta \tilde{\boldsymbol{u}}_p^{\mathrm{T}} \tilde{\boldsymbol{\lambda}}_p \mathrm{d}\Gamma = \int_{\Gamma_u} \delta \left(\sum_{J=1}^{n} \boldsymbol{\phi}_J \hat{\boldsymbol{u}}_J \right)^{\mathrm{T}} \left(\sum_{I=1}^{n_u} \boldsymbol{N}_I \hat{\boldsymbol{\lambda}}_I \right) \mathrm{d}\Gamma$$

5.3 位移边界条件的施加

$$= \sum_{I=1}^{n_u} \sum_{J=1}^{n} \left\{ \delta \hat{\boldsymbol{u}}_J^{\mathrm{T}} \underbrace{\left(\int_{\Gamma_u} \boldsymbol{\phi}_J^{\mathrm{T}} \boldsymbol{N}_I \mathrm{d}\Gamma \right)}_{-\boldsymbol{G}_{IJ}} \hat{\lambda}_I \right\}$$

$$= -\sum_{I=1}^{n_u} \sum_{J=1}^{n} \left\{ \delta \hat{\boldsymbol{u}}_J^{\mathrm{T}} \boldsymbol{G}_{IJ} \hat{\lambda}_I \right\}$$

$$= -\delta \hat{\boldsymbol{u}}^{\mathrm{T}} \boldsymbol{G} \hat{\boldsymbol{\lambda}} \tag{5-44}$$

因此，式 (5-37) 最终写为

$$\delta \hat{\boldsymbol{u}}^{\mathrm{T}} \left[\boldsymbol{K} \hat{\boldsymbol{u}} + \boldsymbol{G} \hat{\boldsymbol{\lambda}} - \boldsymbol{F} \right] + \delta \hat{\boldsymbol{\lambda}}^{\mathrm{T}} \left[\boldsymbol{G}^{\mathrm{T}} \hat{\boldsymbol{u}} - \boldsymbol{Q} \right] = 0 \tag{5-45}$$

由 $\delta \hat{\boldsymbol{u}}$ 和 $\delta \hat{\boldsymbol{\lambda}}$ 的任意性，上式成立需满足：

$$\begin{cases} \boldsymbol{K} \hat{\boldsymbol{u}} + \boldsymbol{G} \hat{\boldsymbol{\lambda}} - \boldsymbol{F} = 0 \\ \boldsymbol{G}^{\mathrm{T}} \hat{\boldsymbol{u}} - \boldsymbol{Q} = 0 \end{cases} \tag{5-46}$$

将其写成整体系统矩阵方程的形式：

$$\underbrace{\begin{bmatrix} \boldsymbol{K} & \boldsymbol{G} \\ \boldsymbol{G}^{\mathrm{T}} & 0 \end{bmatrix}}_{\boldsymbol{K}^G} \underbrace{\begin{Bmatrix} \hat{\boldsymbol{u}} \\ \hat{\boldsymbol{\lambda}} \end{Bmatrix}}_{\boldsymbol{U}^G} = \underbrace{\begin{Bmatrix} \boldsymbol{F} \\ \boldsymbol{Q} \end{Bmatrix}}_{\boldsymbol{F}^G} \tag{5-47}$$

这个系统方程就是 EFG 法的形式，其中总体刚度矩阵 $\boldsymbol{K}^G$ 仍然是一个对称矩阵，但变成一个 $(2n + 2n_u) \times (2n + 2n_u)$ 维数的矩阵，与式 (5-16) 相比，扩大了 $2n_u$ 个维度。

Lagrange 乘子法施加位移边界条件，其重要的一个优点是可保证边界条件的精确施加，但不利的方面是数值实施比较复杂。此外 $\boldsymbol{K}^G$ 被扩展维数以后，通常是非带状和非正定的，当本质/位移边界上的节点数较多时，对系统方程的求解运算将会耗费更多时间，这会降低运算效率。

2. 罚函数法

采用罚函数法施加位移边界条件，将采用如下泛函形式：

$$\delta \Pi = -\int_{\Omega} \delta u_i \left(\sigma_{ij,j} + b_i \right) \mathrm{d}\Omega + \int_{\Gamma_t} \delta u_i \left(n_j \sigma_{ij} - \bar{t}_i \right) \mathrm{d}\Gamma - \alpha \int_{\Gamma_u} \delta u_i \left(u_i - \bar{u}_i \right) \mathrm{d}\Gamma = 0 \tag{5-48}$$

式中，α 为罚参数，上式对应的矩阵方程形式为

$$\int_{\Omega} (\boldsymbol{L} \delta \boldsymbol{u})^{\mathrm{T}} \boldsymbol{D} (\boldsymbol{L} \boldsymbol{u}) \mathrm{d}\Omega - \int_{\Omega} \delta \boldsymbol{u}^{\mathrm{T}} \boldsymbol{b} \mathrm{d}\Omega - \int_{\Gamma_t} \delta \boldsymbol{u}^{\mathrm{T}} \bar{\boldsymbol{t}} \mathrm{d}\Gamma - \alpha \int_{\Gamma_u} \delta \boldsymbol{u}^{\mathrm{T}} (\boldsymbol{u} - \bar{\boldsymbol{u}}) \mathrm{d}\Gamma = 0 \tag{5-49}$$

对于本质边界 $\varGamma_u$ 上的任意一个积分点 x_p,上式最后一项对应写为

$$\alpha \int_{\varGamma_u} \delta \tilde{\boldsymbol{u}}_p^{\mathrm{T}} (\tilde{\boldsymbol{u}}_p - \bar{\boldsymbol{u}}_p) \mathrm{d}\varGamma$$

$$= \alpha \int_{\varGamma_u} \left(\delta \tilde{\boldsymbol{u}}_p^{\mathrm{T}} \tilde{\boldsymbol{u}}_p - \delta \tilde{\boldsymbol{u}}_p^{\mathrm{T}} \bar{\boldsymbol{u}}_p \right) \mathrm{d}\varGamma$$

$$= \alpha \int_{\varGamma_u} \left\{ \delta \left(\sum_{I=1}^{n} \phi_I \hat{\boldsymbol{u}}_I \right)^{\mathrm{T}} \left(\sum_{J=1}^{n} \phi_J \hat{\boldsymbol{u}}_J \right) \right\} \mathrm{d}\varGamma - \alpha \int_{\varGamma_u} \left\{ \delta \left(\sum_{I=1}^{n} \phi_I \hat{\boldsymbol{u}}_I \right)^{\mathrm{T}} \bar{\boldsymbol{u}}_p \right\} \mathrm{d}\varGamma$$

$$= \alpha \sum_{I=1}^{n} \sum_{J=1}^{n} \left\{ \delta \hat{\boldsymbol{u}}_I^{\mathrm{T}} \left(\int_{\varGamma_u} \boldsymbol{\phi}_I^{\mathrm{T}} \boldsymbol{\phi}_J \mathrm{d}\varGamma \right) \hat{\boldsymbol{u}}_J \right\} - \alpha \sum_{I=1}^{n} \left\{ \delta \hat{\boldsymbol{u}}_I^{\mathrm{T}} \left(\int_{\varGamma_u} \boldsymbol{\phi}_I^{\mathrm{T}} \bar{\boldsymbol{u}}_p \mathrm{d}\varGamma \right) \right\}$$

$$= \delta \hat{\boldsymbol{u}}^{\mathrm{T}} \underbrace{\left\{ \alpha \int_{\varGamma_u} \boldsymbol{\phi}_I^{\mathrm{T}} \boldsymbol{\phi}_J \mathrm{d}\varGamma \right\}}_{\boldsymbol{K}_{IJ}^{\alpha}} \hat{\boldsymbol{u}} - \delta \hat{\boldsymbol{u}}^{\mathrm{T}} \underbrace{\left\{ \alpha \int_{\varGamma_u} \boldsymbol{\phi}_I^{\mathrm{T}} \bar{\boldsymbol{u}}_p \mathrm{d}\varGamma \right\}}_{\boldsymbol{F}_I^{\alpha}} \tag{5-50}$$

式中,

$$\boldsymbol{K}_{IJ}^{\alpha} = \alpha \int_{\varGamma_u} \boldsymbol{\phi}_I^{\mathrm{T}} \boldsymbol{\phi}_J \mathrm{d}\varGamma \tag{5-51}$$

$$\boldsymbol{F}_I^{\alpha} = \alpha \int_{\varGamma_u} \boldsymbol{\phi}_I^{\mathrm{T}} \bar{\boldsymbol{u}}_p \mathrm{d}\varGamma \tag{5-52}$$

则最终的系统方程写为

$$[\boldsymbol{K} + \boldsymbol{K}^{\alpha}] \hat{\boldsymbol{u}} = \boldsymbol{F} + \boldsymbol{F}^{\alpha} \tag{5-53}$$

可以看出,总体刚度矩阵 $\boldsymbol{K} + \boldsymbol{K}^{\alpha}$ 将保持原矩阵维数,不会被扩展。

采用罚函数法施加位移边界条件,其优点是算法简单直接,总体刚度矩阵不仅不会扩展维数,而且将是对称的、带状的、正定的。但其不足之处是施加边界条件精度较低,而且易受罚参数取值的影响。

3. 修正变分法

Lagrange 乘子法是一种将各类自变函数引入变分泛函的通用的方法,式 (5-36) 中,如果将 Lagrange 乘子取为本质边界上的面力 t,即

$$\lambda = t \tag{5-54}$$

则式 (5-36) 将变为

$$\delta \varPi = -\int_{\varOmega} \delta u_i \left(\sigma_{ij,j} + b_i \right) \mathrm{d}\varOmega + \int_{\varGamma_t} \delta u_i \left(n_j \sigma_{ij} - \bar{t}_i \right) \mathrm{d}\varGamma$$
$$- \int_{\varGamma_u} \delta t_i \left(u_i - \bar{u}_i \right) \mathrm{d}\varGamma - \int_{\varGamma_u} \delta u_i t_i \mathrm{d}\varGamma = 0 \tag{5-55}$$

5.3 位移边界条件的施加

其对应的矩阵形式方程写为

$$\int_\Omega (L\delta u)^{\mathrm{T}} D (Lu) \mathrm{d}\Omega - \int_\Omega \delta u^{\mathrm{T}} b \mathrm{d}\Omega - \int_{\Gamma_t} \delta u^{\mathrm{T}} \bar{t} \mathrm{d}\Gamma$$
$$- \int_{\Gamma_u} \delta t^{\mathrm{T}} (u - \bar{u}) \mathrm{d}\Gamma - \int_{\Gamma_u} \delta u^{\mathrm{T}} t \mathrm{d}\Gamma = 0 \tag{5-56}$$

对于本质边界 Γ_u 上的任意一个积分点 x_p,上式等号左侧倒数第二项,对应写为

$$\int_{\Gamma_u} \delta \tilde{t}_p^{\mathrm{T}} (\tilde{u}_p - \bar{u}_p) \mathrm{d}\Gamma$$
$$= \int_{\Gamma_u} \delta (nDB)_p^{\mathrm{T}} \tilde{u}_p \mathrm{d}\Gamma - \int_{\Gamma_u} \delta (nDB)_p^{\mathrm{T}} \bar{u}_p \mathrm{d}\Gamma$$
$$= \int_{\Gamma_u} \left(n_p D \sum_{I=1}^n B_I \delta \hat{u}_I \right)^{\mathrm{T}} \left(\sum_{J=1}^n \phi_J \hat{u}_J \right) \mathrm{d}\Gamma - \int_{\Gamma_u} \left(n_p D \sum_{I=1}^n B_I \delta \hat{u}_I \right)^{\mathrm{T}} \bar{u}_p \mathrm{d}\Gamma$$
$$= \int_{\Gamma_u} \left\{ \sum_{I=1}^n \sum_{J=1}^n \left[\delta \hat{u}_I^{\mathrm{T}} (n_p D B_I)^{\mathrm{T}} \phi_J \hat{u}_J \right] \right\} \mathrm{d}\Gamma - \int_{\Gamma_u} \left(n_p D \sum_{I=1}^n B_I \delta \hat{u}_I \right)^{\mathrm{T}} \bar{u}_p \mathrm{d}\Gamma$$
$$= \sum_{I=1}^n \sum_{J=1}^n \left\{ \delta \hat{u}_I^{\mathrm{T}} \underbrace{\left[\int_{\Gamma_u} B_I^{\mathrm{T}} D n_p^{\mathrm{T}} \phi_J \mathrm{d}\Gamma \right]}_{K_{IJ}^{t1}} \hat{u}_J \right\} - \sum_{I=1}^n \left\{ \delta \hat{u}_I^{\mathrm{T}} \underbrace{\left[\int_{\Gamma_u} B_I^{\mathrm{T}} D n_p^{\mathrm{T}} \bar{u}_p \mathrm{d}\Gamma \right]}_{F_I^{tu}} \right\}$$
$$= \delta \hat{u}^{\mathrm{T}} K^{t1} \hat{u} - \delta \hat{u}^{\mathrm{T}} F^{tu} \tag{5-57}$$

式中,

$$K_{IJ}^{t1} = \int_{\Gamma_u} B_I^{\mathrm{T}} D^{\mathrm{T}} n_p^{\mathrm{T}} \phi_J \mathrm{d}\Gamma \tag{5-58}$$

$$F_I^{tu} = \int_{\Gamma_u} B_I^{\mathrm{T}} D^{\mathrm{T}} n_p^{\mathrm{T}} \bar{u}_p \mathrm{d}\Gamma \tag{5-59}$$

式 (5-37) 等号左侧倒数第一项,对应写为

$$\int_{\Gamma_u} \delta \tilde{u}_p^{\mathrm{T}} \tilde{t}_p \mathrm{d}\Gamma = \int_{\Gamma_u} \delta \left(\sum_{J=1}^n \phi_J \hat{u}_J \right)^{\mathrm{T}} \left(n_p D \sum_{I=1}^n B_I \hat{u}_I \right) \mathrm{d}\Gamma$$
$$= \int_{\Gamma_u} \left\{ \sum_{I=1}^n \sum_{J=1}^n \left[\delta \hat{u}_J^{\mathrm{T}} \phi_J^{\mathrm{T}} (n_p D B_I) \hat{u}_I \right] \right\} \mathrm{d}\Gamma$$
$$= \sum_{I=1}^n \sum_{J=1}^n \left\{ \delta \hat{u}_J^{\mathrm{T}} \underbrace{\left[\int_{\Gamma_u} \phi_J^{\mathrm{T}} n_p D B_I \mathrm{d}\Gamma \right]}_{K_{IJ}^{t2}} \hat{u}_I \right\}$$

$$=\delta\hat{\boldsymbol{u}}^{\mathrm{T}}\boldsymbol{K}^{t2}\hat{\boldsymbol{u}} \tag{5-60}$$

式中,

$$\boldsymbol{K}_{IJ}^{t2}=\int_{\varGamma_u}\boldsymbol{\phi}_J^{\mathrm{T}}\boldsymbol{n}_p\boldsymbol{D}\boldsymbol{B}_I\mathrm{d}\varGamma \tag{5-61}$$

则最终得到的总体系统方程为

$$\left[\boldsymbol{K}-\boldsymbol{K}^{t1}-\boldsymbol{K}^{t2}\right]_{(2n\times 2n)}\hat{\boldsymbol{u}}_{(2n\times 1)}=\left[\boldsymbol{F}-\boldsymbol{F}^{tu}\right]_{(2n\times 1)} \tag{5-62}$$

所以,由修正变分法施加本质/位移边界条件,没有增加未知数的数量,总体刚度矩阵不会扩展维数,而且是对称带状矩阵,其数值实现上比 Lagrange 乘子法要简单一些,但其精度比 Lagrange 乘子法低。

除了上面介绍的几种本质边界条件施加方法外,实际上还有很多可供选择的方案,比如有:修正配点法[15],有限元素法[16],Nistche 法[17,18],位移约束方程法[19],边界变换法等[20]。

5.4 数值积分方法

5.4.1 背景网格积分法

背景网格积分是 EFG 法数值积分的基本方法,如图 5-1 所示,其数值实施可分为以下几个步骤:

1) 根据求解域规划的场节点数,确定刚度矩阵 $\boldsymbol{K}$ 和外力矩阵 $\boldsymbol{F}$ 的维数,预赋 0 处理,以备矩阵组装使用。

2) 规划具有规则形状的背景网格,要求背景网格能够覆盖整个求解域,并要求背景区域尽可能小,这是基于保证计算效率的考虑。

3) 对所有网格循环,对于轮序到的任意一个网格 $\varOmega_i$,根据网格的坐标信息,设置网格内的积分点。

4) 对积分点循环,对于轮序到的任意一个积分点 $\boldsymbol{x}_p$,首先判断其是否位于整个求解域内,如果是,则对积分点进行积分运算,并将 $\boldsymbol{K}_{IJ}$ 和 $\boldsymbol{F}_I$ 累加到总体矩阵的对应项上。

5) 对边界专门设置边界型网格,并执行与上述步骤类似的积分运算,以保证边界条件的施加。

背景网格积分通常采用高斯积分法,对任意的场变量 g,在求解域上的区域积分可表示为

$$\int_\varOmega g\mathrm{d}\varOmega=\sum_i^{n_c}\int_{\varOmega_i}g\mathrm{d}\varOmega=\sum_i^{n_c}\sum_p^{n_p}w_p g_p J_p \tag{5-63}$$

式中，n_c 表示背景网格的网格数量；n_p 表示一个网格内且被求解域覆盖的积分点数量；w_p 表示积分点 $\boldsymbol{x}_p$ 的高斯权；J_p 表示积分点 $\boldsymbol{x}_p$ 的 Jacobian 值。标准的面积积分网格通常设置 16 个积分点，如图 5-2(a) 所示。Galerkin 全局弱式法的求解精度也与积分精确性密切相关，为了保证积分精度的需要，也通常将标准网格进一步划分成 4 个子网格，如图 5-2(b) 所示，因而一个背景网格中将有 64 个积分点参与积分运算。

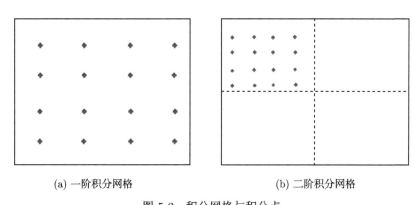

(a) 一阶积分网格   (b) 二阶积分网格

图 5-2 积分网格与积分点

Fig. 5-2 Integration cell and integration points

对于区域边界上的积分，其积分公式可以类似地写为

$$\int_\Gamma g \mathrm{d}\Gamma = \sum_i^{n_c} \int_{\Gamma_i} g \mathrm{d}\Gamma = \sum_i^{n_c} \sum_p^{n_p} w_p g_p J_p \tag{5-64}$$

背景网格积分法的优点是很容易构造积分网格，所使用的积分网格都是标准而形状规则的，但其不利的一面是，对复杂几何结构的求解域问题又会带来数值实施的困难。

5.4.2 有限元积分法

背景网格区域积分时，需要特别注意只对处于求解域内的积分点进行积分，因此需要适时判断积分点与边界信息的相互关系，而且这种操作一般在数值计算的底层。为了避免这种频繁对积分点位置信息的判断，可以考虑在积分网格的规划阶段就进行预先处理。即按照有限元网格划分的方法，对求解域进行更为精确细致的网格划分，如图 5-3 所示。

这样处理可以保证所有积分点自然处于求解域内。但对于复杂的三维问题，有限单元的生成比较困难，而且用有限元网格积分也丧失了无网格法的"不依赖网格"的一个重要优点。

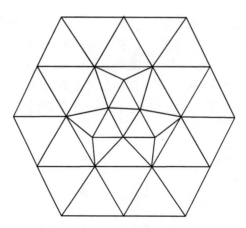

图 5-3 多边形域的网格划分

Fig. 5-3 Meshing a polygonal domain

5.4.3 节点积分法

为了简化积分运算，Beissel 和 Belytschko[21] 采用节点积分方案来近似地进行数值积分计算。节点积分类似于把任意一个场节点 x_I 同时当作一个积分点 x_p 来简单处理，如图 5-4 所示。此时执行计算只需要场节点信息，不需要任何积分网格及积分点信息。

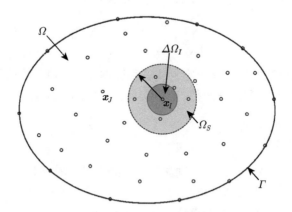

图 5-4 节点积分法图示

Fig. 5-4 Schematics of the node-integration method

对应于式 (5-63)，对整个求解区域的节点积分写为

$$\int_\Omega g \mathrm{d}\Omega = \sum_I^{n-n_\mathrm{B}} g_I \Delta \Omega_I \tag{5-65}$$

式中，n 表示所有的场节点数，n_B 表示边界上的场节点数，则 $n-n_B$ 表示求解域 Ω 内的场节点数；$\Delta\Omega_I$ 表示区域节点 x_I 所代表的区域面积 (三维时，为体积)。同理，对整个边界的节点积分写为

$$\int_\Gamma g \mathrm{d}\Gamma = \sum_I^{n_B} g_I \Delta\Gamma_I \tag{5-66}$$

式中，$\Delta\Gamma_I$ 表示边界节点 x_I 所代表的边界弧长 (三维时，为面积)。

节点积分需要合理确定 $\Delta\Omega_I$ 和 $\Delta\Gamma_I$ 的值，显然需尽可能满足：

$$\begin{cases} \sum_I^{n-n_B} \Delta\Omega_I = \Omega \\ \sum_I^{n_B} \Delta\Gamma_I = \Gamma \end{cases} \tag{5-67}$$

一个简单方法是取

$$\Delta\Omega_I = \frac{A_\Omega}{n-n_B} \tag{5-68}$$

$$\Delta\Gamma_I = \frac{s_\Gamma}{n_B} \tag{5-69}$$

式中，对二维问题，A_Ω 表示整个求解域面积，s_Γ 表示整个边界弧长；对三维问题，A_Ω 表示整个求解域体积，s_Γ 表示整个边界曲面积。

节点积分法数值执行非常简便，计算效率很高，但它是一种非常粗略的积分方案，其数值计算通常是不稳定的 [22]，其求解精度依赖于场节点设置足够细密。需注意到，Wen 等 [23-25] 提出的有限积分法 (FIM) 也属于一种节点积分方法，是一种精确的积分方案，但需要节点规则等距离布置。如果采用规则的积分点布置，并执行积分运算，FIM 可以适应于散乱节点布设问题。

5.4.4 介点积分法

为了克服节点积分法计算不稳定的问题，并吸收其数值执行简单的特点。可以在求解域内及边界上引入一组弱介点 $\{x_p\}_{p=1}^{n_p}$，由其承担积分点的功能执行积分计算，如图 5-5 所示。

介点积分公式对应写为

$$\int_\Omega g \mathrm{d}\Omega = \sum_p^{n_p-n_p^B} g_p \Delta\Omega_p \tag{5-70}$$

$$\int_\Gamma g \mathrm{d}\Gamma = \sum_p^{n_p^B} g_p \Delta\Gamma_p \tag{5-71}$$

式中，$\Delta\Omega_p$ 和 $\Delta\Gamma_p$ 的确定可参照式 (5-68) 和式 (5-69) 执行。

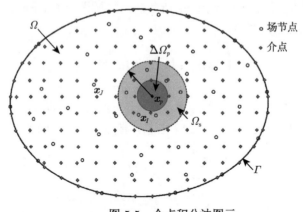

图 5-5 介点积分法图示

Fig. 5-5 Schematics of the intervention-point-integration method

介点积分法同样不需要任何网格，介点的设置方法类似于场节点，其数值执行同样非常简便，计算效率也能够得到保证。从理论上来讲，只要对介点设置合适的密度，例如，对二维问题，$n_p=4n \sim 9n$，则其积分精度是能够得到保证的；数值方法的收敛性将不再完全依赖于场节点的足够致密。

5.5 XFEM 法及其对有限元法的改进

随着无网格法研究的发展，将散点近似的思想引入传统有限元法，则改进后的有限元法有多种特定名称，如单位分解有限元法 (PUFEM)[26]，广义有限元法 (GFEM)[12,27]，扩展有限元法 (XFEM)[11,28-31]。本书用 XFEM 统一指代此类方法。

XFEM 法的构造思想是在传统有限元法的基础中，引入无网格近似函数 (也被称为增强函数)，即单位分解近似的方法 [9,10,26]。如式 (3-114) 所示，XFEM 的单位分解近似函数可写为

$$\tilde{u}(\boldsymbol{x}) = \underbrace{\sum_{\forall I} N_I(\boldsymbol{x}) u_I}_{u^{\mathrm{FE}}} + \underbrace{\sum_{\forall I} \phi_I(\boldsymbol{x}) \Psi_I(\boldsymbol{x}) q_I}_{u^{\mathrm{enr}}} \tag{5-72}$$

等式右边第一项即传统有限元近似函数，第二项即具有散点近似特征的增强函数，N 为有限元形函数，ϕ 为无网格形函数，Ψ 为任意的场量分布函数 (通常取为某种场变量的渐进解函数)，而 u_I 和 q_I 均表示单元节点的自由度。

XFEM 法因为使用了增强近似函数，可以有效地求解一些传统 FEM 很难分析的问题，比如裂纹扩展、位错演化、晶格界面模拟、相界的延伸等。例如，图 5-6

给出细观层面晶格材料的网格离散模型,传统 FEM 要求晶格边界两侧的网格是严格对称的,而且每个独立的晶格区域必须采用独立网格系统离散,这在网格自动生成中是一个非常困难的工作,如果晶格区域比较大,则用于网格划分的时间成本很大,而且通常难以实现。而在 XFEM 法中,网格划分可以无视晶格界面的限制,网格可以贯穿界面,由此可用规则化的网格离散整个区域;而晶格界面可用一组独立的网格进行定义。

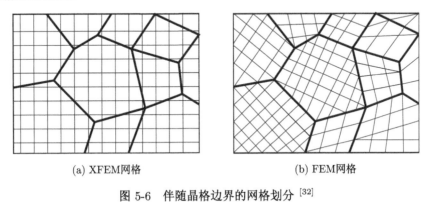

(a) XFEM网格　　　　　　　　(b) FEM网格

图 5-6　伴随晶格边界的网格划分 [32]

Fig. 5-6　Mesh generation with grain boundary[32]

在裂纹扩展问题中,传统 FEM 的网格是不能跨越裂纹边界的,而且网格需随着裂缝的扩展进行适应性的重新规划。而在 XFEM 法中,裂纹线是可以贯穿单元的。因此,裂纹的随意扩展不需要网格重新规划,在此情况下,网格只是一种"背景",而非对实际结构的描述,如图 5-7 所示。

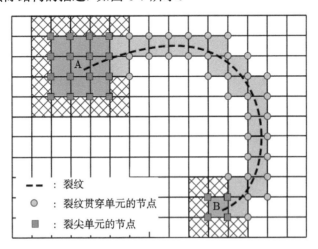

图 5-7　任意裂纹及其增强单元 [32]

Fig. 5-7　An arbitrary crack line with enriched elements[32]

XFEM 有某些方面不依赖于网格的无网格法特性，但其计算框架是严格依赖网格的。因此，XFEM 更应当归类于网格类方法。

5.6 本章小结

式 (5-1) 定义的泛函是最小势能变分原理，依据其他的变分原理或者是一般的变分方法，完全可以发展出多种形式的全局弱式法。但是，式 (5-1) 应当是最为简单的一种形式，因此，这个泛函便成为 Galerkin 弱式的最基本、最普遍的形式。

到目前为止，以 EFG 法为代表的全局弱式方法，是无网格法研究和发展中非常重要的一类方法。全局弱式无网格法是近代无网格法的起源性方法，直接启发了无网格法研究的发展。但是，作为一类无网格法，全局弱式无网格法仍然有一定局限性，因而目前发展的大多数方法，在数值计算中依然依赖于背景网格。虽然这种背景网格可以脱离求解域的限制，采用类似于差分法那样的规则网格，但仍然对网格存在一定依赖性。

参 考 文 献

[1] Nayroles B, Touzot G, Villon P. Generalizing the finite element method: Diffuse approximation and diffuse elements[J]. Computational Mechanics, 1992, 10(5): 307-318.

[2] Belytschko T, Lu Y Y, Gu L. Element-free Galerkin methods[J]. International Journal for Numerical Methods in Engineering, 1994, 37(2): 229-256.

[3] Liu W K, Jun S, Zhang Y F. Reproducing kernel particle methods[J]. International Journal for Numerical Methods in Fluids, 1995, 20(8-9): 1081-1106.

[4] Sukumar N, Moran B, Belytschko T. The natural element method in solid mechanics[J]. International Journal for Numerical Methods in Engineering, 1998, 43: 839-887.

[5] Sukumar N, Moran B, Yu Semenov A, et al. Natural neighbour Galerkin methods[J]. International Journal for Numerical Methods in Engineering, 2001, 50(1): 1-27.

[6] Liu G R, Gu Y T. A point interpolation method for two-dimensional solids[J]. International Journal for Numerical Methods in Engineering, 2001, 50(4): 937-951.

[7] Wendland H. Meshless Galerkin methods using radial basis functions[J]. Mathematics of Computation of the American Mathematical Society, 1999, 68(228): 1521-1531.

[8] Liu G R, Zhang G Y, Gu Y T, et al. A meshfree radial point interpolation method (RPIM) for three-dimensional solids[J]. Computational Mechanics, 2005, 36(6): 421-430.

[9] Duarte C A, Oden J T. An hp adaptive method using clouds[J]. Computer Methods in Applied Mechanics and Engineering, 1996, 139(1-4): 237-262.

[10] Duarte C A, Oden J T. Hp clouds-an hp meshless method[J]. Numerical Methods for Partial Differential Equations, 1996, 12(6): 673-706.

[11] Sukumar N, Moës N, Moran B, et al. Extended finite element method for three-dimensional crack modelling[J]. International Journal for Numerical Methods in Engineering, 2000, 48(11): 1549-1570.

[12] Duarte C A, Babuška I, Oden J T. Generalized finite element methods for three-dimensional structural mechanics problems[J]. Computers & Structures, 2000, 77(2): 215-232.

[13] Shaw A, Roy D. Analyses of wrinkled and slack membranes through an error reproducing mesh-free method[J]. International Journal of Solids and Structures, 2007, 44(11): 3939-3972.

[14] Liu G R, Gu Y T. An Introduction to Meshfree Methods and Their Programming[M]. Singapore: Springer Science & Business Media, 2005.

[15] Zhu T, Atluri S N. A modified collocation method and a penalty formulation for enforcing the essential boundary conditions in the element free Galerkin method[J]. Computational Mechanics, 1998, 21(3): 211-222.

[16] Krongauz Y, Belytschko T. Enforcement of essential boundary conditions in meshless approximations using finite elements[J]. Computer Methods in Applied Mechanics and Engineering, 1996, 131(1-2): 133-145.

[17] Fernández-Méndez S, Huerta A. Imposing essential boundary conditions in mesh-free methods[J]. Computer Methods in Applied Mechanics and Engineering, 2004, 193(12): 1257-1275.

[18] Embar A, Dolbow J, Harari I. Imposing Dirichlet boundary conditions with Nitsche's method and spline-based finite elements[J]. International Journal for Numerical Methods in Engineering, 2010, 83(7): 877-898.

[19] Zhang X, Liu X, Lu M W, et al. Imposition of essential boundary conditions by displacement constraint equations in meshless methods[J]. Communications in Numerical Methods in Engineering, 2001, 17(3): 165-178.

[20] Liu W K, Jun S, Li S, et al. Reproducing kernel particle methods for structural dynamics[J]. International Journal for Numerical Methods in Engineering, 1995, 38(10): 1655-1679.

[21] Beissel S, Belytschko T. Nodal integration of the element-free Galerkin method [J]. Comput. Methods Appl. Mech. Engrg., 1996, 139: 49-74.

[22] Chen J S, Wu C T, Yoon S, et al. A stabilized conforming nodal integration for Galerkin mesh-free methods[J]. International Journal for Numerical Methods in Engineering, 2001, 50(2): 435-466.

[23] Wen P H, Hon Y C, Li M, et al. Finite integration method for partial differential equations[J]. Applied Mathematical Modelling, 2013, 37(24): 10092-10106.

[24] Li M, Hon Y C, Korakianitis T, et al. Finite integration method for nonlocal elastic bar under static and dynamic loads[J]. Engineering Analysis with Boundary Elements, 2013, 37(5): 842-849.

[25] Li M, Chen C S, Hon Y C, et al. Finite integration method for solving multi-dimensional partial differential equations[J]. Applied Mathematical Modelling, 2015, 39(17): 4979-4994.

[26] Melenk J M, Babuška I. The partition of unity finite element method: Basic theory and applications[J]. Computer Methods in Applied Mechanics and Engineering, 1996, 139(1-4): 289-314.

[27] Duarte C A, Hamzeh O N, Liszka T J, et al. A generalized finite element method for the simulation of three-dimensional dynamic crack propagation[J]. Computer Methods in Applied Mechanics and Engineering, 2001, 190(15): 2227-2262.

[28] Sukumar N, Chopp D L, Moës N, et al. Modeling holes and inclusions by level sets in the extended finite-element method[J]. Computer Methods in Applied Mechanics and Engineering, 2001, 190(46): 6183-6200.

[29] Sukumar N, Belytschko T. Arbitrary branched and intersecting cracks with the extended finite element method[J]. Int. J. Numer. Meth. Eng, 2000, 48: 1741-1760.

[30] Moës N, Belytschko T. Extended finite element method for cohesive crack growth[J]. Engineering Fracture Mechanics, 2002, 69(7): 813-833.

[31] Sukumar N, Prévost J H. Modeling quasi-static crack growth with the extended finite element method Part I: Computer implementation[J]. International Journal of Solids and Structures, 2003, 40(26): 7513-7537.

[32] Belytschko T, Gracie R, Ventura G. A review of extended/generalized finite element methods for material modeling[J]. Modelling and Simulation in Materials Science and Engineering, 2009, 17(4): 043001.

第 6 章 无网格局部弱式法

无网格局部弱式法采用加权残值法的子域法进行数值离散，其收敛性解释符合介点原理的推论 1。局部弱式方法的提出，是无网格发展的一大进步。这种方法执行数值积分不再需要背景网格，只需要局部的积分单元即可，使弱式无网格法实现了从 "网格自由" 到 "网格免除" 的转变。美国计算力学科学家 Atluri 及其研究组，对此类方法的提出和发展做出了重要贡献。最初提出的此类特定方法，被称为无网格局部 Petrov-Galerkin (MLPG) 法 [1,2]。这种离散思想与其他近似方法的结合，或者是离散方法的衍生，又形成多种具体的方法。如局部边界积分方程 (LBIE) 法 [3,4]、自然邻接 Petrov-Galerkin (NNPG) 法 [5]、Petrov-Galerkin 自然单元 (PGNE) 法 [6]、局部径向基点插值 (LRPIM) 法 [7,8]、有限球 (FSM) 法 [9]、局部 Kriging (LoKriging) 法 [10] 等。

6.1 Petrov-Galerkin 局部弱式

Petrov-Galerkin 局部弱式法的离散思想如图 6-1 所示。图中，x_I 表示计算 (目标) 节点，Ω_q 表示伴随此计算节点定义的局部积分域，而 x_p 表示此积分域内任意一个积分点 (弱介点)；Ω_S 表示伴随此积分点定义的近似支撑域，而 x_J 表示被此支撑域覆盖到的任意一个场节点。

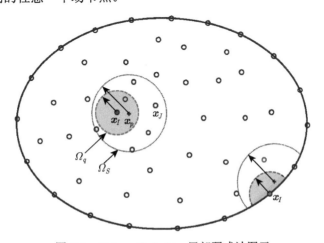

图 6-1　Petrov-Galerkin 局部弱式法图示

Fig. 6-1　Schematics of the local Petrov-Galerkin weak-form methods

以二维弹性静力学问题为例，式 (2-1) 对应的加权残值局部弱式方程写为

$$\int_{\Omega_q} w_q \left(\sigma_{ij,j} + b_i\right) \mathrm{d}\Omega = 0 \qquad (6\text{-}1)$$

式中，w_q 为积分域 Ω_q 上的检验函数 (也可称为权函数)。因 Ω_q 伴随计算节点 x_I 定义，故 w_q 也可用 w_I 表示。上式运用散度定理：

$$\int_{\Omega_q} w_q \sigma_{ij,j} \mathrm{d}\Omega = \int_{\Gamma_q} w_q n_j \sigma_{ij} \mathrm{d}\Gamma - \int_{\Omega_q} w_{q,j} \sigma_{ij} \mathrm{d}\Omega \qquad (6\text{-}2)$$

式中，n_j 表示积分域边界的外法向分量。则式 (6-1) 可推得

$$\int_{\Gamma_q} w_q n_j \sigma_{ij} \mathrm{d}\Gamma - \int_{\Omega_q} w_{q,j} \sigma_{ij} \mathrm{d}\Omega + \int_{\Omega_q} w_q b_i \mathrm{d}\Omega = 0 \qquad (6\text{-}3)$$

如果 Ω_q 与问题域的边界 Γ 相交，如图 6-2 所示。并对边界符号给出如下定义：

$$\partial \Omega_q = \Gamma_q + \Gamma_{qu} + \Gamma_{qt}, \quad \Gamma_{qu} = \Omega_q \cap \Gamma_u, \quad \Gamma_{qt} = \Omega_q \cap \Gamma_t \qquad (6\text{-}4)$$

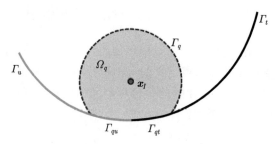

图 6-2 局部积分域 Ω_q 与全局边界 Γ 的交界

Fig. 6-2 The intersection of the local quadrature domain Ω_q and the global boundary Γ

则式 (6-3) 的第一项对应改写为

$$\int_{(\Gamma_q)} w_q n_j \sigma_{ij} \mathrm{d}\Gamma = \int_{\Gamma_q} w_q n_j \sigma_{ij} \mathrm{d}\Gamma + \int_{\Gamma_{qu}} w_q n_j \sigma_{ij} \mathrm{d}\Gamma + \int_{\Gamma_{qt}} w_q n_j \sigma_{ij} \mathrm{d}\Gamma \qquad (6\text{-}5)$$

由 $\{n_j \sigma_{ij} = \bar{t}_i\}_{\Gamma_{qt}}$，并将本质边界条件用罚参数法引入：

$$\int_{\Gamma_q} w_q n_j \sigma_{ij} \mathrm{d}\Gamma + \int_{\Gamma_{qu}} w_q n_j \sigma_{ij} \mathrm{d}\Gamma + \int_{\Gamma_{qt}} w_q \bar{t}_i \mathrm{d}\Gamma - \int_{\Omega_q} w_{q,j} \sigma_{ij} \mathrm{d}\Omega$$

$$+ \int_{\Omega_q} w_q b_i \mathrm{d}\Omega - \alpha \int_{\Gamma_u} w_q \left(u_i - \bar{u}_i\right) \mathrm{d}\Gamma = 0 \qquad (6\text{-}6)$$

则可得到一个局部对称弱式方程 (LSWF)：

$$\int_{\Omega_q} w_{q,j}\sigma_{ij}\mathrm{d}\Omega - \int_{\Gamma_q} w_q n_j \sigma_{ij}\mathrm{d}\Gamma - \int_{\Gamma_{qu}} w_q n_j \sigma_{ij}\mathrm{d}\Gamma + \alpha\int_{\Gamma_u} w_q u_i\mathrm{d}\Gamma$$

$$= \int_{\Omega_q} w_q b_i\mathrm{d}\Omega + \int_{\Gamma_{qt}} w_q \bar{t}_i\mathrm{d}\Gamma + \alpha\int_{\Gamma_u} w_q \bar{u}_i\mathrm{d}\Gamma \tag{6-7}$$

由 $\sigma_{ij} = E_{ijkl}\varepsilon_{ij} = E_{ijkl}u_{i,j}$，对式 (6-3) 的第二项再运用一次散度定理，即

$$\int_{\Omega_q} w_{q,j}\sigma_{ij}\mathrm{d}\Omega = \int_{\Omega_q} w_{q,j} E_{ijkl}u_{i,j}\mathrm{d}\Omega = \int_{\Gamma_q} w_{q,j} n_j E_{ijkl}u_i\mathrm{d}\Gamma - \int_{\Omega_q} w_{q,ij}\sigma_{ij}\mathrm{d}\Omega \tag{6-8}$$

则可得到另外一个局部非对称弱式方程 (LUSWF)：

$$\int_{\Omega_q} w_{q,ij}\sigma_{ij}\mathrm{d}\Omega + \int_{\Gamma_q} w_q n_j \sigma_{ij}\mathrm{d}\Gamma - \int_{\Gamma_q} w_{q,j} n_j E_{ijkl}u_i\mathrm{d}\Gamma$$

$$- \int_{\Gamma_{qt}} w_{q,j} n_j E_{ijkl}u_i\mathrm{d}\Gamma + \int_{\Gamma_{qu}} w_q n_j \sigma_{ij}\mathrm{d}\Gamma$$

$$= -\int_{\Omega_q} w_q b_i\mathrm{d}\Omega - \int_{\Gamma_{qt}} w_q \bar{t}_i\mathrm{d}\Gamma + \int_{\Gamma_{qu}} w_{q,j} n_j E_{ijkl}\bar{u}_i\mathrm{d}\Gamma \tag{6-9}$$

由式 (6-7) 定义的局部对称弱式方程 (LSWF)，和由式 (6-9) 定义的局部非对称弱式方程 (LUSWF)，是 Petrov-Galerkin 局部弱式法的两类标准离散方程。需注意到，LUSWF 方程通过运用两次散度定理，本质边界条件将自动引入。此外，LSWF 中，要求检验函数 w_q 是 C^0 连续的；而在 LUSWF 中，要求检验函数是 C^1 连续的。

在 Petrov-Galerkin 局部弱式法中，通过选用不同的检验函数 w_q，以及采用 LSWF 或 LUSWF 两种离散形式，可以形成多种风格迥异的离散方法，接下来将对其进一步阐述。

6.2 无网格局部 Petrov-Galerkin 法

如果采用 MLS 权函数作为检验函数 w_q，并采用式 (6-7) 定义的 LSWF 方程离散，则对应的局部弱式法被称为无网格局部 Petrov-Galerkin(MLPG) 法。则得到的离散系统方程的一般形式为

$$\boldsymbol{K}\cdot\boldsymbol{U} = \boldsymbol{F} \tag{6-10}$$

式中，系统刚度矩阵 $\boldsymbol{K}$ 和系统荷载矩阵 $\boldsymbol{F}$ 的离散形式定义为

$$\boldsymbol{K}_{IJ} = \int_{\Omega_q} \boldsymbol{V}_p^{\mathrm{T}}\boldsymbol{D}\boldsymbol{B}_J\mathrm{d}\Omega - \int_{\Gamma_q\cup\Gamma_{qu}} w_p \boldsymbol{n}_p\boldsymbol{D}\boldsymbol{B}_J\mathrm{d}\Gamma + \alpha\int_{\Gamma_u} w_p \boldsymbol{N}_J\mathrm{d}\Gamma \tag{6-11}$$

$$F_I = \int_{\Omega_q} w_p \boldsymbol{b}_p \mathrm{d}\Omega + \int_{\Gamma_{qt}} w_p \bar{\boldsymbol{t}}_p \mathrm{d}\Gamma + \alpha \int_{\Gamma_u} w_p \bar{\boldsymbol{u}}_p \mathrm{d}\Gamma \tag{6-12}$$

式中，$\boldsymbol{D}$ 表示由 (2-15a) 或式 (2-15b) 表示的材料常数矩阵；下标 "p" 表示积分点 $\boldsymbol{x}_p$ 处的变量；下标 "J" 表示场节点 $\boldsymbol{x}_J$ 处的变量。对 MLS 权函数有如下定义：

$$w_p = w_p(\boldsymbol{x} - \boldsymbol{x}_I) \equiv w(\boldsymbol{x}_p - \boldsymbol{x}_I) \tag{6-13}$$

并对其他符号定义如下：

$$\boldsymbol{V}_p = \boldsymbol{L}w_p = \begin{bmatrix} \partial w_p/\partial x & 0 \\ 0 & \partial w_p/\partial y \\ \partial w_p/\partial y & \partial w_p/\partial x \end{bmatrix} \tag{6-14}$$

$$\boldsymbol{B}_J = \boldsymbol{L}\boldsymbol{\phi}_J = \begin{bmatrix} \partial \phi_J(\boldsymbol{x},\boldsymbol{x}_p)/\partial x & 0 \\ 0 & \partial \phi_J(\boldsymbol{x},\boldsymbol{x}_p)/\partial y \\ \partial \phi_J(\boldsymbol{x},\boldsymbol{x}_p)/\partial y & \partial \phi_J(\boldsymbol{x},\boldsymbol{x}_p)/\partial x \end{bmatrix} \tag{6-15}$$

$$\boldsymbol{n}_p = \begin{bmatrix} n_x & 0 & n_y \\ 0 & n_y & n_x \end{bmatrix}_p \tag{6-16}$$

$$\boldsymbol{N}_J = \begin{bmatrix} \phi_J(\boldsymbol{x},\boldsymbol{x}_p) & 0 \\ 0 & \phi_J(\boldsymbol{x},\boldsymbol{x}_p) \end{bmatrix} \tag{6-17}$$

$$\boldsymbol{b}_p = \begin{bmatrix} b_x \\ b_y \end{bmatrix}_p, \quad \bar{\boldsymbol{t}}_p = \begin{bmatrix} \bar{t}_x \\ \bar{t}_y \end{bmatrix}_p, \quad \bar{\boldsymbol{u}}_p = \begin{bmatrix} \bar{u}_x \\ \bar{u}_y \end{bmatrix}_p \tag{6-18}$$

须特别说明一点，式 (6-6) 中强行引入本质边界条件在实际应用中可以更为灵活地处理。MLPG 法的局部积分域 Ω_q 是伴随计算节点定义的，因此其离散系方程也是逐节点组装的，每个节点对应于总系方程中的 D 行（D 为问题的维数）。鉴于此，位于本质边界上的计算节点，可以采用直接配点法施加本质边界条件，这是一种更加简单有效的处理方法。即，式 (6-11) 和式 (6-12) 中的带罚参数的位移边界条件项可以取消，对 $\boldsymbol{x}_I \in \Gamma_u$，可以直接用如下离散方程组装在总体系统方程的对应行即可：

$$\begin{cases} \boldsymbol{K}_{IJ} = \bar{\boldsymbol{N}}_J = \begin{bmatrix} \phi_J(\boldsymbol{x},\boldsymbol{x}_I) & 0 \\ 0 & \phi_J(\boldsymbol{x},\boldsymbol{x}_I) \end{bmatrix} \\ \boldsymbol{F}_I = \bar{\boldsymbol{u}}_I \end{cases} \tag{6-19}$$

6.3 阶跃检验函数 MLPG 法

如果采用 Heaviside 阶跃函数作为检验函数 w_q, 即

$$w_q(\boldsymbol{x}) = H(\boldsymbol{x}) = \begin{cases} 1, & \boldsymbol{x} \in (\Omega_q \cup \partial\Omega_q) \\ 0, & \boldsymbol{x} \notin (\Omega_q \cup \partial\Omega_q) \end{cases} \qquad (6\text{-}20)$$

其函数图像如图 6-3(b) 所示。显然 $H(\boldsymbol{x})$ 在积分域 Ω_q 上的分布将是一个单位常数, 其微分将是 "0" 值, 这就意味着在式 (6-7) 定义的 LSWF 中, 等号左边第一个区域积分项将消除, 即有

$$-\int_{\varGamma_q \cup \varGamma_{qu}} n_j \sigma_{ij} \mathrm{d}\varGamma + \alpha \int_{\varGamma_u} u_i \mathrm{d}\varGamma = \int_{\Omega_q} b_i \mathrm{d}\Omega + \int_{\varGamma_{qt}} \bar{t}_i \mathrm{d}\varGamma + \alpha \int_{\varGamma_u} \bar{u}_i \mathrm{d}\varGamma \qquad (6\text{-}21)$$

则其离散系统方程可写为

$$\boldsymbol{K}_{IJ} = -\int_{\varGamma_q \cup \varGamma_{qu}} \boldsymbol{n}_p \boldsymbol{D} \boldsymbol{B}_J \mathrm{d}\varGamma + \alpha \int_{\varGamma_u} \boldsymbol{N}_J \mathrm{d}\varGamma \qquad (6\text{-}22)$$

$$\boldsymbol{F}_I = \int_{\Omega_q} \boldsymbol{b}_p \mathrm{d}\Omega + \int_{\varGamma_{qt}} \bar{\boldsymbol{t}}_p \mathrm{d}\varGamma + \alpha \int_{\varGamma_u} \bar{\boldsymbol{u}}_p \mathrm{d}\varGamma \qquad (6\text{-}23)$$

如果改用配点法施加本质边界条件, 则离散方程对应改写为

$$\boldsymbol{K}_{IJ} = \begin{cases} -\int_{\varGamma_q \cup \varGamma_{qu}} \boldsymbol{n}_p \boldsymbol{D} \boldsymbol{B}_J \mathrm{d}\varGamma, & \boldsymbol{x}_I \notin \varGamma_u \\ \bar{\boldsymbol{N}}_J, & \boldsymbol{x}_I \in \varGamma_u \end{cases} \qquad (6\text{-}24)$$

$$\boldsymbol{F}_I = \begin{cases} \int_{\Omega_q} \boldsymbol{b}_p \mathrm{d}\Omega + \int_{\varGamma_{qt}} \bar{\boldsymbol{t}}_p \mathrm{d}\varGamma, & \boldsymbol{x}_I \notin \varGamma_u \\ \bar{\boldsymbol{u}}_I, & \boldsymbol{x}_I \in \varGamma_u \end{cases} \qquad (6\text{-}25)$$

Atluri 和 Shen[11-13] 也将这种方法用 "MLPG5" 表示。可以看出, 式 (6-24) 和式 (6-25) 相比于式 (6-11) 和式 (6-12) 具有更为简单的形式。而且关键在于: 构造刚度矩阵时, 不再需要执行区域积分运算, 仅需要执行局部边界积分即可。局部积分域上的区域积分点通常要远多于边界积分点。这种改进, 避免了在稠密区域积分点上的近似计算, 可以大幅度地提高计算效率。此外, 这种改进并不会对数值方法的计算精度和稳定性造成明显影响。由此可以说, 阶跃检验函数 MLPG 法 (MLPG5) 是局部弱式方法中最具竞争力的一种方法。

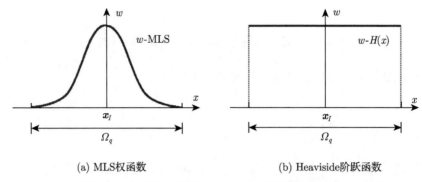

(a) MLS权函数　　　　　　　(b) Heaviside阶跃函数

图 6-3　MLS 权函数与 Heaviside 阶跃函数的对比

Fig. 6-3　Comparison of the MLS weight function and the Heaviside step function

6.4　局部边界积分方程法

如果采用基本解函数 u^* 作为检验函数 $w_q^{[3]}$，要求基本解函数 u^* 满足如下条件：

$$\boldsymbol{L}^{\mathrm{T}}\{\boldsymbol{D}\cdot\boldsymbol{L}[u^*(\boldsymbol{x},\boldsymbol{x}_I)]\}+\delta(\boldsymbol{x},\boldsymbol{x}_I)=0,\quad \boldsymbol{x}\in\Omega_q \tag{6-26}$$

$$u^*(\boldsymbol{x},\boldsymbol{x}_I)=0,\quad \boldsymbol{x}\in\Gamma_q \tag{6-27}$$

并要求满足条件：

$$\Omega_q\cap\Gamma=\varnothing \tag{6-28}$$

即局部积分域与全局边界不相交。

对于二维位势问题，检验函数 u^* 取为

$$u^*(\boldsymbol{x},\boldsymbol{x}_I)=\frac{1}{2\pi}\ln\frac{r_0}{r} \tag{6-29}$$

式中，$r=|\boldsymbol{x}-\boldsymbol{x}_I|$ 表示任意积分点 $\boldsymbol{x}$ 到计算节点 $\boldsymbol{x}_I$ 的距离；r_0 表示局部域 Ω_q 的半径。

对于二维弹性力学问题，检验函数 u^* 可取为式 (2-79) 表示的 Kelvin 解。在式 (6-9) 的 LUSWF 方程中将基本解检验函数 u^* 代入，可得到 LBIE 式：

$$\alpha_I u_I=\int_{\Omega_q}u_i^*b_i\mathrm{d}\Omega-\int_{\partial\Omega_q}u_{i,j}^*n_jE_{ijkl}u_i\mathrm{d}\Gamma+\int_{\Gamma_{qu}\cup\Gamma_{qt}}u_i^*n_j\sigma_{ij}\mathrm{d}\Gamma \tag{6-30}$$

将近似函数代入，可得

$$\alpha_I u_I=\boldsymbol{F}_I'+\sum_{J=1}^N\boldsymbol{K}_{IJ}'\hat{u}_J \tag{6-31}$$

其中，
$$\alpha_I = \begin{cases} 1, & \boldsymbol{x}_I \in \Omega \\ 1/2, & \boldsymbol{x}_I \in \widehat{\varGamma} \\ \theta/2\pi, & \boldsymbol{x}_I \in \hat{\varGamma} \end{cases} \quad (6\text{-}32)$$

式中，$\widehat{\varGamma}$ 表示光滑边界；$\hat{\varGamma}$ 表示非光滑边界。另有

$$\boldsymbol{K}'_{IJ} = \int_{\varGamma_{qu}} u_p^*(\boldsymbol{x},\boldsymbol{x}_I)\,\boldsymbol{n}_p \boldsymbol{D}\boldsymbol{B}_J \mathrm{d}\varGamma - \int_{\varGamma_{qt}\cup\varGamma_q} \phi_J(\boldsymbol{x},\boldsymbol{x}_p)\,\boldsymbol{n}_p \boldsymbol{D}\boldsymbol{B}_p^* \mathrm{d}\varGamma \quad (6\text{-}33)$$

$$\boldsymbol{F}'_I = \int_{\Omega_q} u_p^*(\boldsymbol{x},\boldsymbol{x}_I)\,\boldsymbol{b}_p \mathrm{d}\Omega + \int_{\varGamma_{qt}} u_p^*(\boldsymbol{x},\boldsymbol{x}_I)\,\bar{\boldsymbol{t}}_p \mathrm{d}\varGamma - \int_{\varGamma_{qu}} \boldsymbol{n}_p \boldsymbol{D}\boldsymbol{B}_p^* \bar{\boldsymbol{u}}_p \mathrm{d}\varGamma \quad (6\text{-}34)$$

其中，
$$\boldsymbol{B}_p^* = \boldsymbol{L}u_p^* = \begin{bmatrix} \partial u_p^*(\boldsymbol{x},\boldsymbol{x}_I)/\partial x & 0 \\ 0 & \partial u_p^*(\boldsymbol{x},\boldsymbol{x}_I)/\partial y \\ \partial u_p^*(\boldsymbol{x},\boldsymbol{x}_I)/\partial y & \partial u_p^*(\boldsymbol{x},\boldsymbol{x}_I)/\partial x \end{bmatrix} \quad (6\text{-}35)$$

如果形函数 $\phi(\boldsymbol{x})$ 不具有 Kronecker δ 函数性质，则有

$$\boldsymbol{K}_{IJ} = \begin{cases} -\boldsymbol{K}'_{IJ}, & \boldsymbol{x}_I \in \varGamma_u \\ -\boldsymbol{K}'_{IJ} + \alpha_I \boldsymbol{N}_J, & \boldsymbol{x}_I \notin \varGamma_u \end{cases} \quad (6\text{-}36)$$

$$\boldsymbol{F}_I = \begin{cases} \boldsymbol{F}'_I - \alpha_I \bar{\boldsymbol{u}}_I, & \boldsymbol{x}_I \in \varGamma_u \\ \boldsymbol{F}'_I, & \boldsymbol{x}_I \notin \varGamma_u \end{cases} \quad (6\text{-}37)$$

而如果形函数 $\phi(\boldsymbol{x})$ 具有 Kronecker δ 函数性质，则有

$$\boldsymbol{K}_{IJ} = \begin{cases} -\boldsymbol{K}'_{IJ}, & I \neq J \\ \alpha_I - \boldsymbol{K}'_{IJ}, & I = J \end{cases} \quad (6\text{-}38)$$

$$\boldsymbol{F}_I = \boldsymbol{F}'_I \quad (6\text{-}39)$$

需注意到，LBIE 虽然是基于边界积分方程方法，但有别于全局边界解法。它的场节点分布在这个求解域上，而不仅是问题域的边界上，因此赋予该方法区域求解的特征。

6.5 其他局部弱式离散法

6.5.1 Galerkin 型 MLPG 法

如果在 LSWF 中采用近似函数 $\phi(\boldsymbol{x})$ 作为检验函数，即 $w_q(\boldsymbol{x}) = \phi(\boldsymbol{x})$，则对

应于式 (6-11) 和式 (6-12)，其离散系统方程写为

$$\boldsymbol{K}_{IJ} = \int_{\Omega_q} \boldsymbol{B}_p^{\mathrm{T}} \boldsymbol{D} \boldsymbol{B}_J \mathrm{d}\Omega - \int_{\varGamma_q \cup \varGamma_{qu}} \phi_p(\boldsymbol{x}, \boldsymbol{x}_I) \boldsymbol{n}_p \boldsymbol{D} \boldsymbol{B}_J \mathrm{d}\varGamma + \alpha \int_{\varGamma_u} \phi_p(\boldsymbol{x}, \boldsymbol{x}_I) \boldsymbol{N}_J \mathrm{d}\varGamma \tag{6-40}$$

$$\boldsymbol{F}_I = \int_{\Omega_q} \phi_p(\boldsymbol{x}, \boldsymbol{x}_I) \boldsymbol{b}_p \mathrm{d}\Omega + \int_{\varGamma_{qt}} \phi_p(\boldsymbol{x}, \boldsymbol{x}_I) \bar{\boldsymbol{t}}_p \mathrm{d}\varGamma + \alpha \int_{\varGamma_u} \phi_p(\boldsymbol{x}, \boldsymbol{x}_I) \bar{\boldsymbol{u}}_p \mathrm{d}\varGamma \tag{6-41}$$

式中，

$$\boldsymbol{B}_p = \boldsymbol{L}\phi_p = \begin{bmatrix} \partial \phi_p(\boldsymbol{x}, \boldsymbol{x}_I)/\partial x & 0 \\ 0 & \partial \phi_p(\boldsymbol{x}, \boldsymbol{x}_I)/\partial y \\ \partial \phi_p(\boldsymbol{x}, \boldsymbol{x}_I)/\partial y & \partial \phi_p(\boldsymbol{x}, \boldsymbol{x}_I)/\partial x \end{bmatrix} \tag{6-42}$$

Atluri 和 Shen[13] 也将这种方法用 "MLPG6" 表示。这种方法与 6.2 节介绍的 MLPG 法在计算形式上更为接近，但计算要更复杂一些，因为形函数的计算比权函数的计算要复杂得多。但相比于标准的 MLPG 法，这种方法在计算精度和稳定性上却没有什么明显的优势。

6.5.2 最小二乘 MLPG 法

如果取控制方程的残值函数为检验函数，即

$$w_i = R_i = \sigma_{ij,j} + b_i \tag{6-43}$$

则可以构造一个局部弱式泛函：

$$\varPi = \int_{\Omega_q} \left(\sum_{p=1}^{N_p} (\sigma_{ij,j} + b_i) \right)^2 \mathrm{d}\Omega \tag{6-44}$$

令 $\partial \varPi / \partial \hat{u}_J = 0$，求其驻值，可得到

$$\int_{\Omega_q} \sigma_{ij,j} \left(\sum_{p=1}^{N_p} (\sigma_{ij,j} + b_i) \right) \mathrm{d}\Omega = 0 \tag{6-45}$$

而对边界条件，直接采用配点法施加。则其离散系统方程可写为

$$\boldsymbol{K}_{IJ} = \begin{cases} \int_{\Omega_q} \left(\boldsymbol{L}^{\mathrm{T}} \boldsymbol{D} \boldsymbol{B}_p\right)^{\mathrm{T}} \left(\boldsymbol{L}^{\mathrm{T}} \boldsymbol{D} \boldsymbol{B}_J\right) \mathrm{d}\Omega, & \boldsymbol{x}_I \in \Omega \\ \boldsymbol{n}_I \boldsymbol{D} \bar{\boldsymbol{B}}_J, & \boldsymbol{x}_I \in \varGamma_t \\ \bar{\boldsymbol{N}}_J, & \boldsymbol{x}_I \in \varGamma_u \end{cases} \tag{6-46}$$

$$\boldsymbol{F}_I = \begin{cases} \int_{\Omega_q} \left(\boldsymbol{L}^{\mathrm{T}} \boldsymbol{D} \boldsymbol{B}_p\right)^{\mathrm{T}} \boldsymbol{b}_p \mathrm{d}\Omega, & \boldsymbol{x}_I \in \Omega \\ \bar{\boldsymbol{t}}_I, & \boldsymbol{x}_I \in \varGamma_t \\ \bar{\boldsymbol{u}}_I, & \boldsymbol{x}_I \in \varGamma_u \end{cases} \tag{6-47}$$

式中，

$$\bar{B}_J = L\phi_J(x, x_I) = \begin{bmatrix} \partial\phi_J(x, x_I)/\partial x & 0 \\ 0 & \partial\phi_J(x, x_I)/\partial y \\ \partial\phi_J(x, x_I)/\partial y & \partial\phi_J(x, x_I)/\partial x \end{bmatrix} \quad (6\text{-}48)$$

Atluri 和 Shen[13] 也将这种方法用 "MLPG3" 表示。这种方法在局部域积分中，没有通过弱式转换来降低导数近似的阶数，这对求解精度和稳定性是不利的，求解表现应当会明显差于标准的 MLPG 法。此外，这种方法的数值执行上，也要比标准的 MLPG 复杂。

6.5.3 配点法

如果用 Dirac δ 函数作为检验函数，即 $w_i = \delta(x - x_i)$，则局部弱式将退化为强式的配点法，其离散系统方程写为

$$K_{IJ} = \begin{cases} L^T D \bar{B}_J, & x_I \in \Omega \\ n_I D \bar{B}_J, & x_I \in \Gamma_t \\ \bar{N}_J, & x_I \in \Gamma_u \end{cases} \quad (6\text{-}49)$$

$$F_I = \begin{cases} b_I, & x_I \in \Omega \\ \bar{t}_I, & x_I \in \Gamma_t \\ \bar{u}_I, & x_I \in \Gamma_u \end{cases} \quad (6\text{-}50)$$

将这种方法在此处列出，旨在说明局部弱式可以灵活使用几乎所有的加权残值法，从而使这种离散方法富于变化，形成多种完全不同风格的方法。

6.6 本章小结

本章重点介绍的 MLPG 法、使用阶跃检验函数 MLPG 法 (MLPG5) 和局部边界积分方程法 (LBIE)，这三种方法是局部弱式方法的典型代表，从计算精度、求解效率、数值实施难易程度、对任意问题的适用性等方面综合评价，我们认为使用阶跃检验函数 MLPG 法是最具有优势的一种。

局部弱式方法中，各种加权残值法都很容易应用 [14-21]。通过采用不同的检验函数，使得局部弱式方程可以推演到多种风格迥异的形式，若再辅之以多种近似函数的灵活选用，则局部弱式方法家族会变得足够繁荣。

局部弱式方法的优点主要表现为：

1) 是真正意义上 "无网格" 的，其积分运算完全免除了网格。

2) 是计算精确稳定的方法，对一般问题，通常都能得到高精度的解。即这类方法具有对各种问题较好的普遍适应性。

3) 相比于全局弱式，数值实施要简洁一些，只需对节点循环，然后逐节点组装离散方程。而不像全局弱式那样，需要按网格顺序循环，然后叠加组装系统方程。

但是，局部弱式方法也具有一定的局限性。早期 MLPG 法有计算效率较低的问题，但随着使用阶跃检验函数 MLPG 法的提出，其计算效率问题已显著得到改善。MLPG 法的系统刚度矩阵 K 是一个非对称矩阵，但它是稀疏条带状的，这一问题对其求解效率的影响也非常有限。局部弱式方法最大的问题是对不规则结构的处理，在不规则边界上进行局部积分是其面临的严峻挑战。在 FEM 中，对边界积分的计算准备是在前期网格划分中就规划好的，网格一旦生成，边界上的积分信息便已经准备妥当。而局部弱式方法中，局部域的设置和全局边界的交互信息分散在数值计算的底层，特别是局部积分域与问题边界相交时，需要在计算的底层去精准地判定交互边界区域，然后在这些交互边界上另外设置积分点执行积分运算。因此，对于不规则的几何结构，MLPG 法的边界积分运算是一个非常棘手的问题。

参 考 文 献

[1] Atluri S N, Zhu T. A new meshless local Petrov-Galerkin (MLPG) approach in computational mechanics[J]. Computational Mechanics, 1998, 22(2): 117-127.

[2] Atluri S N, Cho J Y, Kim H G. Analysis of thin beams, using the meshless local Petrov-Galerkin method, with generalized moving least squares interpolations[J]. Computational Mechanics, 1999, 24(5): 334-347.

[3] Zhu T, Zhang J, Atluri S N. A meshless local boundary integral equation (LBIE) method for solving nonlinear problems[J]. Computational Mechanics, 1998, 22(2): 174-186.

[4] Atluri S N, Kim H G, Cho J Y. A critical assessment of the truly meshless local Petrov-Galerkin (MLPG), and local boundary integral equation (LBIE) methods[J]. Computational mechanics, 1999, 24(5): 348-372.

[5] Wang K, Zhou S, Shan G. The natural neighbour Petrov–Galerkin method for elastostatics[J]. International Journal for Numerical Methods in Engineering, 2005, 63(8): 1126-1145.

[6] Cho J R, Lee H W. A Petrov-Galerkin natural element method securing the numerical integration accuracy[J]. Journal of Mechanical Science and Technology, 2006, 20(1): 94-109.

[7] Liu G R, Gu Y T. A local radial point interpolation method (LRPIM) for free vibration analyses of 2-D solids[J]. Journal of Sound and Vibration, 2001, 246(1): 29-46.

[8] Wu Y L, Liu G R. A meshfree formulation of local radial point interpolation method (LRPIM) for incompressible flow simulation[J]. Computational Mechanics, 2003, 30(5-6): 355-365.

[9] De S, Bathe K J. The method of finite spheres[J]. Computational Mechanics, 2000,

25(4): 329-345.

[10] Lam K Y, Wang Q X, Li H. A novel meshless approach–local Kriging (LoKriging) method with two-dimensional structural analysis[J]. Computational Mechanics, 2004, 33(3): 235-244.

[11] Atluri S N, Shen S. The Meshless Local Petrov-Galerkin (MLPG) Method[M]. USA: Tech Science Press(CREST), 2002.

[12] Atluri S N. The Meshless Method (MLPG) for Domain & BIE Discretizations[M]. Forsyth: Tech Science Press, 2004.

[13] Atluri S N, Shen S. The basis of meshless domain discretization: The meshless local Petrov–Galerkin (MLPG) method[J]. Advances in Computational Mathematics, 2005, 23(1-2): 73-93.

[14] Lin H, Atluri S N. Meshless local Petrov-Galerkin(MLPG) method for convection diffusion problems[J]. Computer Modelling in Engineering & Sciences, 2000, 1(2): 45-60.

[15] Lin H, Atluri S N. The meshless local Petrov-Galerkin (MLPG) method for solving incompressible Navier-Stokes equations[J]. CMES-Computer Modeling in Engineering and Sciences, 2001, 2(2): 117-142.

[16] Atluri S N, Zhu T L. A new meshless local Petrov-Galerkin (MLPG) approach to nonlinear problems in computer modeling and simulation[J]. Computer Modeling and Simulation in Engineering, 1998, 3: 187-196.

[17] Atluri S N, Han Z D, Rajendran A M. A new implementation of the meshless finite volume method, through the MLPG "mixed" approach[J]. Computer Modeling in Engineering & Sciences, 2004, 6(6): 491-514.

[18] Gu Y T, Liu G R. A meshless local Petrov-Galerkin (MLPG) method for free and forced vibration analyses for solids[J]. Computational Mechanics, 2001, 27(3): 188-198.

[19] Atluri S N, Cho J Y, Kim H G. Analysis of thin beams, using the meshless local Petrov–Galerkin method, with generalized moving least squares interpolations[J]. Computational Mechanics, 1999, 24(5): 334-347.

[20] Dehghan M, Mirzaei D. The meshless local Petrov–Galerkin (MLPG) method for the generalized two-dimensional non-linear Schrödinger equation[J]. Engineering Analysis with Boundary Elements, 2008, 32(9): 747-756.

[21] Atluri S N, Han Z D, Shen S. Meshless local Petrov-Galerkin (MLPG) approaches for solving the weakly-singular traction & displacement boundary integral equations[J]. Computer Modeling in Engineering and Sciences, 2003, 4(5): 507-518.

第 7 章 配点类无网格法

配点法天然基于散点计算，是真正意义上无网格的。此外，配点法具有数值实施简单、计算效率高，对不规则域问题很容易处理等诸多优势。但配点法也存在一个致命问题，其计算不稳定性是众所周知的。因为这一局限性，限制了配点型无网格法的广泛应用。然而进一步研究表明，配点法的这一局限性是可以有效克服的。配点法基于介点原理加以发展和完善，完全可以成为一类极具竞争力的方法。

实际上，配点法的发展起源很早。比如，早期比较有代表性的方法有广义有限差分法 (GFDM)[1-5]、光滑粒子流体动力学 (SPH) 方法 [6-13] 等。晚近发展的方法有：有限点法 (FPM)[14,15]、Hp 无网格云法 [16,17]、双网格扩散配点 (DGDC) 法 [18]、最小二乘配点 (LSCM) 法 [19-23]、无网格介点 (MIP) 法 [24-26]、无网格全局介点 (MGIP) 法 [27] 等。本章将择要介绍几种代表性的方法。

7.1 有 限 点 法

有限点法 (FPM)[14,15] 当属最为简单的无网格法之一，其离散原理如图 7-1 所示。FPM 采用 MLS 近似方法，任意一个计算节点 x_I 的场变量由式 (3-20) 近似得到。

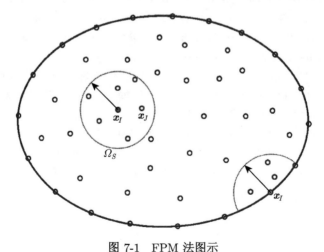

图 7-1　FPM 法图示

Fig. 7-1　Schematics of the FPM

7.1 有限点法

对应于式 (2-1)~ 式 (2-3) 定义的二维弹性力学问题。其离散系统方程的一般形式为

$$\boldsymbol{K}\cdot\boldsymbol{U}=\boldsymbol{F} \tag{7-1}$$

式中，系统刚度矩阵 $\boldsymbol{K}$ 和系统荷载矩阵 $\boldsymbol{F}$ 的离散形式定义为

$$\boldsymbol{K}_{IJ}=\begin{cases} \boldsymbol{L}^{\mathrm{T}}\boldsymbol{D}\boldsymbol{B}_J, & \boldsymbol{x}_I \in \Omega \\ \boldsymbol{n}_I\boldsymbol{D}\boldsymbol{B}_J, & \boldsymbol{x}_I \in \Gamma_t \\ \boldsymbol{N}_J, & \boldsymbol{x}_I \in \Gamma_u \end{cases} \tag{7-2}$$

$$\boldsymbol{F}_I=\begin{cases} \boldsymbol{b}_I, & \boldsymbol{x}_I \in \Omega \\ \bar{\boldsymbol{t}}_I, & \boldsymbol{x}_I \in \Gamma_t \\ \bar{\boldsymbol{u}}_I, & \boldsymbol{x}_I \in \Gamma_u \end{cases} \tag{7-3}$$

上面两式与式 (6-49) 和式 (6-50) 相同，如果材料矩阵 $\boldsymbol{D}$ 采用式 (2-15a) 的平面应力形式，则有

$$\boldsymbol{L}^{\mathrm{T}}\boldsymbol{D}\boldsymbol{B}_J=\frac{E}{1-\nu^2}\cdot\begin{bmatrix} \dfrac{\partial^2\phi_J(\boldsymbol{x},\boldsymbol{x}_I)}{\partial x^2}+\dfrac{1-\nu}{2}\dfrac{\partial^2\phi_J(\boldsymbol{x},\boldsymbol{x}_I)}{\partial y^2} & \dfrac{1+\nu}{2}\dfrac{\partial^2\phi_J(\boldsymbol{x},\boldsymbol{x}_I)}{\partial x\partial y} \\ \dfrac{1+\nu}{2}\dfrac{\partial^2\phi_J(\boldsymbol{x},\boldsymbol{x}_I)}{\partial x\partial y} & \\ \dfrac{\partial^2\phi_J(\boldsymbol{x},\boldsymbol{x}_I)}{\partial y^2}+\dfrac{1-\nu}{2}\dfrac{\partial^2\phi_J(\boldsymbol{x},\boldsymbol{x}_I)}{\partial x^2} & \end{bmatrix} \tag{7-4}$$

$$\boldsymbol{n}_I\boldsymbol{D}\boldsymbol{B}_J=\frac{E}{1-\nu^2}\cdot\begin{bmatrix} n_x\dfrac{\partial\phi_J(\boldsymbol{x},\boldsymbol{x}_I)}{\partial x}+n_y\dfrac{1-\nu}{2}\dfrac{\partial\phi_J(\boldsymbol{x},\boldsymbol{x}_I)}{\partial y} & n_y\dfrac{1-\nu}{2}\dfrac{\partial\phi_J(\boldsymbol{x},\boldsymbol{x}_I)}{\partial x}+n_x\nu\dfrac{\partial\phi_J(\boldsymbol{x},\boldsymbol{x}_I)}{\partial y} \\ n_y\nu\dfrac{\partial\phi_J(\boldsymbol{x},\boldsymbol{x}_I)}{\partial x}+n_x\dfrac{1-\nu}{2}\dfrac{\partial\phi_J(\boldsymbol{x},\boldsymbol{x}_I)}{\partial y} & n_x\dfrac{1-\nu}{2}\dfrac{\partial\phi_J(\boldsymbol{x},\boldsymbol{x}_I)}{\partial x}+n_y\dfrac{\partial\phi_J(\boldsymbol{x},\boldsymbol{x}_I)}{\partial y} \end{bmatrix} \tag{7-5}$$

$$\boldsymbol{N}_J=\begin{bmatrix} \phi_J(\boldsymbol{x},\boldsymbol{x}_I) & 0 \\ 0 & \phi_J(\boldsymbol{x},\boldsymbol{x}_I) \end{bmatrix} \tag{7-6}$$

式中，

$$\boldsymbol{B}_J = \boldsymbol{L}\phi_J(\boldsymbol{x},\boldsymbol{x}_I) = \begin{bmatrix} \dfrac{\partial \phi_J(\boldsymbol{x},\boldsymbol{x}_I)}{\partial x} & 0 \\ 0 & \dfrac{\partial \phi_J(\boldsymbol{x},\boldsymbol{x}_I)}{\partial y} \\ \dfrac{\partial \phi_J(\boldsymbol{x},\boldsymbol{x}_I)}{\partial y} & \dfrac{\partial \phi_J(\boldsymbol{x},\boldsymbol{x}_I)}{\partial x} \end{bmatrix} \tag{7-7}$$

另，

$$\boldsymbol{b}_I = \begin{bmatrix} b_x \\ b_y \end{bmatrix}_I, \quad \bar{\boldsymbol{t}}_I = \begin{bmatrix} \bar{t}_x \\ \bar{t}_y \end{bmatrix}_I, \quad \bar{\boldsymbol{u}}_I = \begin{bmatrix} \bar{u}_x \\ \bar{u}_y \end{bmatrix}_I \tag{7-8}$$

为了提高 FPM 的求解稳定性，Oñate 等[28]建议对式 (2-1) 定义的控制方程和式 (2-2) 定义的自然边界条件增加稳定项，并改写成如下的形式：

$$\frac{\partial \sigma_{ij}}{\partial x_j} + b_i - \frac{1}{2} h_k \frac{\partial}{\partial x_k}\left(\frac{\partial \sigma_{ij}}{\partial x_j} + b_i\right) = 0, \quad 在\Omega内 \tag{7-9}$$

$$n_j \sigma_{ij} - \bar{t}_i - \frac{1}{2} h_k n_k \left(\frac{\partial \sigma_{ij}}{\partial x_j} + b_i\right) = 0, \quad 在\Gamma_t 上 \tag{7-10}$$

式中，i, j, k 表示三维问题的三个维度分量；h_k 表示稳定项的影响参数。其对应的离散方程此处不再详细给出。需注意到，这样处理后提高了对近似函数连续性的要求，要求近似函数 ϕ 至少是 C^3 连续的。那么，会有一个问题，其稳定性的增强究竟是近似函数 ϕ 连续性提高的原因？还是稳定项发挥了作用？对此有必要进一步研究。

7.2 双网格扩散配点法

就我们的理解，Breitkopf 等建议的双网格扩散配点 (double grid diffuse collocation, DGDC) 法[18]，实际上也属于"介点原理"在配点法中的一种应用。在该法中，问题域不仅需要场节点离散，还需引入全局设置的"赋值点"(evaluation point)，这种赋值点属于介点的定义。DGDC 法的数值离散如图 7-2 所示。

在 DGDC 法中，对域内计算点 $\boldsymbol{x}_I$ 处的场变量的二阶微分近似 $u^h_{I,ij}$ 使用广义有限差分法 (GFDM) 和 MLS 结合的方法进行阐述。此处，为了便于理解和表述上的简洁性，我们改用一般的 MLS 近似进行表述。如图 7-2 所示，对场变量的微分近似需要两个"网格"，一个网格是以目标节点 $\boldsymbol{x}_I$ 为中心定义的 Ω_I，另一个网格是以介点 $\boldsymbol{x}_p$(被 Ω_I 覆盖到的任意一个) 为中心定义的 Ω_p。当然，此处的"网格"

实际上就是局部支撑域,这两个网格的使用,就是该方法名称中"双网格"所包含的意思。

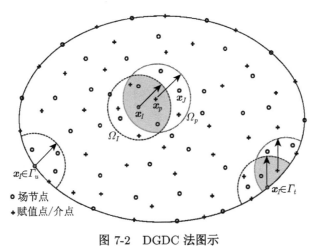

图 7-2 DGDC 法图示

Fig. 7-2 Schematics of the DGDC method

DGDC 法的二阶微分近似使用所谓的"两步双网格近似"法,如图 7-3 所示。若要获得域内计算点 x_I 处场变量的二阶微分近似 $u_{I,ij}^h$,第一步首先需要用邻近场节点位移变量来近似介点 x_p 处的应变量,即

$$\varepsilon_p^h \equiv u_{p,i}^h = \sum_{J=1}^{N_J} \varphi_{J,i}(\boldsymbol{x}, \boldsymbol{x}_p)\hat{u}_J, \quad \boldsymbol{x}_J \in \Omega_p \tag{7-11}$$

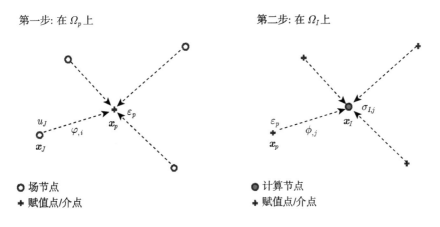

图 7-3 两步双网格近似

Fig. 7-3 Two steps and double grid approach

接下来是第二步，计算点 x_I 处应力的微分状态 $\sigma_{I,j}^h$ 用邻近介点 x_p 处的应变状态来近似得到，即

$$\sigma_{I,j}^h = E_{ijkl} \sum_{p=1}^{N_p} \phi_{p,j}(\boldsymbol{x}, \boldsymbol{x}_I) \varepsilon_p^h, \quad \boldsymbol{x}_p \in \Omega_I \tag{7-12}$$

需注意到，式 (7-11) 和式 (7-12) 中所使用的均为 MLS 近似法 (当然也可以是其他的近似方法)，但由于这两步近似在 Ω_p 和 Ω_I 两个不同的近似空间上进行，所以，这里用 φ 和 ϕ 两种符号来对不同空间的近似函数加以区分。

此外，自然边界 Γ_t 上计算点 x_I 处的应力状态近似函数 σ_I^h 写为

$$\sigma_I^h = E_{ijkl} \sum_{p=1}^{N_p} \phi_p(\boldsymbol{x}, \boldsymbol{x}_I) \varepsilon_p^h, \quad \boldsymbol{x}_I \in \Gamma_t, \boldsymbol{x}_p \in \Omega_I \tag{7-13}$$

而本质边界 Γ_u 上计算点 x_I 处的位移状态近似函数 u_I^h 可以通过邻近场节点位移来表达，不必要使用两步双网格近似，即

$$u_I^h = \sum_{J=1}^{N_J} \phi_J(\boldsymbol{x}, \boldsymbol{x}_I) \hat{u}_J, \quad \boldsymbol{x}_I \in \Gamma_u \tag{7-14}$$

接下来我们可以给出 DGDC 法的离散方程，对应于式 (2-1)~ 式 (2-3) 定义的二维弹性力学问题。其离散系统方程的一般形式亦如式 (7-1)，则 DGDC 法的系统刚度矩阵 $\boldsymbol{K}$ 和系统荷载矩阵 $\boldsymbol{F}$ 的离散形式定义为

$$\boldsymbol{K}_{IJ} = \begin{cases} \sum_{p=1}^{N_p} \boldsymbol{B}_p^{\mathrm{T}} \boldsymbol{D} \boldsymbol{B}_J, & \boldsymbol{x}_I \in \Omega \\ \sum_{p=1}^{N_p} \phi_p(\boldsymbol{x}, \boldsymbol{x}_I) \boldsymbol{n}_I \boldsymbol{D} \boldsymbol{B}_J, & \boldsymbol{x}_I \in \Gamma_t \\ \boldsymbol{N}_J, & \boldsymbol{x}_I \in \Gamma_u \end{cases} \tag{7-15}$$

$$\boldsymbol{F}_I = \begin{cases} \boldsymbol{b}_I, & \boldsymbol{x}_I \in \Omega \\ \bar{\boldsymbol{t}}_I, & \boldsymbol{x}_I \in \Gamma_t \\ \bar{\boldsymbol{u}}_I, & \boldsymbol{x}_I \in \Gamma_u \end{cases} \tag{7-16}$$

式中,

$$\boldsymbol{B}_p = \boldsymbol{L}\phi_p(\boldsymbol{x}, \boldsymbol{x}_I) = \begin{bmatrix} \dfrac{\partial \phi_p(\boldsymbol{x}, \boldsymbol{x}_I)}{\partial x} & 0 \\ 0 & \dfrac{\partial \phi_p(\boldsymbol{x}, \boldsymbol{x}_I)}{\partial y} \\ \dfrac{\partial \phi_p(\boldsymbol{x}, \boldsymbol{x}_I)}{\partial y} & \dfrac{\partial \phi_p(\boldsymbol{x}, \boldsymbol{x}_I)}{\partial x} \end{bmatrix} \quad (7\text{-}17)$$

$$\boldsymbol{B}_J = \boldsymbol{L}\varphi_J(\boldsymbol{x}, \boldsymbol{x}_p) = \begin{bmatrix} \dfrac{\partial \varphi_J(\boldsymbol{x}, \boldsymbol{x}_p)}{\partial x} & 0 \\ 0 & \dfrac{\partial \varphi_J(\boldsymbol{x}, \boldsymbol{x}_p)}{\partial y} \\ \dfrac{\partial \varphi_J(\boldsymbol{x}, \boldsymbol{x}_p)}{\partial y} & \dfrac{\partial \varphi_J(\boldsymbol{x}, \boldsymbol{x}_p)}{\partial x} \end{bmatrix} \quad (7\text{-}18)$$

$$\boldsymbol{n}_I = \begin{bmatrix} n_x & 0 & n_y \\ 0 & n_y & n_x \end{bmatrix}_I \quad (7\text{-}19)$$

$$\boldsymbol{N}_J = \begin{bmatrix} \phi_J(\boldsymbol{x}, \boldsymbol{x}_I) & 0 \\ 0 & \phi_J(\boldsymbol{x}, \boldsymbol{x}_I) \end{bmatrix} \quad (7\text{-}20)$$

综上,DGDC 法采用两步双网格近似技术后,只需要执行一阶导数近似,而传统配点法 (如 FPM) 需要二阶导数近似。也就是说,传统配点法要求近似函数 C^2 连续,而 DGDC 法仅需要近似函数 C^1 连续。这种转换,对数值求解的精度和稳定性均会有明显改善。此外,DGDC 法因降低了近似函数连续性的要求,因此 MLS 可以使用一次基函数计算,所需的支撑点数将显著降低,从而可以显著提高求解效率。

7.3 最小二乘配点法

张雄等发展的最小二乘配点法 (LSCM)[19,20] 也可视为介点原理在配点法中的另一种应用。LSCM 中,问题域在 n 个场节点离散的基础上,在域内又引入 n_a 个"辅助点"(介点),辅助点集 $\{\boldsymbol{x}_p\}_{p=1}^{n_a} \subset \Omega$。控制方程的残值不仅要求在场节点上消除,也同时要求在域内的辅助点上消除。计算节点,或计算辅助点的场变量由场节点信息执行近似。如图 7-4 所示。

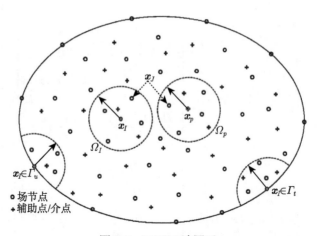

图 7-4 LSCM 法图示
Fig. 7-4 Schematics of the LSCM method

场节点的数量为

$$n = n_d + n_t + n_u \tag{7-21}$$

式中，n_d 为域内场节点数量；n_t 为自然边界 Γ_t 上的场节点数量；n_u 为本质边界 Γ_u 上的场节点数量。

令所有域内节点和域内辅助点满足平衡方程式 (2-1)，自然边界 Γ_t 上的节点满足自然边界条件式 (2-2)，本质边界 Γ_u 上的节点满足本质边界条件式 (2-3)，则可得到如下两个方程：

$$\bm{K}_{11}\bm{u}_1 + \bm{K}_{12}\bm{u}_2 = \bm{F}_1 \tag{7-22}$$

$$\bm{K}_{21}\bm{u}_1 + \bm{K}_{22}\bm{u}_2 = \bm{F}_2 \tag{7-23}$$

式中，$\bm{u}_1$ 为边界 Γ_t 和 Γ_u 上 $n_t + n_u$ 个节点的待求位移向量；$\bm{u}_2$ 为域内 n_d 个节点的待求位移向量。其中，式 (7-22) 对应于边界条件，式 (7-23) 对应于平衡方程。令边界条件严格满足，由式 (7-22) 可得

$$\bm{u}_1 = \bm{K}_{11}^{-1}(\bm{F}_1 - \bm{K}_{12}\bm{u}_2) \tag{7-24}$$

将其代入式 (7-23)，可得

$$\bm{K}\bm{u}_2 = \bm{F} \tag{7-25}$$

式中，

$$\begin{cases} \bm{K} = \bm{K}_{22} - \bm{K}_{21}\bm{K}_{11}^{-1}\bm{K}_{12} \\ \bm{F} = \bm{F}_2 - \bm{K}_{21}\bm{K}_{11}^{-1}\bm{F}_1 \end{cases} \tag{7-26}$$

7.3 最小二乘配点法

式 (7-25) 的方程组中，有 $2n_d$ 个未知数，有 $2(n+n_a)$ 个方程，即方程数多于未知数，不可直接求解。对其残值向量 $\boldsymbol{R} = \boldsymbol{K}\boldsymbol{u}_2 - \boldsymbol{F}$ 构造欧几里得 (Euclidean) 范数：

$$\|\boldsymbol{R}\|_2^2 = \|\boldsymbol{K}\boldsymbol{u}_2 - \boldsymbol{F}\|_2^2 \tag{7-27}$$

命

$$\frac{\partial \|\boldsymbol{R}\|_2^2}{\partial \boldsymbol{u}_2} = 0 \tag{7-28}$$

求该残值向量的极小值，可得到

$$\boldsymbol{K}^{\mathrm{T}}\boldsymbol{K}\boldsymbol{u}_2 = \boldsymbol{K}^{\mathrm{T}}\boldsymbol{F} \tag{7-29}$$

由上式即可对 $\boldsymbol{u}_2$ 求解。

此外，还可采用一种直接构造出可解系统方程的方法，需要构造一个最小二乘加权残值泛函，定义如下：

$$\begin{aligned} \Pi &= \sum_{\boldsymbol{x}_n \cup \boldsymbol{x}_p} \frac{\partial \left\{R\left(u^h\right)\right\}^2}{\partial \hat{u}} \\ &= \sum_{(\boldsymbol{x}_n \cup \boldsymbol{x}_p) \in \Omega} \frac{\partial \left(\sigma_{ij,j} - b_i\right)^2}{\partial \hat{u}} + \sum_{\boldsymbol{x}_n \in \Gamma_t} \lambda_t \frac{\partial \left(n_i \cdot \sigma_{ij} - \bar{t}_i\right)^2}{\partial \hat{u}} + \sum_{\boldsymbol{x}_n \in \Gamma_u} \lambda_u \frac{\partial \left(u_i - \bar{u}_i\right)^2}{\partial \hat{u}} \end{aligned} \tag{7-30}$$

式中，$\boldsymbol{x}_n$ 表示场节点；$\boldsymbol{x}_p$ 表示介点 (辅助点)；λ_t 和 λ_u 为保证本质边界条件的罚参数。命该泛函为 0，即

$$\Pi = 0 \tag{7-31}$$

则域内节点和介点必然满足平衡方程，自然边界上的节点必然满足自然边界条件，本质边界上的节点必然满足本质边界条件。并可得到一个离散系统方程：

$$\boldsymbol{K} \cdot \boldsymbol{U} = \boldsymbol{F} \tag{7-32}$$

式中，

$$\boldsymbol{K} = \sum_{(\boldsymbol{x}_n \cup \boldsymbol{x}_p) \in \Omega} \boldsymbol{H}^{\mathrm{T}}\boldsymbol{H} + \sum_{\boldsymbol{x}_n \in \Gamma_t} \lambda_t \boldsymbol{Q}^{\mathrm{T}}\boldsymbol{Q} + \sum_{\boldsymbol{x}_n \in \Gamma_u} \lambda_u \boldsymbol{N}^{\mathrm{T}}\boldsymbol{N} \tag{7-33}$$

$$\boldsymbol{F} = \sum_{(\boldsymbol{x}_n \cup \boldsymbol{x}_p) \in \Omega} \boldsymbol{H}^{\mathrm{T}}\boldsymbol{b} + \sum_{\boldsymbol{x}_n \in \Gamma_t} \lambda_t \boldsymbol{Q}^{\mathrm{T}}\bar{\boldsymbol{t}} + \sum_{\boldsymbol{x}_n \in \Gamma_u} \lambda_u \boldsymbol{N}^{\mathrm{T}}\bar{\boldsymbol{u}} \tag{7-34}$$

另，

$$\begin{cases} \boldsymbol{H} = [\boldsymbol{H}_1, \boldsymbol{H}_2, \cdots, \boldsymbol{H}_J, \cdots, \boldsymbol{H}_n] \\ \boldsymbol{H}_J = \boldsymbol{L}^{\mathrm{T}} \boldsymbol{D} \boldsymbol{L}(\phi_J) = \dfrac{E}{1-\nu^2} \begin{bmatrix} \dfrac{\partial^2 \phi_J}{\partial x^2} + \dfrac{1-\nu}{2} \dfrac{\partial^2 \phi_J}{\partial y^2} & \dfrac{1+\nu}{2} \dfrac{\partial^2 \phi_J}{\partial x \partial y} \\ \dfrac{1+\nu}{2} \dfrac{\partial^2 \phi_J}{\partial x \partial y} & \dfrac{\partial^2 \phi_J}{\partial y^2} + \dfrac{1-\nu}{2} \dfrac{\partial^2 \phi_J}{\partial x^2} \end{bmatrix} \end{cases} \tag{7-35}$$

$$\begin{cases} \boldsymbol{Q} = [\boldsymbol{Q}_1, \boldsymbol{Q}_2, \cdots, \boldsymbol{Q}_J, \cdots, \boldsymbol{Q}_n] \\ \boldsymbol{Q}_J = \boldsymbol{n}_I \boldsymbol{D} \boldsymbol{L}(\phi_J) = \dfrac{E}{1-\nu^2} \begin{bmatrix} n_x \dfrac{\partial \phi_J}{\partial x} + n_y \dfrac{1-\nu}{2} \dfrac{\partial \phi_J}{\partial y} & n_y \dfrac{1-\nu}{2} \dfrac{\partial \phi_J}{\partial x} + n_x \nu \dfrac{\partial \phi_J}{\partial y} \\ n_y \nu \dfrac{\partial \phi_J}{\partial x} + n_x \dfrac{1-\nu}{2} \dfrac{\partial \phi_J}{\partial y} & n_x \dfrac{1-\nu}{2} \dfrac{\partial \phi_J}{\partial x} + n_y \dfrac{\partial \phi_J}{\partial y} \end{bmatrix} \end{cases} \tag{7-36}$$

$$\begin{cases} \boldsymbol{N} = [\boldsymbol{N}_1, \boldsymbol{N}_2, \cdots, \boldsymbol{N}_J, \cdots, \boldsymbol{N}_n] \\ \boldsymbol{N}_J = \begin{bmatrix} \phi_J & 0 \\ 0 & \phi_J \end{bmatrix} \end{cases} \tag{7-37}$$

可见,其整体刚度矩阵的组织是逐点叠加式的,与全局弱式法相类似。张雄等[21]将这种构造泛函直接得到离散系统方程的方法,称为加权最小二乘无网格 (MWLS) 法,罚参数 λ_t 和 λ_u 对数值求解有直接影响,通常建议取为

$$\lambda_t = 10^5, \quad \lambda_u = \lambda_t \left(\dfrac{E}{1-\nu^2} \right)^2 \tag{7-38}$$

最小二乘配点法 (LSCM) 或加权最小二乘无网格 (MWLS) 法,通过引入辅助点来满足平衡方程,增加了对场节点自由度的约束,明显增强了一般配点法的求解稳定性[19-23]。此外,张雄等还进一步建议,平衡方程的残值不仅要求在域内节点和介点上消除,而且还要求在边界节点上消除。对于这一建议,客观上会增加离散方程的数量,进一步加强了对节点自由度的约束,或许对求解稳定性有一定帮助。但是,这种处理有悖于固体力学的基本理论,因为平衡方程是对区域内的要求,而非边界上的,即便这种处理在经验上有效,对其所发挥的作用尚需进一步研究。

7.4 无网格介点法

无网格介点 (MIP) 法是一种计算性能优秀的配点方法,是介点原理在配点法中的更合理应用[24-26]。MIP 法的关键策略是采用了一种所谓的局部介点近似技术,其近似策略如图 7-5 所示。对于计算节点 $\boldsymbol{x}_I$,以其为中心定义一个局部域 Ω_{L}^I,该局部域尺度根据周边场节点信息限定,然后在此局部域内插入介点集 $\{\boldsymbol{x}_p\}$,计

算节点 x_I 的场信息，通过介点集 $\{x_p\}$ 来近似。对于任意一个介点 x_p，以其为中心定义一个覆盖域 Ω_C^p，此介点的场信息由覆盖域内的场节点集 $\{x_J\}$ 来近似。

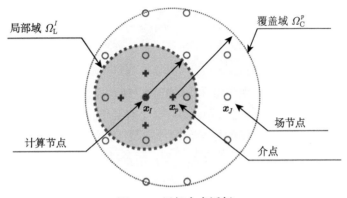

图 7-5 局部介点近似

Fig. 7-5 Local intervention-point approximation

需特别注意，MIP 法须使用 MLSc[29] 来执行进行。因为局部域 Ω_L^I 的尺度可能控制得非常小，如果使用高次基函数，足够局部的域上介点间距将是致密的，这种情况下标准的 MLS 将变得不稳定。

在局部域 Ω_L^I 上，计算节点 x_I 的场变量通过介点集 $\{x_p\}$ 执行近似，即

$$u_I^h(\boldsymbol{x}) = \sum_{p=1}^{N_p} \varphi_p(\boldsymbol{x}, \boldsymbol{x}_I) \hat{u}_p, \quad 在 \Omega_L^I 上 \tag{7-39}$$

在覆盖域 Ω_C^p 上，介点的场变量通过场节点集 $\{x_J\}$ 执行近似，即

$$\hat{u}_p \approx u_p^h(\boldsymbol{x}) = \sum_{J=1}^{N_J} \phi_J(\boldsymbol{x}, \boldsymbol{x}_p) \hat{u}_J, \quad 在 \Omega_C^p 上 \tag{7-40}$$

需注意到，以上两个近似公式分别取自不同的近似空间，故在此处采用 φ 和 ϕ 两种形函数记号。将式 (7-40) 代入式 (7-39)，则可消掉 $\hat{u}_p$，即有

$$u_I^h(\boldsymbol{x}) \approx \sum_{p=1}^{N_p} \sum_{J=1}^{N_J} \varphi_p(\boldsymbol{x}, \boldsymbol{x}_I) \phi_J(\boldsymbol{x}, \boldsymbol{x}_p) \hat{u}_J \tag{7-41}$$

对计算节点场变量 u_I 的一阶导数近似建议为

$$\left. \begin{array}{l} u_{I,i}^h(\boldsymbol{x}) = \sum_{p=1}^{N_p} \varphi_{p,i}(\boldsymbol{x}, \boldsymbol{x}_I) \hat{u}_p \\ \hat{u}_p \approx u_p^h(\boldsymbol{x}) = \sum_{J=1}^{N_J} \phi_J(\boldsymbol{x}, \boldsymbol{x}_p) \hat{u}_J \end{array} \right\} \Rightarrow u_{I,i}^h(\boldsymbol{x}) \approx \sum_{p=1}^{N_p} \sum_{J=1}^{N_J} \varphi_{p,i}(\boldsymbol{x}, \boldsymbol{x}_I) \phi_J(\boldsymbol{x}, \boldsymbol{x}_p) \hat{u}_J$$

$$\tag{7-42}$$

对 u_I 的一阶导数近似还有另外一种选择方案, 即

$$\left.\begin{aligned} u_{I,i}^h(\boldsymbol{x}) &= \sum_{p=1}^{N_p} \varphi_p(\boldsymbol{x},\boldsymbol{x}_I)\,\hat{u}_{p,i} \\ \hat{u}_{p,i} &\approx u_{p,i}^h(\boldsymbol{x}) = \sum_{J=1}^{N_J} \phi_{J,i}(\boldsymbol{x},\boldsymbol{x}_p)\,\hat{u}_J \end{aligned}\right\} \Rightarrow u_{I,i}^h(\boldsymbol{x}) \approx \sum_{p=1}^{N_p}\sum_{J=1}^{N_J} \varphi_p(\boldsymbol{x},\boldsymbol{x}_I)\,\phi_{J,i}(\boldsymbol{x},\boldsymbol{x}_p)\,\hat{u}_J \tag{7-43}$$

因为局部域 Ω_L^I 尺度要小于覆盖域 Ω_C^p 的尺度, 理论上讲, 局部域上的场变量近似精度更好, 优先在局部域上执行导数近似, 会更有利于控制导数近似的精度。因此, 一般问题中, 式 (7-42) 是建议的标准一阶导数近似方法。

对计算节点场变量 u_I 的二阶导数近似, 也会有两种形式。其一为

$$\left.\begin{aligned} u_{I,ij}^h(\boldsymbol{x}) &= \sum_{p=1}^{N_p} \varphi_{p,ij}(\boldsymbol{x},\boldsymbol{x}_I)\,\hat{u}_p \\ \hat{u}_p &\approx u_p^h(\boldsymbol{x}) = \sum_{J=1}^{N_J} \phi_J(\boldsymbol{x},\boldsymbol{x}_p)\,\hat{u}_J \end{aligned}\right\} \Rightarrow u_{I,ij}^h(\boldsymbol{x}) \approx \sum_{p=1}^{N_p}\sum_{J=1}^{N_J} \varphi_{p,ij}(\boldsymbol{x},\boldsymbol{x}_I)\,\phi_J(\boldsymbol{x},\boldsymbol{x}_p)\,\hat{u}_J \tag{7-44}$$

显然, 上式的二阶导数近似全部在局部域 Ω_L^I 上完成, 可以把这种二阶导数近似格式记为 "L2C0" 型, L 表示局部域, 其后的数字 2 表示在该域上的导数近似阶数; C 表示覆盖域, 其后的数字 0 表示在该域上的导数近似阶数。

另外一种二阶导数近似公式可以写为

$$\left.\begin{aligned} u_{I,ij}^h(\boldsymbol{x}) &= \sum_{p=1}^{N_p} \varphi_{p,i}(\boldsymbol{x},\boldsymbol{x}_I)\,\hat{u}_{p,j} \\ \hat{u}_{p,j} &\approx u_{p,j}^h(\boldsymbol{x}) = \sum_{J=1}^{N_J} \phi_{J,j}(\boldsymbol{x},\boldsymbol{x}_p)\,\hat{u}_J \end{aligned}\right\} \Rightarrow u_{I,ij}^h(\boldsymbol{x}) \approx \sum_{p=1}^{N_p}\sum_{J=1}^{N_J} \varphi_{p,i}(\boldsymbol{x},\boldsymbol{x}_I)\,\phi_{J,j}(\boldsymbol{x},\boldsymbol{x}_p)\,\hat{u}_J \tag{7-45}$$

显然, 上式的二阶导数近似在局部域 Ω_L^I 上分配一阶导数近似, 在覆盖域上也分配一阶导数近似。对应地, 可以把这种二阶导数近似格式记为 "L1C1" 型。

这两种二阶导数近似格式各有优势, L2C0 型使得导数近似在局部小域上完成, 理论上更容易控制导数近似精度。而 L1C1 型则将二阶导数近似进行了分配, 降低了对近似函数连续性的要求; 比如, L2C0 中对近似函数 φ 要求是 C^2 连续的, 而 L1C1 中对其只要求 C^1 连续。

对更一般的情况, 或者是对高于二阶 ($d \geqslant 2$) 的导数近似, 计算节点 u_I 的 $d\,(d \geqslant 2)$ 阶导数近似公式可以写为

$$\left\{u_I^h(\boldsymbol{x})\right\}^{(d)} = \sum_{p=1}^{N_p}\sum_{J=1}^{N_J} \varphi_p^{(d-\alpha)}(\boldsymbol{x},\boldsymbol{x}_I)\,\phi_J^{(\alpha)}(\boldsymbol{x},\boldsymbol{x}_p)\,\hat{u}_J, \quad \alpha \leqslant d/2 \tag{7-46}$$

7.4 无网格介点法

综上所述,可以把局部介点近似方法的特点归结为:

1) 随计算节点 x_I 定义的局部域 Ω_L^I 具有自由的尺度,其大小很容易被锁定。在数值计算中,可以把就近的第 3 个节点距离 $\min(r_3)$ 作为它的参考半径。这样一来,Ω_L^I 具有了 FEM 中分片单元的概念,为计算节点赋予一个严格意义上的局部空间。而覆盖域 Ω_C^p 则发挥与周边场节点建立联系的链接作用。

2) 局部域 Ω_L^I 中介点集 $\{x_p\}$ 是计算底层临时借用的位置信息,该插入点集的数量是自由的,只要在满足近似计算连续性要求的条件下,近似函数的基函数 $p(x)$ 的项数和次数的选择是相对自由的,即所谓的 "p-适应性"。

3) 既然介点集 $\{x_p\}$ 的使用是灵活的,那么可以根据形函数的分布特点更合理地布置这些介点。通过使用标准模板的方式,把介点集 $\{x_p\}$ 摆放在关键的位置上 (形函数及其导数图像的峰值位置),并合理控制其使用数量。这个标准模板在计算底层随时调用,随计算节点定义,大小根据周边节点信息缩放。也就是说,这些介点可以通过精密的设计,保证数值计算的精度和效率,并有效实现二者的平衡 [30]。

4) 计算节点 u_I 的 d 阶导数近似可以自由地分配在两个形函数 φ 和 ϕ 之上。此处,可以把这种 d 阶导数在两个形函数上的自由分配机制称为 "d-适应性"。这样一来,赋予数值方法更多的离散方案,即数值计算的灵活性。

5) 对目标节点上场变量的近似需要使用两个形函数 φ 和 ϕ。这两个形函数可以采用同一种近似方法,也可采用不同的近似方法。因此,很容易实现近似方法的耦合,这种特性可能在一些特殊问题中有一定的应用潜力。比如,SPH 法具有对粒子物理特性可直接描述的优势,但其导数运算适应能力较差,很容易出现不稳定性,如果使用耦合近似方案,即

$$u_{I,i}^h(\boldsymbol{x}) \approx \sum_{p=1}^{N_p}\sum_{J=1}^{N_J} \varphi_{p,i}^{\mathrm{MLSc}} \phi_J^{\mathrm{SPH}} \hat{u}_J \tag{7-47}$$

可以作为一种应对挑战的方案。

上述局部介点近似方法的的 5 个特征,实际上就是 MIP 法所具有的优势和特点。局部介点近似适用于配点技术的方法,是 "h-适应性" 的,加之其具有 "p-适应性" 和 "d-适应性"。因此,MIP 法将具有 "h-,p-,d-适应性"。

MIP 法采用局部介点近似方法,主要是通过控制导数近似的精度和稳定性,来提高数值方法的计算精度和稳定性。因此在数值离散中,只需对问题域内节点和自然边界上的节点使用该近似方法。而在本质边界上的节点因不涉及导数近似,可以直接近似即可。因此,MIP 法的离散原理如图 7-6 所示。

对二维弹性力学问题,令所有域内节点满足平衡方程 (2-1),自然边界 Γ_t 上的节点满足自然边界条件 (2-2),本质边界 Γ_u 上的节点满足本质边界条件 (2-3),则

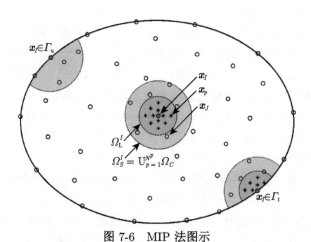

图 7-6 MIP 法图示
Fig. 7-6 Schematics of the MIP method

可得到 MIP 法的离散系统方程：

$$\boldsymbol{K} \cdot \boldsymbol{U} = \boldsymbol{F} \tag{7-48}$$

其中，

$$\boldsymbol{K}_{IJ} = \begin{cases} \begin{cases} \sum\limits_{p=1}^{N_p} \boldsymbol{A}_p \phi_J(\boldsymbol{x}, \boldsymbol{x}_p), & \text{L2C0}, \quad \boldsymbol{x}_I \in \Omega \\ \sum\limits_{p=1}^{N_p} \boldsymbol{B}_p^{\mathrm{T}} \boldsymbol{D} \boldsymbol{B}_J, & \text{L1C1}, \quad \boldsymbol{x}_I \in \Omega \end{cases} \\ \sum\limits_{p=1}^{N_p} \boldsymbol{n}_I \boldsymbol{D} \boldsymbol{B}_p \phi_J(\boldsymbol{x}, \boldsymbol{x}_p), & \boldsymbol{x}_I \in \Gamma_t \\ \boldsymbol{N}_J, & \boldsymbol{x}_I \in \Gamma_u \end{cases} \tag{7-49}$$

$$\boldsymbol{F}_I = \begin{cases} \boldsymbol{b}_I, & \boldsymbol{x}_I \in \Omega \\ \bar{\boldsymbol{t}}_I, & \boldsymbol{x}_I \in \Gamma_t \\ \bar{\boldsymbol{u}}_I, & \boldsymbol{x}_I \in \Gamma_u \end{cases} \tag{7-50}$$

式中，$\boldsymbol{D}$ 为由式 (2-15a) 定义的材料常数矩阵。其他符号定义为

$$\begin{aligned}
\boldsymbol{A}_p &= \boldsymbol{L}^{\mathrm{T}} \boldsymbol{D} \boldsymbol{L} \varphi_p(\boldsymbol{x}, \boldsymbol{x}_I) \\
&= \frac{E}{1-\nu^2} \begin{bmatrix} \dfrac{\partial^2 \varphi_p(\boldsymbol{x}, \boldsymbol{x}_I)}{\partial x^2} + \dfrac{1-\nu}{2} \dfrac{\partial^2 \varphi_p(\boldsymbol{x}, \boldsymbol{x}_I)}{\partial y^2} & \dfrac{1+\nu}{2} \dfrac{\partial^2 \varphi_p(\boldsymbol{x}, \boldsymbol{x}_I)}{\partial x \partial y} \\ \dfrac{1+\nu}{2} \dfrac{\partial^2 \varphi_p(\boldsymbol{x}, \boldsymbol{x}_I)}{\partial x \partial y} & \dfrac{\partial^2 \varphi_p(\boldsymbol{x})}{\partial y^2} + \dfrac{1-\nu}{2} \dfrac{\partial^2 \varphi_p(\boldsymbol{x}, \boldsymbol{x}_I)}{\partial x^2} \end{bmatrix}
\end{aligned} \tag{7-51}$$

$$\boldsymbol{n}_I = \begin{bmatrix} n_x & 0 & n_y \\ 0 & n_y & n_x \end{bmatrix}_I \tag{7-52}$$

$$\boldsymbol{N}_J = \begin{bmatrix} \phi_J(\boldsymbol{x}, \boldsymbol{x}_I) & 0 \\ 0 & \phi_J(\boldsymbol{x}, \boldsymbol{x}_I) \end{bmatrix} \tag{7-53}$$

$$\boldsymbol{B}_p = \boldsymbol{L}\varphi_p(\boldsymbol{x}, \boldsymbol{x}_I) = \begin{bmatrix} \dfrac{\partial \varphi_p(\boldsymbol{x}, \boldsymbol{x}_I)}{\partial x} & 0 \\ 0 & \dfrac{\partial \varphi_p(\boldsymbol{x}, \boldsymbol{x}_I)}{\partial y} \\ \dfrac{\partial \varphi_p(\boldsymbol{x}, \boldsymbol{x}_I)}{\partial y} & \dfrac{\partial \varphi_p(\boldsymbol{x}, \boldsymbol{x}_I)}{\partial x} \end{bmatrix} \tag{7-54}$$

$$\boldsymbol{B}_J = \boldsymbol{L}\phi_J(\boldsymbol{x}, \boldsymbol{x}_p) = \begin{bmatrix} \dfrac{\partial \phi_J(\boldsymbol{x}, \boldsymbol{x}_p)}{\partial x} & 0 \\ 0 & \dfrac{\partial \phi_J(\boldsymbol{x}, \boldsymbol{x}_p)}{\partial y} \\ \dfrac{\partial \phi_J(\boldsymbol{x}, \boldsymbol{x}_p)}{\partial y} & \dfrac{\partial \phi_J(\boldsymbol{x}, \boldsymbol{x}_p)}{\partial x} \end{bmatrix} \tag{7-55}$$

$$\boldsymbol{b}_I = \begin{bmatrix} b_x \\ b_y \end{bmatrix}_I, \quad \bar{\boldsymbol{t}}_I = \begin{bmatrix} \bar{t}_x \\ \bar{t}_y \end{bmatrix}_I, \quad \bar{\boldsymbol{u}}_I = \begin{bmatrix} \bar{u}_x \\ \bar{u}_y \end{bmatrix}_I \tag{7-56}$$

式中，E 和 ν 分别为弹性模量和泊松比。

MIP 法的总体刚度矩阵组装如 FPM 一样，非常简单直接。总体刚度矩阵 $\boldsymbol{K}$ 的行用计算节点的序号 "I" 索引，其列用支撑节点序号 "J" 索引，只需将离散矩阵 $\boldsymbol{K}_{IJ}$ 装配到对应位置即可，如下式所示：

$$\begin{matrix} & J \rightarrow \\ I \downarrow & \begin{bmatrix} \boldsymbol{K}_{11} & \cdots & \boldsymbol{K}_{1J} & \cdots & \boldsymbol{K}_{1n} \\ & & \vdots & & \\ \boldsymbol{K}_{I1} & \cdots & \boldsymbol{K}_{IJ} & \cdots & \boldsymbol{K}_{In} \\ & & \vdots & & \\ \boldsymbol{K}_{n1} & \cdots & \boldsymbol{K}_{nJ} & \cdots & \boldsymbol{K}_{nn} \end{bmatrix} \cdot \begin{bmatrix} \boldsymbol{U}_1 \\ \vdots \\ \boldsymbol{U}_I \\ \vdots \\ \boldsymbol{U}_n \end{bmatrix} = \begin{bmatrix} \boldsymbol{F}_1 \\ \vdots \\ \boldsymbol{F}_I \\ \vdots \\ \boldsymbol{F}_n \end{bmatrix} \end{matrix} \tag{7-57}$$

此外，MIP 法的离散方程很容易从原始平衡方程得到，计算逻辑非常简洁，在实际应用和数值实施中很容易实现。接下来，通过一些数值算例，进一步演示和验证 MIP 法的计算效果。

算例 1 带圆孔的无限板问题

考虑一带圆孔的无限大板，在 x 方向受集度为 q 的均布拉伸荷载，如图 7-7 所示。利用结构的对称性，取边长为 b 的右上方 1/4 板的有限结构建模，如图 7-8 所示。计算中采用无量纲化处理 (或者默认为均采用标准国际单位),计算参数取为:圆孔半径 $a=1$,方板长度 $b=5$,荷载强度 $q=1$,弹性模量 $E=10^3$,泊松比 $\nu=0.3$。该问题存在如下解析解 [31]:

$$\begin{cases} u_r = \dfrac{q}{4\eta}\left[r\left(\dfrac{\kappa-1}{2}+\cos 2\theta\right)+\dfrac{a^2}{r}[1+(1+\kappa)\cos 2\theta]-\dfrac{a^4}{r^3}\cos 2\theta\right] \\ u_\theta = \dfrac{q}{4\eta}\left[(1-\kappa)\dfrac{a^2}{r}-r-\dfrac{a^4}{r^3}\right]\sin 2\theta \end{cases} \quad (7\text{-}58)$$

$$\begin{cases} \sigma_x(x,y) = q\left[1-\dfrac{a^2}{r^2}\left(\dfrac{3}{2}\cos 2\theta+\cos 4\theta\right)+\dfrac{3a^4}{2r^4}\cos 4\theta\right] \\ \sigma_y(x,y) = -q\left[\dfrac{a^2}{r^2}\left(\dfrac{1}{2}\cos 2\theta-\cos 4\theta\right)+\dfrac{3a^4}{2r^4}\cos 4\theta\right] \\ \tau_{xy}(x,y) = -q\left[\dfrac{a^2}{r^2}\left(\dfrac{1}{2}\sin 2\theta+\sin 4\theta\right)-\dfrac{3a^4}{2r^4}\sin 4\theta\right] \end{cases} \quad (7\text{-}59)$$

式中,

$$\eta = \dfrac{E}{2(1+\nu)}, \quad \kappa = 3-4\nu \quad (7\text{-}60)$$

问题的数值计算模型如图 7-8 所示,外力边界条件使用解析解的结果代入进行计算。数值计算时的场节点离散方案如图 7-9(a) 所示,板中的 x 方向应力计算结果如图 7-9(b) 所示。

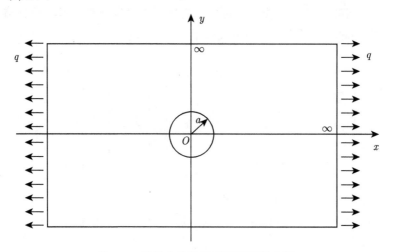

图 7-7 受单向拉伸的带圆孔无限大板

Fig. 7-7 An infinite plate with a circle hole forced by uniaxial tensile load

7.4 无网格介点法

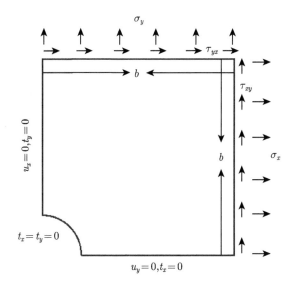

图 7-8　带圆孔无限大板问题的计算模型

Fig. 7-8　Numerical model for the problem of infinite plate with circle hole

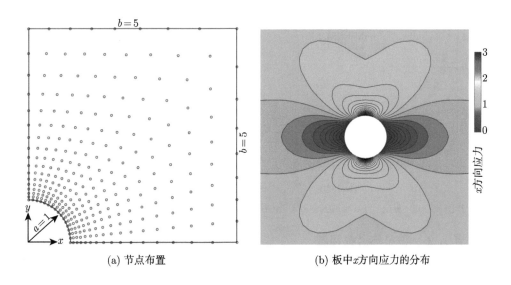

(a) 节点布置　　　　　　　　(b) 板中 x 方向应力的分布

图 7-9　带圆孔无限板问题的离散模型和求解结果

Fig. 7-9　Discrete model and numerical results of the problem of infinite plate with a hole

选用 $x=0, y>0$ 的路径作为计算结果的输出和分析路径，并定义如下的平均 L2 误差范数 (error norm)：

$$e(\xi) = \frac{1}{N}\sqrt{\sum_{I=1}^{N}(\xi_I^{\text{num}} - \xi_I^{\text{exa}})^2 \bigg/ \sum_{I=1}^{N}(\xi_I^{\text{exa}})^2} \tag{7-61}$$

式中, ξ_I^{num} 表示数值计算得到的值; ξ_I^{exa} 表示解析得到的精确值; N 表示输出路径上考察的计算点数量。图 7-10 中给出了 MIP 法的两种计算格式 (L2C0 和 L1C1) 以及标准配点法 (FPM) 的应力数值解计算结果比较, 从图 7-10(a) 的计算结果可以看出, MIP 的两种计算格式求解精度要明显高于普通配点法。并对平均节点间距给出如下定义:

$$h_a = \sqrt{S_\Omega}\bigg/\left(\sqrt{N}-1\right) \tag{7-62}$$

式中, S_Ω 表示问题域面积; N 表示场节点总数量。图 7-10(b) 给出了数值法的求解误差比较, 并据此分析其收敛性。从图中可以看出, MIP 法的收敛率线总是位于直接配点法的下面, 说明 MIP 法相比于普通配点法, 具有更高的求解精度和更好的收敛性。计算结果初步表明, MIP 法的两种求解格式中, L2C0 求解格式的计算精度和收敛性通常要优于 L1C1 求解格式。

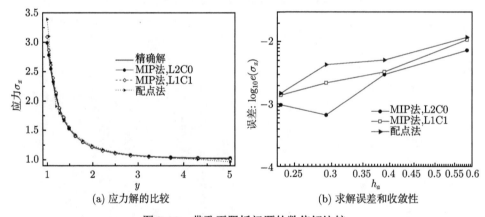

(a) 应力解的比较　　(b) 求解误差和收敛性

图 7-10　带孔无限板问题的数值解比较

Fig. 7-10　Comparing the numerical solutions for the problem of infinite plate with a hole

算例 2　轴对称薄板问题

考虑一周边简支圆板, 圆板上表面施加集度为 q, 指向 z 轴负方向的均布荷载, 如图 7-11(a) 所示。该问题具有轴对称性, 可以简化为平面问题求解分析, 即求解域的离散只需选取一个对称面即可, 如图中给出的任意的轴对称面 S。其数值计算模型如图 7-11(b) 所示, 该问题实际上对数值求解有一定挑战性, 其对称轴位置只有单向约束, 而在全部边界上只有外边底脚一个点处有全位移约束。对计算参数进行量纲归一化处理 (或默认为国际标准单位), 参数取为: 板的半径长度 $a=2$, 板厚 $2c=0.12$, 弹性模量 $E=7\times 10^5$, 泊松比 $\nu=1/3$, 表面荷载强度 $q=100$。

7.4 无网格介点法

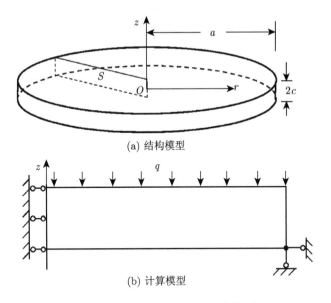

(a) 结构模型

(b) 计算模型

图 7-11 轴对称薄板问题计算模型

Fig. 7-11 Numerical model for the axisymmetric plate problem

在柱坐标系中的轴对称方程写为

$$\begin{cases} \dfrac{1}{r}\dfrac{\partial(r\sigma_r)}{\partial r} - \dfrac{\sigma_\theta}{r} + \dfrac{\partial \tau_{rz}}{\partial z} + b_r = 0 \\ \dfrac{\partial \sigma_z}{\partial z} + \dfrac{1}{r}\dfrac{\partial(r\tau_{rz})}{\partial r} + b_z = 0 \end{cases} \tag{7-63}$$

该问题存在如下解析解[32]：

$$\begin{cases} u_z = \dfrac{3q(1-\nu^2)}{128Ec^3}\left[r^4 - \dfrac{3+\nu}{1+\nu}2a^2r^2 + \dfrac{5+\nu}{1+\nu}a^4\right] \\ \sigma_r = q\left[\dfrac{2+\nu}{8}\dfrac{z^3}{c^3} - \dfrac{3(3+\nu)}{32}\dfrac{r^2z}{c^3} - \dfrac{3}{8}\dfrac{2+\nu}{5}\dfrac{z}{c} + \dfrac{3(3+\nu)}{32}\dfrac{a^2z}{c^3}\right] \end{cases} \tag{7-64}$$

MIP 法求解该问题采用 L1C1 型离散方案，数值计算采用规则节点离散方案，并在 r 和 z 方向取不同的节点间距，且有 $h_r=10h_z$，如图 7-12 所示。当 $h_r=0.1$ 时，板底的位移和应力解如图 7-13 所示，并与 MLPG 法 (MLPG5)[33] 和一般配点法 (FPM)[14] 的数值计算结果进行比较。可见，MIP 法与 MLPG 法的求解精确度较高，而配点法的求解结果严重偏离精确解。

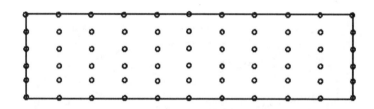

图 7-12　轴对称薄板问题的节点布置

Fig. 7-12　Nodal arrangement for the axisymmetric plate problem

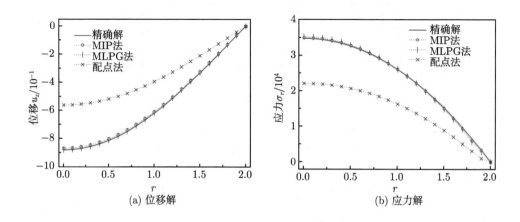

图 7-13　轴对称薄板问题的数值解

Fig. 7-13　Numerical solutions for the axisymmetric plate problem

在此计算路径上，对这 3 种方法的求解误差和数值收敛性进一步比较，如图 7-14 所示。显然，MIP 法与 MLPG 法的精度和收敛性要明显地优于配点法。MIP 法与 MLPG 法相比较，在节点稀疏时，MIP 法的精度和收敛性优于 MLPG 法，整体上，MLPG 法的收敛性优于 MIP 法。

进一步比较 3 种方法的计算效率，如表 7-1 所示。显然，配点法的计算效率最高，MIP 法次之，而 MLPG 法计算效率在三者中最低。其中，MIP 法与 MLPG 法作为计算精度可接受的两种方法，其效率比较是我们更应当关心的。需特别注意，此处采用的 MLPG 法，是 6.3 节所介绍的采用了阶跃检验函数的 MLPG 法，即 MLPG5，该法的计算效率相比于传统的 MLPG 法 (MLPG1)，已经得到大幅提高。很显然，MIP 法的计算效率要远高于 MLPG 法，在考察的节点数范围内，MIP 法的计算耗时仅为 MLPG 法的 17% 左右。也就是说，对这个算例而言，MIP 法的求解精度与 MLPG 法相当，但其计算效率远高于 MLPG 法。

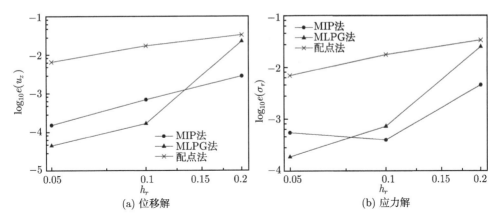

图 7-14 轴对称薄板问题的求解误差

Fig. 7-14 Numerical errors for the axisymmetric plate problem

表 7-1 轴对称薄板问题的数值计算时间比较

Table 7-1 Solution-time costs for the axisymmetric plate problem

Number of field nodes	Time cost for numerical solution /s		
	配点法	MIP 法	MLPG 法
147	0.084188	0.936725	4.481785
273	0.140762	1.864536	10.101940
533	0.319651	3.963082	23.046047
793	0.534845	6.238015	37.716791

算例 3 泊松方程问题

考虑 4.6 节算例 4 给出的泊松方程求解问题,采用不规则节点离散方案,如图 7-15 所示。

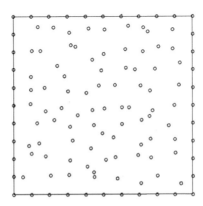

图 7-15 泊松方程问题的节点布置

Fig. 7-15 Nodal arrangement for the Poisson's equation problem

MIP 法和配点法求解该问题的数值收敛性给出，如图 7-16 所示。显然，对于这一随机布点问题，MIP 法表现为更高的求解精度，并具有良好的收敛性。

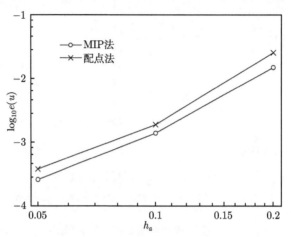

图 7-16　泊松方程问题的求解误差

Fig. 7-16　Numerical errors for the Poisson's equation problem

算例 4　Helmholtz 方程问题

MIP 法无需执行积分运算，对边界条件的处理非常简单直接，因此很容易处理具有不规则几何形状的边界问题。考虑一个 Helmholtz 方程问题，其域边界由外旋轮线函数确定，即

$$\rho(\theta) = \sqrt{(a+b)^2 + 1 - 2(a+b)\cos(a\theta/b)} \tag{7-65}$$

式中，形状参数取 $a=3, b=1$。给定的 Helmholtz 方程为

$$\frac{\partial^2 u}{\partial x^2} + \frac{\partial^2 u}{\partial y^2} - 2u = 0 \tag{7-66}$$

其解析解为

$$u = \exp(x+y) \tag{7-67}$$

外边界采用本质边界条件，使用上式的精确解直接代入。采用的节点离散方案如图 7-17 所示，在问题域上得到数值解，如图 7-18 所示。

另取以原点为圆心，$r=1$ 的圆形路径，该路径上的求解结果如图 7-19 所示。将 MIP 法与普通配点法的计算收敛性进行比较，如图 7-20 所示。显然，MIP 法的计算精度和收敛性要明显优于配点法，而且采用"L2C0"型计算格式的 MIP 法具有最佳的收敛性。

7.4 无网格介点法

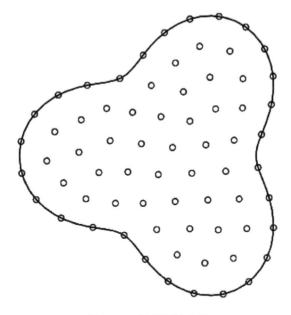

图 7-17 节点离散方案

Fig. 7-17 Nodal arrangement for the domain

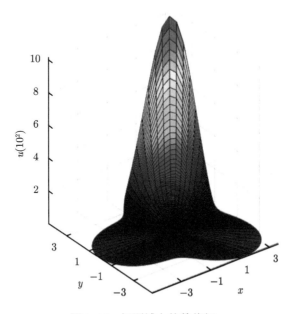

图 7-18 问题域上的数值解

Fig. 7-18 Numerical solution on the domain

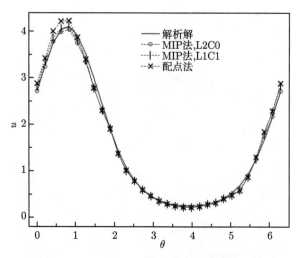

图 7-19 Helmholtz 方程问题数值解的比较

Fig. 7-19 Comparision of the numerical solutions for the Helmholtz equation problem

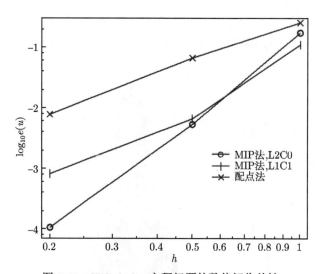

图 7-20 Helmholtz 方程问题的数值解收敛性

Fig. 7-20 Convergences for the Helmholtz equation problem

算例 5 欧拉梁问题

MIP 法具有 p-适应性，这意味着该方法在求解高阶偏微分方程问题中有理论上的可能性，为了验证这一设想，故引入该算例进行初步检验。受均布荷载作用的欧拉梁问题如图 7-21 所示。该问题是一个四阶微分方程问题，其平衡方程写为

$$EI \cdot \frac{\partial^4 u(x)}{\partial x^4} = q(x) \tag{7-68}$$

7.4 无网格介点法

式中，$u(x)$ 表示欧拉梁的挠度函数；EI 表示梁的弯曲劲度。该问题有如下解析解：

$$u(x) = -\frac{qx}{24EI}\left(L^3 - 2Lx^2 + x^3\right) \quad (7\text{-}69)$$

式中的计算参数取值为 $EI=10^5$, $q=1$, $L=100$。其边界条件为 $u_{x=L}(x)=0$，$\partial u/\partial x|_{x=0} = -\dfrac{qL^3}{24EI}$。另，引入求解控制条件 $\partial u/\partial x|_{x=L/2}=0$。

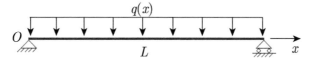

图 7-21 欧拉梁

Fig. 7-21 The Euler beam

本书求解该问题使用 51 个节点进行一维离散。对于该四阶微分方程问题，利用 MIP 的 d 适应性，可以有更多的计算格式，此处考虑 L2C2, L3C1 和 L4C0 三种求解格式，计算结果如图 7-22 所示。可以看出，本书方法能对该问题给出较为合理的解答。而对这种高阶微分方程问题，通常方法是不能直接进行求解的。如果采用配点法直接求解这个问题，无论节点如何加密，其数值解都是严重发散而不收敛的。

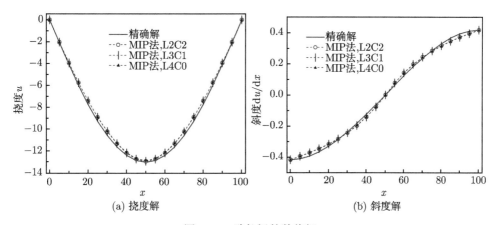

图 7-22 欧拉梁的数值解

Fig. 7-22 Numerical solutions for the Euler-beam problem

经过以上数值验证，表明 MIP 法不仅具有计算简单、效率高、精度高的优点，而且对多类求解问题具有"广泛适用"的潜在优势。此外，对 MIP 法而言，合理设

计局部介点模板是保证数值方法精确、稳定和高效的重要环节。建议将距计算节点最近的第 k 个点作为参考点，两点间距 $r_k = \|\boldsymbol{x}_k - \boldsymbol{x}_I\|$ 作为局部域 Ω_L^I 半径尺度的参考值，其中 k 的取值建议为

$$k = 2D - 1 \tag{7-70}$$

式中，D 为问题的维度数。则局部域 Ω_L^I 半径尺度取值为

$$r_\mathrm{L} = \alpha_\mathrm{L} r_k \tag{7-71}$$

式中，α_L 为尺度参数，建议取为 [30]

$$\alpha_\mathrm{L} = [1.3, 1.5] \tag{7-72}$$

在单位半径为 1 的介点模板上，介点取值建议为

$$r_p = \begin{bmatrix} 0, & 0.25, & 0.5 \end{bmatrix} \tag{7-73}$$

该径向点集间隔 $\pi/4$ 角度布设一周即形成单位介点模板。

7.5 无网格全局介点法

无网格全局介点 (MGIP) 法中 [27]，用 n 个场节点离散问题域，并引入 n_p 个介点，分别布置在域内和边界上，且要求 $n_p > n$，如图 7-23 所示。这些介点是隐性的，仅在构造离散方程时使用，求解结果是映射在场节点上，无需依靠介点来表达。

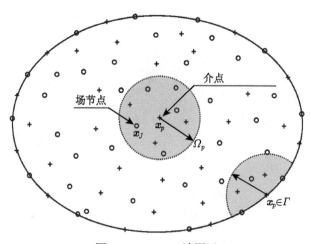

图 7-23 MGIP 法图示

Fig. 7-23 Schematics of the MGIP method

7.5 无网格全局介点法

MGIP 法要求控制方程和边界条件方程的残值在介点集 $\{\boldsymbol{x}_p\}_{p=1}^{n_p}$ 上消除, 并在介点集上构造如下加权残值泛函:

$$\begin{aligned}&\Pi_3\left(\delta u, \delta u_{,i}, \delta u_{,ij}\right)\\&=\sum_{\boldsymbol{x}_p\in\Omega}\delta u_{,ij}\left(\sigma_{ij,i}-b_i\right)+\sum_{\boldsymbol{x}_p\in\Gamma_t}\lambda_t\delta u_{,i}\left(n_i\cdot\sigma_{ij}-\bar{t}_i\right)+\sum_{\boldsymbol{x}_p\in\Gamma_u}\lambda_u\cdot\delta u\left(u_i-\bar{u}_i\right)\end{aligned} \tag{7-74}$$

该泛函基于有限点的广义变分法, 显然与 LSCM 的泛函式 (7-30) 不同。式中, λ_t 和 λ_u 为保证本质边界条件的罚参数。命该泛函为 0, 即

$$\Pi_3=0 \tag{7-75}$$

则域内介点必然满足平衡方程, 自然边界上的介点必然满足自然边界条件, 本质边界上的介点必然满足本质边界条件。并可得到一个离散系统方程:

$$\boldsymbol{K}\cdot\boldsymbol{U}=\boldsymbol{F} \tag{7-76}$$

式中,

$$\boldsymbol{K}=\sum_{\boldsymbol{x}_p\in\Omega}\bar{\boldsymbol{H}}^\mathrm{T}\boldsymbol{H}+\sum_{\boldsymbol{x}_p\in\Gamma_t}\lambda_t\overline{\boldsymbol{Q}}^\mathrm{T}\boldsymbol{Q}+\sum_{\boldsymbol{x}_p\in\Gamma_u}\lambda_u\boldsymbol{N}^\mathrm{T}\boldsymbol{N} \tag{7-77}$$

$$\boldsymbol{F}=\sum_{\boldsymbol{x}_p\in\Omega}\bar{\boldsymbol{H}}^\mathrm{T}\boldsymbol{b}+\sum_{\boldsymbol{x}_p\in\Gamma_t}\lambda_t\overline{\boldsymbol{Q}}^\mathrm{T}\bar{\boldsymbol{t}}+\sum_{\boldsymbol{x}_p\in\Gamma_u}\lambda_u\boldsymbol{N}^\mathrm{T}\bar{\boldsymbol{u}} \tag{7-78}$$

需注意到这个离散方程也有别于 MWLS 法。其中,

$$\begin{cases}\overline{\boldsymbol{H}}=\left[\overline{\boldsymbol{H}}_1,\overline{\boldsymbol{H}}_2,\cdots,\overline{\boldsymbol{H}}_J,\cdots,\overline{\boldsymbol{H}}_n\right]\\[6pt]\overline{\boldsymbol{H}}_J=\boldsymbol{L}^\mathrm{T}\boldsymbol{L}\left(\phi_J\right)=\begin{bmatrix}\dfrac{\partial^2\phi_J}{\partial x^2}+\dfrac{\partial^2\phi_J}{\partial y^2} & \dfrac{\partial^2\phi_J}{\partial x\partial y}\\[8pt]\dfrac{\partial^2\phi_J}{\partial x\partial y} & \dfrac{\partial^2\phi_J}{\partial x^2}+\dfrac{\partial^2\phi_J}{\partial y^2}\end{bmatrix}\end{cases} \tag{7-79}$$

$$\begin{cases}\boldsymbol{H}=\left[\boldsymbol{H}_1,\boldsymbol{H}_2,\cdots,\boldsymbol{H}_J,\cdots,\boldsymbol{H}_n\right]\\[6pt]\boldsymbol{H}_J=\boldsymbol{L}^\mathrm{T}\boldsymbol{D}\boldsymbol{L}\left(\phi_J\right)=\dfrac{E}{1-\nu^2}\begin{bmatrix}\dfrac{\partial^2\phi_J}{\partial x^2}+\dfrac{1-\nu}{2}\dfrac{\partial^2\phi_J}{\partial y^2} & \dfrac{1+\nu}{2}\dfrac{\partial^2\phi_J}{\partial x\partial y}\\[8pt]\dfrac{1+\nu}{2}\dfrac{\partial^2\phi_J}{\partial x\partial y} & \dfrac{\partial^2\phi_J}{\partial y^2}+\dfrac{1-\nu}{2}\dfrac{\partial^2\phi_J}{\partial x^2}\end{bmatrix}\end{cases} \tag{7-80}$$

$$\begin{cases} \boldsymbol{N} = [\boldsymbol{N}_1, \boldsymbol{N}_2, \cdots, \boldsymbol{N}_J, \cdots, \boldsymbol{N}_n] \\ \boldsymbol{N}_J = \begin{bmatrix} \phi_J & 0 \\ 0 & \phi_J \end{bmatrix} \end{cases} \quad (7\text{-}81)$$

$$\begin{cases} \overline{\boldsymbol{Q}} = [\overline{\boldsymbol{Q}}_1, \overline{\boldsymbol{Q}}_2, \cdots, \overline{\boldsymbol{Q}}_J, \cdots, \overline{\boldsymbol{Q}}_n] \\ \overline{\boldsymbol{Q}}_J = \boldsymbol{n}_p \boldsymbol{L}(\phi_J) = \begin{bmatrix} n_x \dfrac{\partial \phi_J}{\partial x} + n_y \dfrac{\partial \phi_J}{\partial y} & n_y \dfrac{\partial \phi_J}{\partial x} \\ n_x \dfrac{\partial \phi_J}{\partial y} & n_x \dfrac{\partial \phi_J}{\partial x} + n_y \dfrac{\partial \phi_J}{\partial y} \end{bmatrix} \end{cases} \quad (7\text{-}82)$$

$$\begin{cases} \boldsymbol{Q} = [\boldsymbol{Q}_1, \boldsymbol{Q}_2, \cdots, \boldsymbol{Q}_J, \cdots, \boldsymbol{Q}_n] \\ \boldsymbol{Q}_J = \boldsymbol{n}_I \boldsymbol{D} \boldsymbol{L}(\phi_J) = \dfrac{E}{1-\nu^2} \begin{bmatrix} n_x \dfrac{\partial \phi_J}{\partial x} + n_y \dfrac{1-\nu}{2}\dfrac{\partial \phi_J}{\partial y} & n_y \dfrac{1-\nu}{2}\dfrac{\partial \phi_J}{\partial x} + n_x \nu \dfrac{\partial \phi_J}{\partial y} \\ n_y \nu \dfrac{\partial \phi_J}{\partial x} + n_x \dfrac{1-\nu}{2}\dfrac{\partial \phi_J}{\partial y} & n_x \dfrac{1-\nu}{2}\dfrac{\partial \phi_J}{\partial x} + n_y \dfrac{\partial \phi_J}{\partial y} \end{bmatrix} \end{cases}$$
$$(7\text{-}83)$$

MGIP 法与 MWLS 法的离散方程和数值实施类似,但其刚度方程不再是"二乘"的形式。其中,罚参数 λ_t 和 λ_u 建议取为

$$\lambda_t = 10^5, \quad \lambda_u = \lambda_t \left(\dfrac{E}{1-\nu^2} \right) \quad (7\text{-}84)$$

在 MGIP 中,$\overline{\boldsymbol{H}}$ 和 $\overline{\boldsymbol{Q}}$ 中不包含材料常数矩阵。因此,对系统刚度矩阵 $\boldsymbol{K}$ 的数量级而言,MWLS 的数量级约为 E^2,而 MGIP 方法数量级仅约为 E。通常而言,弹性模量 E 是远大于形函数值的一个大数,而从求解的原理上讲,MGIP 法的刚度矩阵数量级比 MWLS 的要显著降低,这对提高计算精度通常是有帮助的。此外,通过一些数值试验表明:过多使用介点并不会对求解精度有明显帮助,反而会直接影响计算效率。因此,对二维问题而言,介点数量 n_p 的取值建议为

$$n_p \approx 2n \quad (7\text{-}85)$$

式中,n 为场节点数量。

算例 1 泊松方程问题

考虑 4.6 节算例 4 给出的泊松方程求解问题。MGIP 法求解该问题采用 121 个不规则分布的场节点布点方案,并在域背景上引入 169 个规则分布的介点,如图 7-24 所示。

MGIP 法的求解结果及其与精确解的比较如图 7-25 所示。同时列入比较的还有 MLPG5 法和配点法 (FPM) 的计算结果。配点法求解该问题时,场节点分布与

图 7-24 一致,只是不需要用到介点。计算结果可看出 MGIP 法能对该问题给出较为精确的求解,该法与 MLPG 法的求解精度明显高于配点法。

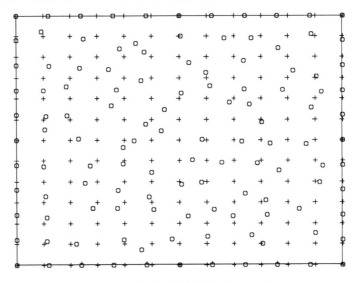

图 7-24 泊松方程问题的节点布置

Fig. 7-24 Nodal arrangement for the Poisson's equation problem

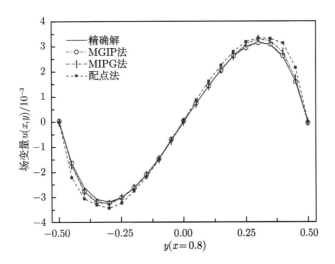

图 7-25 泊松方程问题的数值解

Fig. 7-25 Numerical solutions for the Poisson's equation problem

为了便于对 MGIP 法的求解效率有一个更为直接的判断,结合这个算例,设置一组疏密不同的节点离散方案,并将本书方法与 MLPG5 法[33]的计算机求解运算时间列于表 7-2。可以看出,本书方法的计算效率远高于 MLPG 法。

表 7-2 数值计算耗时 (单位: s)

Table 7-2 Time costs for numerical solution /s

Number of field nodes	MGIP 法	MLPG 法
36	0.024429	3.013041
64	0.031879	5.963321
121	0.063674	12.216659
256	0.179510	27.378109

算例 2 悬臂梁问题

考虑 4.6 节算例 3 给出的悬臂梁问题。问题域用 186 个节点离散,并引入 344 个背景介点,如图 7-26 所示。

图 7-26 悬臂梁问题布点方案

Fig. 7-26 Distribution of nodes and intervention points for the cantilever-beam problem

图 7-27 给出了板中 ($y = 0$) 沿板长路径上的数值解与精确解的比较计算结果,其中图 (a) 为竖向位移解,图 (b) 为 x 方向的应力解。在比较计算中,配点法的节点布置完全与图 7-26 的节点布置相同,MWLS 法的求解流程参照 MGIP 法执行,仅改变了相应的计算公式。由计算结果可以看出,本书方法的求解精度要明显高于普通配点法,而且相比于 MWLS 法也有一定优势。

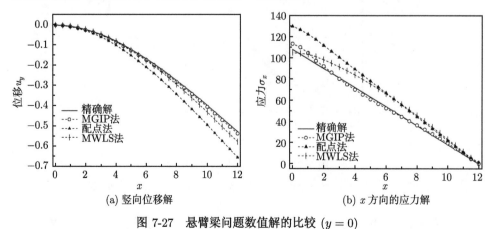

图 7-27 悬臂梁问题数值解的比较 ($y = 0$)

Fig. 7-27 Comparing the numerical solutions for the cantilever beam problem ($y = 0$)

算例 3　带圆孔的无限板问题

考虑 7.4 节算例 1 给出的带圆孔的无限板问题。问题域用 169 个节点离散，并引入 361 个背景介点，如图 7-28 所示。

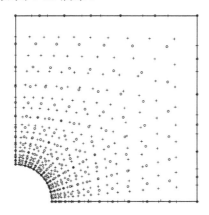

图 7-28　带圆孔无限板问题的布点方案

Fig. 7-28　Distribution of nodes and intervention points for the problem of infinite plate with a hole

将 $x = 0$ 沿 y 分布的线作为分析路径，其上径向位移和 x 方向的应力计算结果如图 7-29 所示。图中给出了几种方法的比较计算结果，列为对比数值计算的有一般配点法 (FPM) 和 EFG 法，其场节点布点方案与图 7-28 中的场节点布设完全一致。从计算结果可以看出，MGIP 法的求解精度更接近于 EFG 法。而 MGIP 法作为一种配点型的方法，在接近圆孔处的应力集中区域，其计算精度要明显高于普通配点法。

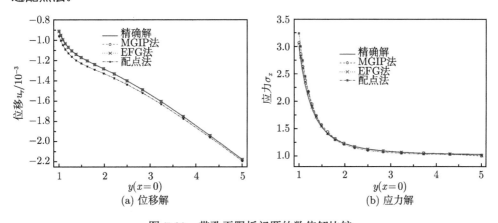

图 7-29　带孔无限板问题的数值解比较

Fig. 7-29　Comparing the numerical solutions for the problem of infinite plate with a hole

无网格全局介点 (MGIP) 法通过引入超过场节点数量的背景介点，增强了对场节点自由度的约束；通过基于介点的广义变分法求解，是寻求残值泛函驻立值的一种有效方法。MGIP 法的系统刚度矩阵 K 是稀疏带状对称矩阵，在大规模求解中有利于保证求解效率。MGIP 法作为一种配点型的方法，与一般配点法相比，通常计算精度更高，计算稳定性更好。MGIP 法属于强式方法，其计算执行与配点法类似，不需要构造背景网格或积分单元，无须执行积分运算，数值实施比弱式方法更为简洁，效率也更高。

7.6 本章小结

除了计算稳定性这一问题之外，配点法几乎具备所有无网格法所提倡的优点。配点法计算效率很高，可以完全免除网格的使用，数值实施简单，并容易处理不规则结构问题。在一般问题中，只要场节点足够细密，配点法通常能够收敛于精确解。但是，在场节点较为稀疏，或场函数光滑性较差的情况下，配点法很容易产生不稳定计算。

本章重点介绍的双网格扩散配点 (DGDC) 法、最小二乘配点法 (LSCM)、无网格介点 (MIP) 法和无网格全局介点 (MGIP) 法等，都是对传统配点法进行稳定化的改进方法。本质上，这几种方法都是介点原理的应用范例。其中，DGDC 法和 MIP 法是介点原理推论 3 的应用；而 LSCM 法和 MGIP 法属于介点原理推论 2 的应用。相比于一般配点法，这几种方法经介点辅助后，均加速了收敛性，并提高了稳定性。这几种方法中，通过数值实施简洁性、计算稳定性和计算的灵活性等综合比较，我们认为 MIP 法是更具竞争力的一种方法。

计算方法作为一种解决实际问题的工具，有效性、简洁性和灵活性是同等重要的几个考虑因素。逐步改进和完善后的配点类方法完全有可能在计算有效性上与弱式方法相媲美，而在简洁性和灵活性上则更胜一筹。因此，对无网格法的完善和发展而言，配点类方法的持续进步尤其值得关注。

参 考 文 献

[1] Girault V. Theory of a GDM on irregular net works[J]. SIAM J. Num. Anal., 1974, 11: 260-282.

[2] Perrone N, Kao R. A general finite difference method for arbitrary meshes[J]. Computers & Structures, 1975, 5(1): 45-57.

[3] Liszka T, Orkisz J. Finite Difference Method for Arbitrary Irregular Meshes in Nonlinear Problems of Applied Mechanics[M]. San Francisco: IV SMiRt, 1977.

[4] Liszka T, Orkisz J. The finite difference method at arbitrary irregular grids and its

application in applied mechanics [J]. Comput. Struct., 1980, 11: 83-95.

[5] Liszka T. An interpolation method for an irregular net of nodes[J]. International Journal for Numerical Methods in Engineering, 1984, 20(9): 1599-1612.

[6] Gingold R A, Monaghan J J. Smoothed particle hydrodynamics: theory and application to non-spherical stars[J]. Monthly Notices of The Royal Astronomical Society, 1977, 181(3): 375-389.

[7] Monaghan J J. Smoothed particle hydrodynamics[J]. Annual Review of Astronomy and Astrophysics, 1992, 30(1): 543-574.

[8] Lucy L B. A numerical approach to the testing of the fission hypothesis [J]. The A. Stron J., 1977, 8(12): 1013-1024.

[9] Morris J P, Fox P J, Zhu Y. Modeling low Reynolds number incompressible flows using SPH[J]. Journal of Computational Physics, 1997, 136(1): 214-226.

[10] Colagrossi A, Landrini M. Numerical simulation of interfacial flows by smoothed particle hydrodynamics[J]. Journal of Computational Physics, 2003, 191(2): 448-475.

[11] Randles P W, Libersky L D. Smoothed particle hydrodynamics: some recent improvements and applications[J]. Computer Methods in Applied Mechanics and Engineering, 1996, 139(1-4): 375-408.

[12] Liu M B, Liu G R. Smoothed particle hydrodynamics (SPH): An overview and recent developments[J]. Archives of Computational Methods in Engineering, 2010, 17(1): 25-76.

[13] Liu G R, Liu M B. Smoothed Particle Hydrodynamics: A Meshfree Particle Method[M]. Singapore: World Scientific, 2003.

[14] Onate E, Idelsohn S, Zienkiewicz O C, et al. A finite point method in computational mechanics. Applications to convective transport and fluid flow[J]. International Journal for Numerical Methods in Engineering, 1996, 39(22): 3839-3866.

[15] Onate E, Idelsohn S, Zienkiewicz O C, et al. A stabilized finite point method for analysis of fluid mechanics problems[J]. Computer Methods in Applied Mechanics and Engineering, 1996, 139(1-4): 315-346.

[16] Duarte C A, Oden J T. An hp adaptive method using clouds[J]. Computer Methods in Applied Mechanics and Engineering, 1996, 139(1-4): 237-262.

[17] Duarte C A, Oden J T. Hp clouds-an hp meshless method[J]. Numerical Methods for Partial Differential Equations, 1996, 12(6): 673-706.

[18] Breitkopf P, Touzot G, Villon P. Double grid diffuse collocation method[J]. Computational Mechanics, 2000, 25(2): 199-206.

[19] Zhang X, Liu X H, Song K Z, et al. Least-squares collocation meshless method[J]. International Journal for Numerical Methods in Engineering, 2001, 51(9): 1089-1100.

[20] Park S H, Youn S K. The least-squares meshfree method[J]. International Journal for Numerical Methods in Engineering, 2001, 52(9): 997-1012.

[21] 张雄, 胡炜, 潘小飞, 等. 加权最小二乘无网格法 [J]. 力学学报, 2003, 35(4): 425-430.
[22] Liu G R, Kee B B T, Chun L. A stabilized least-squares radial point collocation method (LS-RPCM) for adaptive analysis[J]. Computer Methods in Applied Mechanics and Engineering, 2006, 195(37): 4843-4861.
[23] Kee B B T, Liu G R, Lu C. A regularized least-squares radial point collocation method (RLS-RPCM) for adaptive analysis[J]. Computational Mechanics, 2007, 40(5): 837-853.
[24] Yang J J, Zheng J L. Intervention-point principle of meshless method[J]. Chinese Science Bulletin, 2013, 58(4-5): 478-485.
[25] 杨建军, 郑健龙. 无网格法介点原理 [J]. 科学通报, 2012, 57(26): 2456-2462.
[26] 杨建军, 郑健龙. 无网格介点法：一种具有 h,p,d 适应性的无网格法 [J]. 应用数学和力学, 2016, 37(10): 1013-1025.
[27] 杨建军, 郑健龙. 无网格全局介点法 [J]. 应用力学学报, 2017, 34(5): 956-962.
[28] Oñate E, Perazzo F, Miquel J. A finite point method for elasticity problems[J]. Computers & Structures, 2001, 79(22): 2151-2163.
[29] 杨建军, 郑健龙. 移动最小二乘法的近似稳定性 [J]. 应用数学学报, 2012, 35(4): 637-648.
[30] 李承城, 王聪, 杨建军. 无网格介点法计算参数分析 [J], 力学季刊, 2017, 38(3): 567-578.
[31] Timoshenko S P, Goodier J N. Theory of Elasticity[M]. Beijing: Tsinghua University Press, 2004.
[32] 徐秉业，王建学. 弹性力学 [M]. 北京: 清华大学出版社, 2007.
[33] Atluri S N, Shen S. The basis of meshless domain discretization: The meshless local Petrov–Galerkin (MLPG) method[J]. Advances in Computational Mathematics, 2005, 23(1-2): 73-93.

第 8 章 边界型无网格法

无网格法散点近似的策略，很容易应用于传统的边界元法 (BEM)，可消除 BEM 对网格的依赖性，进而发展出一类边界型无网格法。根据建立系统方程时是否需使用积分运算，边界型无网格法通常可分为弱式和强式两大类别。列入同一大类的方法，或者采用了不同的积分策略，或者使用不同的近似函数，或者对基本解使用不同的替代解，或者对基本解的奇异性做了某种特殊的技术处理，便形成各具特点的多种具体方法。

可归类为弱式方法的有：边界节点法 (BNM)[1,2]、边界云法 (BCM)[3]、杂交边界节点法 (HBNM)[4]、边界面法 (BFM)[5]、边界无单元法 (BEFM)[6]、边界径向基点插值法 (BRPIM)[7]、插值型边界无单元法 (IBEFM)[8]、零场边界积分方程法[9] 和边界分布源方法 (BDS)[10] 等。

可归类为强式方法的有基本解方法 (MFS)[11-18]、边界配点法 (BCoM)[19-22]、边界粒子法 (BPM)[23]、边界点法 (BKM)[24,25]、改良基本解法 (MMFS)[26-30]、奇异边界法 (SBM)[31,32] 和广义基本解法 (GMFS) [33] 等。

8.1 边界节点法

Mukherjee 等 [1,2] 将 MLS 近似与 BEM 相结合，发展了边界节点法 (BNM)。BNM 使用节点离散问题域边界，并在域边界上设置背景网格执行积分运算，如图 8-1 所示。

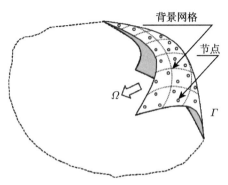

图 8-1 边界节点及其背景网格

Fig. 8-1 Boundary nodes with background mesh

对线弹性力学问题，边界积分方程的正则形式写为

$$\int_\Gamma \{t_i u_{ij}^*(P,Q) - t_{ij}^*(P,Q)[u_i(Q) - u_i(P)]\}\mathrm{d}S_Q = 0 \tag{8-1}$$

式中，位移基本解 u_{ij}^* 和面力基本解 t_{ij}^* 分别由式 (2-79) 和式 (2-80) 给出；P 和 Q 分别代表源点和场点。将位移和面力的未知变量用 MLS 近似，即

$$u_i = \sum_{J=1}^{N} \phi_J \hat{u}_J, \quad t_i = \sum_{J=1}^{N} \phi_J \hat{t}_J \tag{8-2}$$

并代入式 (8-1)，得到

$$\int_\Gamma \left\{ u_{ij}^*(P,Q) \sum_{J=1}^{N_Q} \phi_J(Q) \hat{t}_J - t_{ij}^*(P,Q) \left[\sum_{J=1}^{N_Q} \phi_J(Q) \hat{u}_J - \sum_{J=1}^{N_P} \phi_J(P) \hat{u}_J \right] \right\} \mathrm{d}S_Q = 0 \tag{8-3}$$

式中，N_Q 和 N_P 分别表示场点 Q 和源点 P 的局部支撑域中覆盖的场节点数量。由上式即可得到一组离散方程：

$$\boldsymbol{A}\hat{\boldsymbol{u}} + \boldsymbol{B}\hat{\boldsymbol{t}} = 0 \tag{8-4}$$

将近似函数式 (8-2) 代入式 (2-70) 和式 (2-71) 表示的自然边界条件和本质边界条件，又可得到两组离散方程：

$$\begin{cases} \boldsymbol{H}\hat{\boldsymbol{u}} = \bar{\boldsymbol{u}} \\ \boldsymbol{H}\hat{\boldsymbol{t}} = \bar{\boldsymbol{t}} \end{cases} \tag{8-5}$$

将式 (8-4) 和式 (8-5) 联立，便得到问题的整体离散系统方程，并可确定边界上所有节点的虚拟位移 $\hat{\boldsymbol{u}}$ 和虚拟面力 $\hat{\boldsymbol{t}}$。然后，边界节点上的真实位移和面力由式 (8-2) 计算，而域内点的位移和应力通过边界积分方程求得。

8.2 杂交边界节点法

为了避免 BNM 的背景网格积分，可采用局部弱式法类似的局部积分策略，由此张见明等[4] 发展了消除背景网格的杂交边界节点法 (HBNM)，如图 8-2 所示。图中，Ω_q^J 为节点 s_J 对应的局部积分域。

8.2 杂交边界节点法

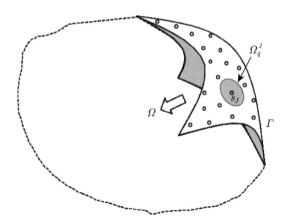

图 8-2 边界节点及其局部积分域

Fig. 8-2 Boundary nodes with a local quadrature domain

基于式 (2-56) 给出的最小势能原理，对二维弹性力学问题定义一个泛函

$$\Pi = \frac{1}{2}\int_\Omega u_{i,j} E_{ijkl} u_{k,l} \mathrm{d}\Omega - \int_\Omega u_i b_i \mathrm{d}\Omega - \int_{\Gamma_t} u_i \bar{t}_i \mathrm{d}\Gamma \tag{8-6}$$

如果不计体力，并将本质边界条件引入上式，将得到一个修正的泛函

$$\Pi_{AB} = \frac{1}{2}\int_\Omega u_{i,j} E_{ijkl} u_{k,l} \mathrm{d}\Omega - \int_\Gamma \tilde{t}_i (u_i - \tilde{u}_i) \mathrm{d}\Gamma - \int_{\Gamma_t} \tilde{u}_i \bar{t}_i \mathrm{d}\Gamma \tag{8-7}$$

式中，对 $\tilde{u}$ 和 $\tilde{t}$ 有如下定义：

$$\begin{cases} u = \tilde{u}, & \in \Gamma \\ t = \tilde{t}, & \in \Gamma \end{cases} \tag{8-8}$$

取泛函 Π_{AB} 的变分，可得到

$$\delta\Pi_{AB} = -\int_\Omega \sigma_{ij,j} \delta u_i \mathrm{d}\Omega + \int_\Gamma (t_i - \tilde{t}_i) \delta u_i \mathrm{d}\Gamma - \int_\Gamma (u_i - \tilde{u}_i) \delta\tilde{t}_i \mathrm{d}\Gamma - \int_{\Gamma_t} (\tilde{t}_i - \bar{t}_i) \delta\tilde{u}_i \mathrm{d}\Gamma \tag{8-9}$$

令

$$\delta\Pi_{AB} = 0 \tag{8-10}$$

寻求该泛函的驻值，则有

$$-\int_\Omega \sigma_{ij,j} \delta u_i \mathrm{d}\Omega + \int_\Gamma (t_i - \tilde{t}_i) \delta u_i \mathrm{d}\Gamma = 0 \tag{8-11}$$

$$\int_\Gamma (u_i - \tilde{u}_i) \delta\tilde{t}_i \mathrm{d}\Gamma = 0 \tag{8-12}$$

$$\int_{\Gamma_t} \left(\tilde{t}_i - \bar{t}_i \right) \delta \tilde{u}_i \mathrm{d}\Gamma = 0 \tag{8-13}$$

假如自然边界条件可以合理施加,则式 (8-13) 将是满足的。考虑到变分方程在任意的局部积分域上均满足,则可得

$$-\int_{\Omega_q} \sigma_{ij,j} v_i \mathrm{d}\Omega + \int_{\Gamma_q \cup \Gamma_s} \left(t_i - \tilde{t}_i \right) v_i \mathrm{d}\Gamma = 0 \tag{8-14}$$

$$\int_{\Gamma_q \cup \Gamma_s} \left(u_i - \tilde{u}_i \right) v_i \mathrm{d}\Gamma = 0 \tag{8-15}$$

式中,v 为检验函数。需注意到上两式中对边界积分项包含局部域的边界 Γ_q 和局部域与全局边界相交的部分 $\Gamma_s = \Omega_q \cap \Gamma$,如图 8-3 所示。如果采用合适的检验函数 v,比如 MLPG1 所使用的 MLS 权函数,其在局部域的边界 Γ_q 上,有 $v \equiv 0$,则在此边界上的积分可以免除。

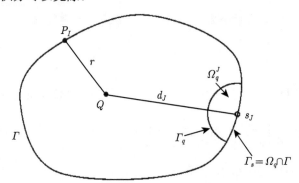

图 8-3 节点 s_J 的局部积分域

Fig. 8-3 The local quadrature domain for the node s_J

对于试探函数 $\tilde{u}$ 和 $\tilde{t}$,采用 MLS 近似:

$$\tilde{u}_i = \sum_{I=1}^{N} \phi_I \hat{u}_I, \quad \tilde{t}_i = \sum_{I=1}^{N} \phi_I \hat{t}_I \tag{8-16}$$

而域内场函数 u 和法向面力函数 t 则采用基本解插值的方式近似:

$$u_i = \sum_{I=1}^{N} u_I^* a_I, \quad t_i = \sum_{I=1}^{N} t_I^* a_I \tag{8-17}$$

式中,u_I^* 和 t_I^* 分别为 I 节点处的位移基本解和面力基本解;a_I 为节点参数。将式 (8-16) 和式 (8-17) 分别代入式 (8-14) 和式 (8-15),并须注意到式 (8-14) 的第一项为 0 值,则得到离散系统方程:

$$\boldsymbol{U}\boldsymbol{a} = \boldsymbol{H}\hat{\boldsymbol{u}} \tag{8-18}$$

8.2 杂交边界节点法

$$Ta = H\hat{t} \tag{8-19}$$

其中，

$$\begin{cases} U_{IJ} = \int_{\Gamma_s} u_I^* v_J(Q) \, d\Gamma = 0 \\ u_I^* = \begin{bmatrix} u_{11}^* & u_{12}^* \\ u_{21}^* & u_{22}^* \end{bmatrix}_I \end{cases} \tag{8-20}$$

$$\begin{cases} T_{IJ} = \int_{\Gamma_s} t_I^* v_J(Q) \, d\Gamma = 0 \\ t_I^* = \begin{bmatrix} t_{11}^* & t_{12}^* \\ t_{21}^* & t_{22}^* \end{bmatrix}_I \end{cases} \tag{8-21}$$

$$\begin{cases} H_{IJ} = \int_{\Gamma_s} N_I v_J(Q) \, d\Gamma = 0 \\ N_I = \begin{bmatrix} \phi_I(s) & 0 \\ 0 & \phi_I(s) \end{bmatrix} \end{cases} \tag{8-22}$$

由式 (8-18) 可解得向量 a：

$$a = U^{-1} H \hat{u} \tag{8-23}$$

将其代入式 (8-19) 得

$$TU^{-1} H \hat{u} = H \hat{t} \tag{8-24}$$

计算中首先利用边界条件通过变换的方法确定本质边界上的虚拟位移 $\hat{u}$ 和虚拟面力 $\hat{t}$

$$\hat{u}_I = \sum_{J=1}^{N} R_{IJ} \tilde{u}_J = \sum_{J=1}^{N} R_{IJ} \bar{u}_J \tag{8-25}$$

$$\hat{t}_I = \sum_{J=1}^{N} R_{IJ} \tilde{t}_J = \sum_{J=1}^{N} R_{IJ} \bar{t}_J \tag{8-26}$$

其中，

$$R_{IJ} = [\phi_J(\zeta_I)]^{-1} \tag{8-27}$$

式中，ζ_I 为边界节点 I 的曲线坐标。然后，将通过边界条件确定的节点参数代入方程 (8-24) 中，可解出其他节点参数。求出 $\hat{u}$ 和 $\hat{t}$ 后，再通过式 (8-23) 确定 a。然后，边界上的位移和面力由式 (8-16) 计算，而域内点的位移由式 (8-17) 的第一个式子插值计算。只要得到域内任意处的位移解，则域内的应力很容易通过 MLS 导数近似得到。需要注意的是，HBNM 的位移不是由边界积分方程来计算的，这一点明显有别于通常 BEM 法的计算规则。

HBNM 在远离边界的域内点上精度很高,而在靠近边界的域内点上误差较大,即存在"边界效应"。这是由于基本解作为插值函数,在靠近边界处产生了奇异性。实际上,HBNM 只是利用基本解作为插值函数,源点是否位于边界上并不重要。因此,一种改进措施就是将源点 P_I 移至域外,即正则化处理,如图 8-4 所示。正则化源点位置由下式确定:

$$P_I = x_I + n(x_I) \cdot h \cdot S_f \tag{8-28}$$

式中,h 为节点平均间距;n 为节点 x_I 处的外法向方向;S_f 为源点偏移量参数。这个偏移量参数对求解效果有重要影响,若取值太小,则源点靠近边界,仍然会出现奇异性问题;若取值较大,则源点偏移量相应变大,系统方程将呈现出病态,也会导致数值解失真。只有 S_f 的取值合理时,才可以有效消除边界效应,并保证获得精确结果。数值试验表明,S_f 的临界值与节点布置有关,而与边界条件无关,一个建议的合理取值区间为 $S_f = [3, 6]$。这种修正后的 HBNM 也被称为正则杂交边界节点法 (RHBNM)[34]。

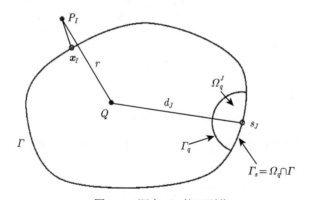

图 8-4 源点 P_I 的正则化

Fig. 8-4 Regualrization of the source point P_I

此外,张见明等也改用局部表面域来代替局部球面域,这个局部表面域伴随节点局部定义,并取在问题域边界表面上,然后对表面规则局部域采取映射的方法进行更为简单的标准参数化积分。这样处理后的方法避免了检验函数 v 如何合理选择的问题,而且积分计算也更加简单。这种方法被称为边界面法 (boundary face method, BFM)[5]。

8.3 基本解方法

与弱式边界元法相对应,还有一类发展较早的强式类边界元法。其场函数由基本解函数构造,其近似函数满足平衡方程,故只需在边界上配点并要求满足边界条

8.3 基本解方法

件。这类将配点技术与基本解近似相结合的边界解法称为基本解方法 (method of fundamental solutions, MFS)[11-15]。MFS 无须积分计算,其边界上的积分网格是自然免除的,因此这种方法也属于无网格方法。MFS 采用问题的基本解函数来近似场函数的思想于 20 世纪 60 年代起源,最早由 Kupradze 等 [16-18] 提出。

MFS 对场函数的近似由一组基本解函数的线性组合来表示:

$$u^h(\boldsymbol{x}) = \sum_{J=1}^{N} a_J \psi(\boldsymbol{x}, \boldsymbol{s}_J), \quad \boldsymbol{x} \in \bar{\Omega} \tag{8-29}$$

式中,$\bar{\Omega}$ 为问题的闭区域,即 $\bar{\Omega} = \Omega \cup \Gamma$;$N$ 为边界上的节点数 (同时也是源点数),a_J 为源点 $\boldsymbol{s}_J$ 对应的待定强度系数;$\psi(\boldsymbol{x}, \boldsymbol{s}_J)$ 即特定问题的基本解函数,该函数满足问题域内的控制方程。因此,可对边界上的节点由式 (8-29) 采用配点法直接满足边界条件,对 N 个边界节点可得到 N 个线性方程组,从而可解得源点集 $\{\boldsymbol{s}\}$ 的系数向量 $\boldsymbol{a}$。然后,由式 (8-29) 可计算域内任意点上的场量。

基本解函数 $\psi(\boldsymbol{x}, \boldsymbol{s}_J)$ 是边界解法的立足点之一。先定义问题域上 (包含边界) 任意点 $\boldsymbol{x}$ 到边界源点 $\boldsymbol{s}_J$ 的空间距离为

$$r = \|\boldsymbol{x} - \boldsymbol{s}_J\|_2 \tag{8-30}$$

若函数 $\psi(r)$ 对任意一个微分算子 L 满足:

$$L\{\psi(r)\} = \delta(r) = \begin{cases} \infty, & r = 0 \\ 0, & r \neq 0 \end{cases} \tag{8-31}$$

则称 $\psi(r)$ 为微分算子 L 的基本解函数,即该函数是齐次微分方程

$$Lu = 0 \tag{8-32}$$

的基本解。表 8-1 列出一些常见微分算子的基本解 [35]。

与边界元法一样,MFS 也选用满足问题控制方程的基本解作为插值函数。与边界元方法不同的是,MFS 将源点布置在问题域外的虚假边界上,以求克服基本解的源点奇异性问题,从而避免了复杂的奇异积分。

MFS 的数值实施如图 8-5 所示。实际上,如果真实边界上设置了 N 个节点 $\{\boldsymbol{x}\}$,则虚拟边界上也应当设置同等数量的源点 $\{\boldsymbol{s}\}$,而且源点集应当对应地布置在节点集的外法向上 (非严格要求),以保证基于源点构造的方程数与节点未知数相同,从而使得到的线性方程组可解。

MFS 的数值实现非常简单,而且研究发现其收敛速度甚至比传统的 BEM 还要快 [35]。但是,亦如 8.2 节对 RHBNM 源点合理偏移问题的讨论,MFS 也存在虚拟边界偏移量如何合理选择的问题,偏移量过小会出现源点奇异性无法消除的问

题,而偏移量过大又会导致插值矩阵的病态性。此外,在数值计算中还发现,当边界节点离散超过一定密度时,也会引起插值矩阵的病态性。

表 8-1 几种常见微分算子的基本解

Table 8-1 Fundamental solutions for some differential operators

L	二维	三维
Δ	$\dfrac{-1}{2\pi}\ln(r)$	$\dfrac{1}{4\pi r}$
$\Delta + \lambda^2$	$\dfrac{1}{2\pi}Y_0(\lambda r)$	$\dfrac{\cos(\lambda r)}{4\pi r}$
$\Delta - \lambda^2$	$\dfrac{1}{2\pi}K_0(\lambda r)$	$\dfrac{\exp(-\lambda r)}{4\pi r}$
$\beta\Delta + \nu\cdot\nabla - \lambda^2$	$\begin{cases}\dfrac{1}{2\pi}K_0(\mu r)\exp\left(\dfrac{-\nu\cdot r}{2\beta}\right)\\ \mu = \sqrt{\left(\dfrac{\|\nu\|}{2\beta}\right)^2 + \dfrac{\lambda}{\beta}}\end{cases}$	$\dfrac{\exp(-\mu r)}{4\pi r}\exp\left(\dfrac{-\nu\cdot r}{2\beta}\right)$
∇^4 或 Δ^2	$\dfrac{-r^2}{8\pi}\ln(r)$	$\dfrac{r}{8\pi}$
$\nabla^4 + \kappa^2$	$\text{Kei}(r\sqrt{\kappa}) + \text{Ber}(r\sqrt{\kappa})$	$\text{Kei}_{(\frac{3}{2})}(r\sqrt{\kappa}) + \text{Ber}_{(\frac{3}{2})}(r\sqrt{\kappa})$
$\nabla^4 - \lambda^4$	$\dfrac{Y_0(\lambda r) + K_0(\lambda r)}{2\pi}$	$\dfrac{\exp(-\lambda r) + \cos(\lambda r)}{4\pi r}$
$\nabla^4 - \lambda^2\nabla^2$	$\dfrac{K_0(\lambda r) + \ln(r)}{2\pi\lambda^2}$	$\dfrac{\exp(-\lambda r) + 1}{4\pi\lambda^2 r}$

注:Δ 表示 Laplace 算子,∇ 表示梯度算子;
λ 为波数,β 为扩散系数,ν 为速度向量,κ 为刚度;
Y_0 表示 0 阶第二类 Bessel 函数,K_0 表示 0 阶第二类修正 Bessel 函数,Kei 表示第二类修正 Kelvin 函数,Ber 表示第一类 Kelvin 函数。

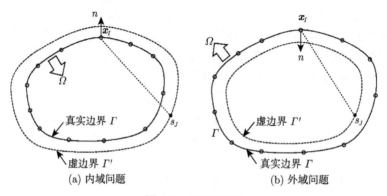

(a) 内域问题 (b) 外域问题

图 8-5 MFS 图示

Fig. 8-5 Schematics of MFS

为了避免引入虚边界，从而导致虚边界与真实边界偏移的难题。有一种解决办法是采用合适的数学手段来消除基本解的奇异性。比如，Young 等[26-30] 提出的修正基本解方法 (modified method of fundamental solutions, MMFS)，该法选用单层或双层势基本解作为近似基函数，引入零场积分方程，采用加减去奇异技术，消去了源点的奇异性，从而得到了非奇异的场量近似函数。

8.4 边界点法

为了避免 MFS 中的虚假边界和基本解的源点奇异性，陈文等[24,25,36-42] 提出了仅需在边界上配点离散的边界点法 (boundary knot method, BKM)。

BKM 的主要特点是使用了非奇异的径向基函数通解来代替奇异基本解，这个径向基函数通解要求满足控制方程。对特定问题，只要找到了这个 "通解"，BKM 的数值实现就变得非常简单。类似于配点法，BKM 只需在边界上设置一组离散节点，不需要积分运算，因此也不需要任何的积分网格或积分子域，如图 8-6 所示。

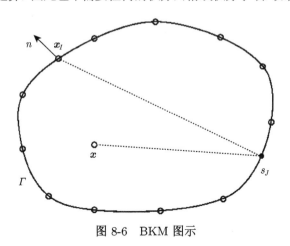

图 8-6 BKM 图示

Fig. 8-6 Schematics of BKM

假如通解存在，则 BKM 对场函数的近似由一组非奇异径向基函数通解的线性组合来表示，有

$$u^h(\boldsymbol{x}) = \sum_{J=1}^{N} a_J \varphi(\boldsymbol{x}, \boldsymbol{s}_J) \tag{8-33}$$

式中，$\varphi(r)$ 为非奇异径向基通解函数；$\{a_J\}$ 为待定系数；$\{\boldsymbol{x}\} \in \varGamma \cup \varOmega$ 表示边界上的节点或域内的计算点；$\{\boldsymbol{s}\} \in \varGamma$ 表示域边界上的源点；N 表示边界节点数量。在实际计算中，可以把边界离散节点 $\{\boldsymbol{x}\} \in \varGamma$ 同时作为边界上的源点 $\{\boldsymbol{s}\}$。

BKM 的数值实施非常简单，但其真正的困难在于寻找对应问题的通解函数 $\varphi(r)$。对于弹性力学所对应的拉普拉斯 (Laplace) 方程问题，非奇异径向基通解为常数，换句话说，这个通解函数并不存在。表 8-2 列出几种常见微分算子的通解[35]。

表 8-2 几种常见微分算子的通解

Table 8-2 General solutions for some differential operator

L	二维	三维
$\Delta + \lambda^2$	$\dfrac{1}{2\pi} J_0(\lambda r)$	$\dfrac{\sin(\lambda r)}{4\pi r}$
$\Delta - \lambda^2$	$\dfrac{1}{2\pi} I_0(\lambda r)$	$\dfrac{\sinh(\lambda r)}{4\pi r}$
$\beta \Delta + \nu \cdot \nabla - \lambda^2$	$\begin{cases} \dfrac{1}{2\pi} I_0(\mu r) \exp\left(\dfrac{-\nu \cdot r}{2\beta}\right) \\ \mu = \sqrt{\left(\dfrac{\|\nu\|}{2\beta}\right)^2 + \dfrac{\lambda}{\beta}} \end{cases}$	$\dfrac{\sinh(\mu r)}{4\pi r} \exp\left(\dfrac{-\nu \cdot r}{2\beta}\right)$
$\nabla^4 + \kappa^2$	$\text{Bei}(r\sqrt{\kappa}) + \text{Ber}(r\sqrt{\kappa})$	$\text{Bei}_{(\frac{3}{2})}(r\sqrt{\kappa}) + \text{Ber}_{(\frac{3}{2})}(r\sqrt{\kappa})$
$\nabla^4 - \lambda^4$	$\dfrac{J_0(\lambda r) + I_0(\lambda r)}{2\pi}$	$\dfrac{\sinh(\lambda r) + \sin(\lambda r)}{4\pi r}$
$\nabla^4 - \lambda^2 \nabla^2$	$\dfrac{I_0(\lambda r) + 1}{2\pi \lambda^2}$	$\dfrac{\sinh(\lambda r) + r}{4\pi \lambda^2 r}$

注：Δ 表示 Laplace 算子，∇ 表示梯度算子；
λ 为波数，β 为扩散系数，ν 为速度向量，κ 为刚度；
J_0 表示 0 阶第一类 Bessel 函数，I_0 表示 0 阶第一类修正 Bessel 函数，Bei 表示第二类修正 Kelvin 函数，Ber 表示第一类 Kelvin 函数。

鉴于很多实际问题难以找到通解，一种可能的解决办法是采用"虚基本解"来代替通解。即，只取基本解的非奇异项，而忽略其奇异项，边界配点法 (boundary collocation method, BCoM)[19,20] 就采用了这样的技巧，使用虚部核函数 (imaginary-part kernel function)，即对基本解为复数变量的问题 (如 Helmholtz 方程问题)，只取其非奇异的虚数部分，来代替式 (8-30) 中的通解函数 $\varphi(r)$ 去逼近场量。

8.5 奇异边界法

为了应对 MFS 中基本解的源点奇异性问题，陈文等[31,32] 发展了奇异边界法 (singular boundary method, SBM)。该法无须再另外设置虚拟边界，而采用引入源点强度因子的概念来消除源点处基本解的奇异性。

SBM 的离散与图 8-5 类似，边界上的节点集 $\{\boldsymbol{x}\}_\Gamma$ 和源点集 $\{\boldsymbol{s}\}$ 取自同一点集空间。类似于 MFS，SBM 的场函数近似由一组基本解函数的线性组合来表示，但在源点 $\boldsymbol{s}_J$ 与边界配点 $\boldsymbol{x}_I$ 重合时，引入强度因子 U^0，即

$$u_I^h(\boldsymbol{x}) = \sum_{\substack{J=1 \\ J \neq I}}^{N} a_J \psi_I(\boldsymbol{x}, \boldsymbol{s}_J) + a_I U_{JJ}^0 \tag{8-34}$$

式中，$\psi(\boldsymbol{x}, \boldsymbol{s}_J)$ 即基本解函数。在数值计算中，需要对场量的边界外法向导数近似，而对应的强度因子又不可直接由 U^0 计算得到。因此，在近似中，还需对应地引入场量的边界外法向导数强度因子 Q^0：

$$\frac{\partial u_I^h(\boldsymbol{x})}{\partial n} = \sum_{\substack{J=1 \\ J \neq I}}^{N} a_J \psi_{I,n}(\boldsymbol{x}, \boldsymbol{s}_J) + a_I Q_{JJ}^0 \tag{8-35}$$

式中，n 为源点 $\boldsymbol{s}_J$ 处的单位外法向向量。

假如，强度因子 U^0 和 Q^0 已经获得，则 SBM 可对边界节点直接配点满足边界条件，便可求得系数向量 $\boldsymbol{a}$。然后再由式 (8-34) 反求任意点 $\boldsymbol{x}$ 处的场量。

SBM 数值实现的关键是确定强度因子 U^0 和 Q^0。一种解决方法被称为"纯反插值技术"[43,44]，其策略大致可以描述为：首先选取一个满足问题控制方程的样本解，并根据原问题的边界条件类型构造该样本解的无限域问题，在域内设置样本点，并基于源点集和样本点集构造出一组标准的基本解插值方程组；然后基于边界节点集 $\{\boldsymbol{x}\}_\Gamma$ 和源点集构造样本解和无限域问题对应的含源点强度因子的 SBM 插值方程组；联立这两个方程组即可求得源点处的强度因子 U^0 和 Q^0。从理论上讲，源点强度因子与样本解和样本点的选取无关，只与源点位置和基本解有关。进一步研究发现，对三维问题，样本解和样本点的选取对计算结果的精度影响较为敏感。为此，Gu 等 [45,46] 将式 (8-35) 引入边界积分方程，采用加减去奇异技术消除强度因子 Q^0，而强度因子 U^0 仍采用纯反插值技术获得。

8.6 边界分布源方法

为了应对 MFS 中基本解的源点奇异性问题，Liu[10] 提出边界分布源方法 (BDS)。其场量近似取为

$$u^h(\boldsymbol{x}) = \sum_{J=1}^{N} a_J \int_{\Omega_q^J} \psi(\boldsymbol{x}, \boldsymbol{s}_p) \, d\Omega, \quad \boldsymbol{s}_p \in \Omega_q^J, \boldsymbol{x} \in \Omega - \bigcup_{J=1}^{N} \Omega_q^J \tag{8-36}$$

式中，N 为边界上的源点数量；$\psi(\boldsymbol{x}, \boldsymbol{s}_p)$ 为基本解函数；a_J 为源点 $\boldsymbol{s}_J$ 对应的待定强度系数；Ω_q^J 为伴随源点 $\boldsymbol{s}_J$ 定义的分布源积分域；$\boldsymbol{s}_p$ 为域内一个分布源点，也可称为弱介点。一个伴随边界源点 $\boldsymbol{s}_J$ 定义的圆形域分布源如图 8-7 所示。

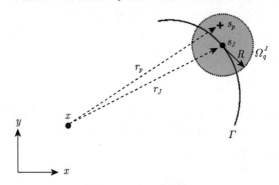

图 8-7 圆形域分布源

Fig. 8-7 Distributed source on a circular domain

理论上讲，对二维问题，分布源区域 Ω_q^J 可以取为过源点的边界切向或法向线段，或者是一个封闭源点的圆环，或者是一个任意形状的面域。在实际应用中，主要考虑到便于得到奇异积分，而采用合适的域形状。

本质上讲，BDS 将 MFS 中基本解的源点奇异性问题，转换成了一个奇异积分问题。而采用合理的奇异积分，可以有效消除直接的奇异性计算。虽然对任意的基本解函数，其奇异积分仍然是一个棘手问题。但对于一些形式简单的基本解问题，其奇异积分可能较为容易实现。比如，对于 Laplace 算子势问题，其基本解函数写为

$$\psi(\boldsymbol{x}, \boldsymbol{s}_p) = \frac{-1}{2\pi}\ln(r_p) \tag{8-37}$$

则圆形域分布源上的奇异积分的解析结果为

$$\int_{\Omega_q^J}\psi(\boldsymbol{x}, \boldsymbol{s}_p)\,\mathrm{d}\Omega = \begin{cases} \dfrac{-R^2}{2}\ln(r_J), & r_J > R \\ \dfrac{-R^2}{2}\ln(R) + \dfrac{R^2-(r_J)^2}{4}, & r_J \leqslant R \end{cases} \tag{8-38}$$

对于导数边界条件，即 NBC 的施加，需执行场量的导数近似，即

$$\frac{\partial u^h(\boldsymbol{x})}{\partial n} = \sum_{J=1}^{N} a_J \int_{\Omega_q^J} \frac{\partial \psi(\boldsymbol{x}, \boldsymbol{s}_p)}{\partial n}\mathrm{d}\Omega \tag{8-39}$$

式中，基本解关于边界外法向导数的奇异积分，在 $r_J > R$ 的情况下容易解析得到

$$\int_{\Omega_q^J}\frac{\partial \psi(\boldsymbol{x}, \boldsymbol{s}_p)}{\partial n}\mathrm{d}\Omega = \frac{-R^2}{2r_J}\frac{\partial r_J}{\partial n}, \quad r_J > R \tag{8-40}$$

然而对 $r_J \leqslant R$ 的情况，该奇异积分却不能解析得到。

寻求任意情况下基本解关于边界外法向导数的奇异积分，首先将该奇异积分记为

$$Q(\boldsymbol{x}, \boldsymbol{s}_J) = \int_{\Omega_q^J} \frac{\partial \psi(\boldsymbol{x}, \boldsymbol{s}_p)}{\partial n} \mathrm{d}\Omega \tag{8-41}$$

一种特别的技巧是利用势函数外法向梯度在域边界上积分为 0 的条件 [47]：

$$\int_{\partial \Omega} \frac{\partial u^h(\boldsymbol{x})}{\partial n} \mathrm{d}\Gamma = \sum_{J=1}^{N} a_J \int_{\partial \Omega} Q(\boldsymbol{x}, \boldsymbol{s}_J) \mathrm{d}\Gamma = 0 \tag{8-42}$$

上式对任意的边界条件成立，则有

$$\int_{\partial \Omega} Q(\boldsymbol{x}, \boldsymbol{s}_J) \mathrm{d}\Gamma = 0 \tag{8-43}$$

则该奇异积分可由下式计算：

$$Q(\boldsymbol{x}_I, \boldsymbol{s}_J) \approx \frac{-1}{L_I} \sum_{\substack{J=1 \\ J \neq I}}^{N} L_J Q(\boldsymbol{x}_J, \boldsymbol{s}_I) \tag{8-44}$$

式中，L_I 为第 I 个节点对应的边界曲线长度，实际计算中可用相邻前后两个边界节点径向距离的 1/2 近似表示。对三维问题，L_I 则表示第 I 个节点对应的边界面积。显然，上式可用于计算 $r_J \leqslant R$ 情况的奇异积分，以作为对式 (8-40) 的补充。

自此，BDS 后续的计算只需边界配点满足边界条件，从而解得系数向量 $\boldsymbol{a}$。然后域内任意点的场量可由式 (8-36) 直接计算得到。

BDS 需要对分布源进行积分运算，属于弱式类的边界方法，但其分布源形状可以自由灵活选取，无须考虑与问题边界的交集信息，可能获得解析的奇异积分。因此，在数值实现上相比其他弱式法更为简单，如果可用解析的奇异积分，则与强式方法相类似。如果奇异积分无法解析得到，可以采用数值积分的方法求得，只不过执行上要复杂一些。另外需特别注意一个问题，对域内点的场量计算，须特别注意式 (8-36) 中的 $\boldsymbol{x} \in \Omega - \bigcup_{J=1}^{N} \Omega_q^J$ 条件，即在靠近边界与分布源域重合的区域，其场量计算将会是不精确的。一种可以考虑的调整方案可将分布源完全设置在域外，但其会带来奇异积分无法解析获得的问题，需要类似于弱式方法那样直接计算奇异积分。

8.7 广义基本解法

为了改善 MFS 的计算稳定性，进而为其虚边界的合理设置难题提供一种可选择的解决方案，本书作者在 MFS 的基础上，提出了广义基本解法 (generalized method of fundamental solution, GMFS)[33]。GMFS 采用基本解的双线性组合来构造场函数的近似，即

$$u^h(\boldsymbol{x}) = \sum_{J=1}^{N} \sum_{p=1}^{N_p} a_J \psi_p^J(\boldsymbol{x},\boldsymbol{s}), \quad \{\boldsymbol{s}_p\} \in \Omega_J \tag{8-45}$$

式中，$\{\boldsymbol{s}_p\} \notin \bar{\Omega}$ 为边界源点 $\boldsymbol{x}_J$ 的"介点弥散"(intervention-point diffuse, IPD)；N_p 表示 IPD 的介点数量；$\Omega_J \notin \bar{\Omega}$ 表示伴随 $\boldsymbol{x}_J$ 定义 (以其为中心) 的局部弥散域。需注意到，式中为基本解函数 ψ 增加一个上标符号 "J"，以表示与边界源点 $\boldsymbol{x}_J$ 的对应关系。我们也可把这种双线性基本解近似称为"广义基本解近似 (GFSA)"，如图 8-8(a) 所示。

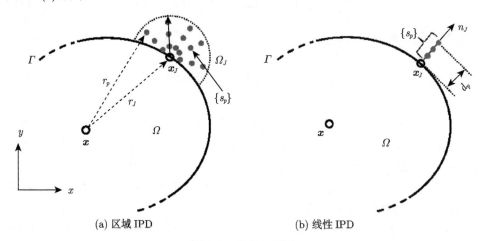

(a) 区域 IPD　　　　　　　　　　(b) 线性 IPD

图 8-8　GFSA 图示

Fig. 8-8　Schematics of the GFSA

实际上，弥散域 Ω_J 可以是任意的形状。基于效率的考虑，我们完全可以线性规划 IPD，比如 $\{\boldsymbol{s}_p\} \in n_J$，如图 8-8(b) 所示。其中，$n_J$ 为边界在 $\boldsymbol{x}_J$ 处的外法向，而 d_p 为任意一个介点的对源偏移，即

$$d_p = \|\boldsymbol{s}_p - \boldsymbol{x}_J\|_2, \quad p = 1, 2, \cdots, N_p \tag{8-46}$$

对于 $\{\boldsymbol{s}_p\} \in n_J$ 的 IPD 方案，建议 $N_p \geqslant 5$，并为 $\{d_p\}$ 初步设定为

$$\{d_p\} = (0.3,\ 0.5,\ 0.7,\ 0.9,\ 1.2) \cdot \bar{R}, \quad \text{外边界} \tag{8-47}$$

$$\{d_p\} = (0.1,\ 0.3,\ 0.5,\ 0.7,\ 0.9) \cdot \bar{R}, \quad \text{内边界} \tag{8-48}$$

8.7 广义基本解法

其中，$\bar{R}$ 为边界尺度的规范系数，定义为

$$\bar{R} = \frac{1}{\sqrt{D}} \left\| \frac{\max(x_i) - \min(x_i)}{2} \right\|_2 \tag{8-49}$$

式中，下标 "i" 表示坐标分量；D 表示问题维数。无特别说明时，上述选择在后续算例中默认设置。

显然，在 $\{s_p\} \in n_J$ 方案中，当 $N_p=1$ 时，GMFS 与 MFS 将完全等价。因此，我们用 "广义" 一词来表征这种新的方法。接下来给出 GMFS 的数值实施方法，不失一般性，考虑一个势问题：

$$\nabla^2 u(\boldsymbol{x}) = 0, \quad \boldsymbol{x} \in \bar{\Omega} \tag{8-50}$$

边界条件为

$$u(\boldsymbol{x}) = \bar{u}(\boldsymbol{x}), \quad \boldsymbol{x} \in \Gamma_u \tag{8-51}$$

$$u_{,n}(\boldsymbol{x}) \equiv \frac{\partial u}{\partial n}(\boldsymbol{x}) = \bar{q}(\boldsymbol{x}), \quad \boldsymbol{x} \in \Gamma_t \tag{8-52}$$

注意到，控制方程中采用的是 Laplace 算子，$\nabla^2 \equiv \Delta$，问题的基本解函数 ψ 见表 8-1。GMFS 的数值实施如图 8-9(a) 所示。以示比较，MFS 如图 8-9(b) 所示。本质上讲，GMFS 的 IPD 是伴随边界源点定义的，而虚边界的概念并不存在。

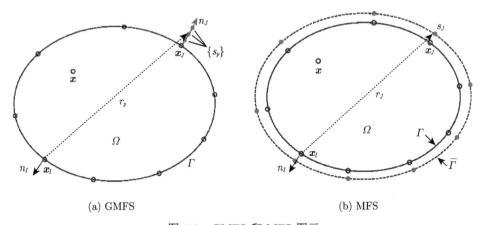

(a) GMFS   (b) MFS

图 8-9 GMFS 和 MFS 图示

Fig. 8-9 Schematics of the GMFS and the MFS

从 MFS 的角度审视，由 GFSA 构造的系统方程是超约束或超静定的。因此，我们首先考虑用变分法求解。基于边界节点，构造如下变分泛函：

$$\delta \Pi_2 = \sum_{x_I \in \Gamma_u} \delta u(u - \bar{u}) + \sum_{x_I \in \Gamma_t} \delta u_{,n}(u_{,n} - \bar{q}) \tag{8-53}$$

命 $\delta \Pi_2 = 0$，则式 (8-51) 和式 (8-52) 定义的边界条件将被满足。将 (8-45) 代入，便可得到 GMFS1 型的系统方程：

$$\bar{K}a = \bar{F} \tag{8-54}$$

其中，

$$\bar{K} = K^{\mathrm{T}}K, \quad \bar{F} = K^{\mathrm{T}}F \tag{8-55}$$

式中，K 和 F 定义为

$$K_{IJ} = \begin{cases} \sum_{p=1}^{N_p} \psi_p^J(x_I, s), & x_I \in \Gamma_u \\ \sum_{p=1}^{N_p} \dfrac{\partial \psi_p^J(x_I, s)}{\partial n}, & x_I \in \Gamma_t \end{cases} \tag{8-56}$$

$$F_I = \begin{cases} \bar{u}(x_I), & x_I \in \Gamma_u \\ \bar{q}(x_I), & x_I \in \Gamma_t \end{cases} \tag{8-57}$$

此外，也可对边界节点直接配点满足边界条件，则可得到另外一类 GMFS2 型的系统方程：

$$Ka = F \tag{8-58}$$

式中，K 和 F 如式 (8-56) 和式 (8-57) 定义。

通过式 (8-54) 或式 (8-58)，则边界节点的待定强度系数 $a = [a_1, a_2, \cdots, a_N]^{\mathrm{T}}$ 可被解得。由此，通过式 (8-45)，任意测量点 $x \in \bar{\Omega}$ 处的场变量可计算得到。接下来将通过几个算例，检验 GMFS 相比于 MFS 的优势及有效性。若无特别说明，算例中的尺度或变量单位均采用国际单位，示例中不再特别标注单位。

算例 1 一个圆域的 Dirichlet 边界条件 (BCs) 问题

一个半径为 $R=1$ 的圆域，作用 Dirichlet BCs。其解析解假定为

$$u(x, y) = \cos(x) \cosh(y) + \sin(x) \sinh(y) \tag{8-59}$$

在初始计算时，用 20 个节点离散域边界，并采用规则布点和随机布点两种方案，如图 8-10 所示。采用规则布点的求解结果比较如图 8-11 所示。

8.7 广义基本解法

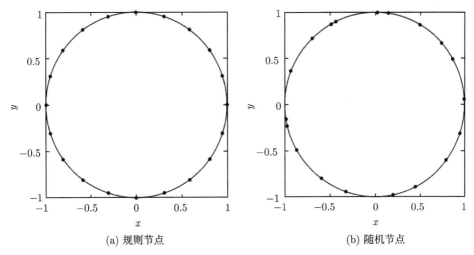

(a) 规则节点　　　　　　　　　(b) 随机节点

图 8-10　圆域问题的边界节点布置

Fig. 8-10　Nodal distribution for the circular domain problem

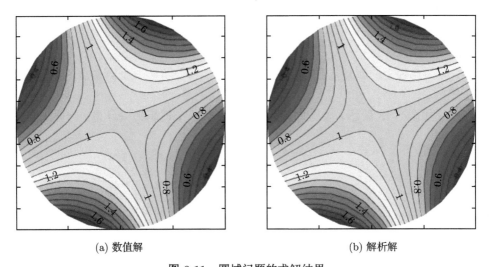

(a) 数值解　　　　　　　　　(b) 解析解

图 8-11　圆域问题的求解结果

Fig. 8-11　Solution results for the circular domain problem

通过试算,可以确定,$d=0.07\bar{R}$ 是 MFS 的一个不合理虚边界偏移量 (坏偏移)。将此坏偏移植入到 GMFS 的偏距集合中,如 $\{d_p\}=\{0.07, 0.5, 0.7, 0.9, 1.2\}\bar{R}$。然后选取一个半径为 $P=0.99R$ 的圆形分析路径,路径上设置 80 个测量点 (取值点)。分析路径上的结果如图 8-12 所示,显然,在坏偏移存在的情况下,GMFS 仍能给出合理结果,而 MFS 的解却严重失真。

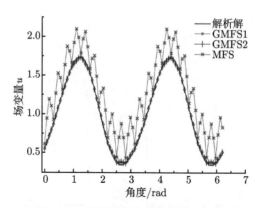

图 8-12 分析路径上的求解结果比较

Fig. 8-12 Comparing numerical solutions on an analytical path

接下来分析求解误差。另设置一个半径为 $P=0.5R$ 的圆形分析路径，路径上均匀设置 80 个测量点。对比分析中，GMFS 的 $\{d_p\}$ 采用默认设置，而 MFS 的偏移 d 取其中值，即有 $d = (\max\{d_p\} - \min\{d_p\})/2$。并定义如下误差范数：

$$Er = \frac{1}{m}\sqrt{\sum_{k=1}^{m}(\xi_k^{\text{num}} - \xi_k^{\text{ana}})^2 \Big/ \sum_{k=1}^{m}(\xi_k^{\text{ana}})^2} \tag{8-60}$$

式中，m 为路径上的测量点数量；ξ_k^{num} 和 ξ_k^{ana} 分别表示第 k 个测量点的数值解和解析解。

结果如图 8-13 所示。本图中因为 MFS 取到一个理想的偏移量，因此 GMFS

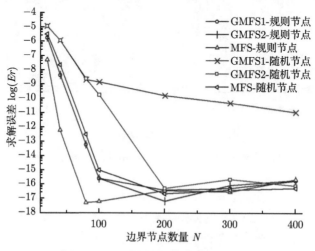

图 8-13 圆域问题的数值解误差比较

Fig. 8-13 Comparing numerical errors for the circular domain problem

8.7 广义基本解法

与其相比并无特别优势。但这个结果可以说明，GMFS 同样可以得到高精度和良好收敛的结果。此外，结果表明，GMFS1 的精度和收敛性，要差于 GMFS2，这与原来的预期有偏差。

算例 2　一个外旋轮线域上的结合边界问题

一个外旋轮线边界定义为

$$\rho(\theta) = \sqrt{(a+b)^2 + 1 - 2(a+b)\cos(a\theta/b)} \tag{8-61}$$

其形状参数取为 $a=3$, $b=1$。在 $0 \leqslant \theta \leqslant \pi$ 的边界上施加 Dirichlet BCs，在另外的 $\pi < \theta < 2\pi$ 边界上施加 Neumann BCs。域上的解析解假定为

$$u(x,y) = \exp(x)\cos(y) \tag{8-62}$$

初始计算中，在边界上设置 30 个等角度离散的节点，如图 8-14 所示。由 GMFS1 计算得到的结果，及其与解析解的比较如图 8-15 所示。

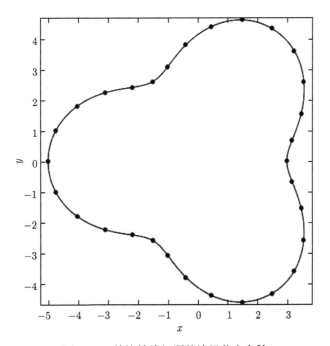

图 8-14　外旋轮线问题的边界节点离散

Fig. 8-14　Nodal distribution for the epitrochoid domain problem

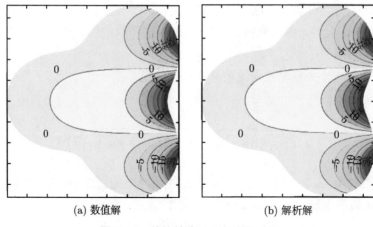

(a) 数值解　　　　　　　　(b) 解析解

图 8-15　外旋轮线问题的求解结果

Fig. 8-15　Solution results for the epitrochoid domain problem

离散方程的系统矩阵条件数也是影响数值方法计算稳定性的一个重要指标，接下来对此进行一个比较计算。计算中，MFS 的偏移量取 GMFS 的偏移集的最大值，以保证二者的最大偏移值相同。计算结果如图 8-16 所示。显然，总体而言 GMFS 的系统矩阵条件数总是比 MFS 的要小。换句话说，包含最大偏移量相同的情况下，GMFS 的稳定性要比 MFS 的要好。

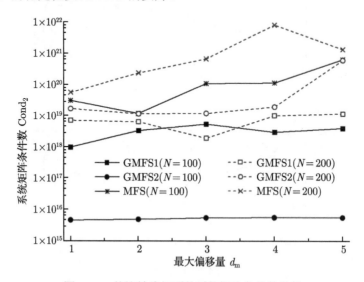

图 8-16　外旋轮线问题的系统矩阵条件数比较

Fig. 8-16　Comparing condition numbers of the system matrix for the epitrochoid domain problem

8.7 广义基本解法

接下来比较求解误差。选取 $P=0.9\rho$ 的一条分析路径，并等角度设置 80 个测量点。MFS 的偏移 d 取 GMFS 的偏移集中值，计算结果如图 8-17 所示。可以看出，在边界节点较密时，MFS 中有明显的误差线呈上扬的不稳定现象，而 GMFS2 更表现为保收敛性的特点。同时，本算例再次表明，GMFS1 的精度和收敛性不及 GMFS2。

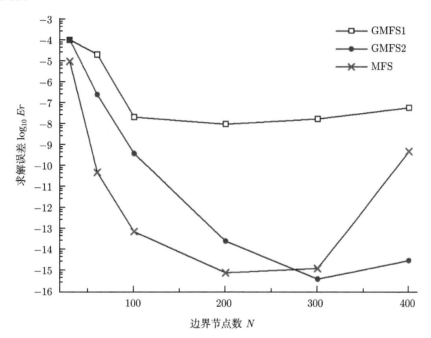

图 8-17 外旋轮线问题的求解误差比较

Fig. 8-17 Comparing numerical errors for the epitrochoid domain problem

算例 3 一个具有复边界的不规则边界问题

一个问题域内设置 $r=0.2$ 的内圆边界，圆心位于 $(x,y)=(0.5,0.5)$。问题的不规则外边界定义为

$$\rho(\theta) = \exp(\sin\theta)\sin^2 2\theta + \exp(\cos\theta)\cos^2 2\theta \tag{8-63}$$

内圆边界施加 Dirichlet BCs，不规则外边界施加 Neumann BCs，并假设问题的解析解为

$$u(x,y) = \ln\sqrt{(x-0.5)^2 + (y-0.5)^2} \tag{8-64}$$

初始计算中，外边界用 30 个节点离散，内边界用 15 个节点离散，如图 8-18 所示。由 GMFS1 计算得到的结果，及其与解析解的比较如图 8-19 所示。

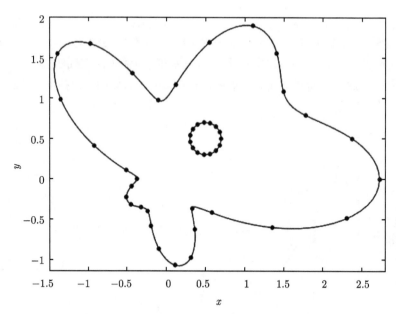

图 8-18　不规则复边界问题的边界节点布置

Fig. 8-18　Nodal distribution for the amoeba-like domain problem

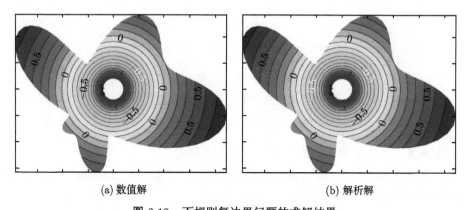

(a) 数值解　　　　　　　　　　　(b) 解析解

图 8-19　不规则复边界问题的求解结果

Fig. 8-19　Solution results for the amoeba-like domain problem

接下来，我们继续验证数值计算的稳定性问题。首先，我们选取一个 $P=0.5\rho$ 的分析路径，并在路径上设置 80 个测量点。通过试算，我们可以确定：$d=0.001$ 是 MFS 的一个坏偏移。然后，我们将此坏偏移植入到 GMFS 的偏移集中。然后，我们取路径上的计算结果进行比较，如图 8-20 所示。显然，当同样使用到一个坏偏移的情况下，MFS 给出了不精确的计算结果，而 GMFS 却能给出较为精确的结果。由此，进一步验证了 GMFS 比 MFS 更具计算稳定性的效果。

8.7 广义基本解法

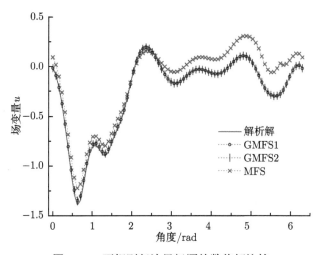

图 8-20　不规则复边界问题的数值解比较

Fig. 8-20　Comparing numerical solutions for the amoeba-like domain problem

最后，进一步考察数值解误差及收敛性。基于前述所设定的 $P=0.5\rho$ 分析路径，并约定 MFS 的偏移取 GMFS 偏移集的中值，结果如图 8-21 所示。可见，固然 MFS 过早地达到了高精度状态 (跳跃收敛)，但 GMFS2 所表现的才是理想的收敛状态。同时，GMFS1 的精度和收敛性不及 GMFS2。

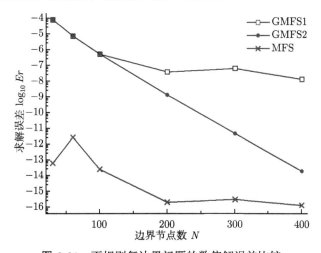

图 8-21　不规则复边界问题的数值解误差比较

Fig. 8-21　Comparing numerical errors for the amoeba-like domain problem

算例 4　一个双调和问题

基于已有的研究工作，对于双调和问题，若采用非调和边界条件，则 MFS 求

解会变得高度敏感而脆弱，需要特别的数值技巧来寻求合理解答[48-52]。双调和问题定义为

$$\nabla^4 u(\boldsymbol{x}) \equiv \nabla^2 (\nabla^2 u) = 0, \quad \boldsymbol{x} \in \Omega \tag{8-65}$$

施加 Robin BCs：

$$u(\boldsymbol{x}) = \bar{u}(\boldsymbol{x}), \quad \nabla^2 u(\boldsymbol{x}) = \bar{Q}(\boldsymbol{x}), \quad \boldsymbol{x} \in \Gamma \tag{8-66}$$

或

$$u(\boldsymbol{x}) = \bar{u}(\boldsymbol{x}), \quad u_{,n}(\boldsymbol{x}) = \bar{q}(\boldsymbol{x}), \quad \boldsymbol{x} \in \Gamma \tag{8-67}$$

假定一个问题的解析边界条件为

$$\bar{u}(x, y) = x^2 y^3 \tag{8-68}$$

以及 $\bar{Q} = \nabla^2 \bar{u}$，或 $\bar{q} = \bar{u}_{,n}$ 对应考虑。须注意到，该边界条件并不满足控制式 (8-65)，因此称其为"非调和"边界条件。对于该问题，域上不存在解析解，因此只可基于一个"最大误差原理"通过边界上的数值解来检验数值方法的有效性[48,49]。

因双调和问题是一个高阶 PDEs，故对应于式 (8-45)，其场近似需要特别处理[48,50]

$$u^h(\boldsymbol{x}) = \sum_{J=1}^{N} \sum_{p=1}^{N_p} \{ a_J \psi_p^J(\boldsymbol{x}, \boldsymbol{s}) + b_J \bar{\psi}_p^J(\boldsymbol{x}, \boldsymbol{s}) \} \tag{8-69}$$

式中，ψ 为 Laplace 算子（所对应的齐次方程）的基本解；$\bar{\psi}$ 为双调和 (biharmonic) 算子的基本解，并有

$$\bar{\psi}(r) = \begin{cases} \dfrac{-r^2}{8\pi} \ln(r), & 2D \\ \dfrac{r}{8\pi}, & 3D \end{cases} \tag{8-70}$$

具体算法介绍此处从略。针对本算例，我们考虑采用一个 L 形域，并使用 40 个边界节点离散，如图 8-22 所示。其实，这样的结构看似简单，但对一些边界解法是存在挑战性的，因其存在尖锐的角点和凹点。在本算例计算中，对 GMFS，我们只考虑 GMFS2 算法。在边界上设置一个周界沿走分析路径，定义为：从左底角点算起，沿边界按逆时针方向行进。在此路径上，设置 160 个数值测量点。采用式 (8-66) 定义的 Robin BCs，介点偏移集取为 $\{d_p\}=(0.02\sim0.1)\bar{R}$，并对其使用 0.01 的弥散间距，则计算结果如图 8-23 所示，可见 GMFS 能够给出精确的解答。

8.7 广义基本解法

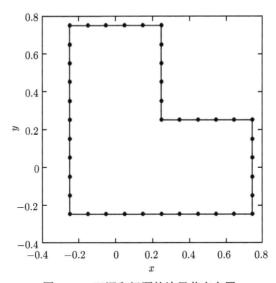

图 8-22 双调和问题的边界节点布置

Fig. 8-22 Nodal distribution for the biharmonic problem

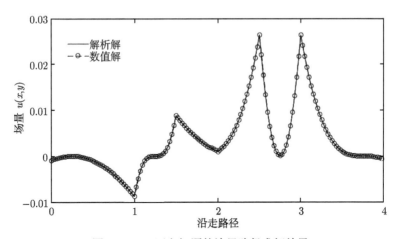

图 8-23 双调和问题的边界路径求解结果

Fig. 8-23 Solutions on the analytical path for the biharmonic problem

接下来，我们通过对 GMFS 设置不同的$\{d_p\}$域，通过计算求解误差来分析其稳定性效果。比较计算中，MFS 的源偏移取 GMFS 偏移集的中值。对于式 (8-66) 的边界条件，IPD 采用 0.01 的间距弥散。而对于式 (8-67) 的边界条件，IPD 采用 0.001 的间距弥散。计算结果见表 8-3 和表 8-4。显然，GMFS 的稳定性和精确性均明显优于 MFS。此外，还可看到一个有趣的现象，在一个合理的宽取值域内，GMFS 表现出一种随$\{d_p\}$值域增加而收敛 (弥散尺度收敛性) 的特点。这可能是 GMFS 所具有的另一种优势。

表 8-3 GMFS 和 MFS 的求解误差比较：式 (8-66) 边界条件

Tab.8-3 Comparing of numerical errors (Er) for GMFS and MFS: Eq.(8-66) BCs

Range of $\{d_p\}$	GMFS	MFS
$(0.01\sim0.1)\bar{R}$	1.0777×10^{-4}	1.1608×10^{-4}
$(0.01\sim0.2)\bar{R}$	9.6753×10^{-5}	1.0908×10^{-4}
$(0.01\sim0.3)\bar{R}$	8.9717×10^{-5}	1.0606×10^{-4}
$(0.01\sim0.4)\bar{R}$	8.4983×10^{-5}	collapsed (warning)
$(0.01\sim0.5)\bar{R}$	8.1330×10^{-5}	9.7136×10^{-4}

表 8-4 GMFS 和 MFS 的求解误差比较：式 (8-67) 边界条件

Tab.8-4 Comparing of numerical errors (Er) for GMFS and MFS: Eq.(8-67) BCs

Range of $\{d_p\}$	GMFS	MFS
$(0.001\sim0.1)\bar{R}$	1.4784×10^{-4}	4.6663×10^{-4}
$(0.001\sim0.2)\bar{R}$	1.2072×10^{-4}	1.5907×10^{-4}
$(0.001\sim0.3)\bar{R}$	1.0564×10^{-4}	2.2991×10^{-4}
$(0.001\sim0.4)\bar{R}$	9.6537×10^{-5}	1.0964 (warning)
$(0.001\sim0.5)\bar{R}$	9.0233×10^{-5}	6.3016×10^{-4}

通过上述几个算例的初步验证，我们有理由得出这样一个初步结论：GMFS 通过一种简便易行的改进手段，可以有效提高 MFS 的计算稳定性，表现为良好的计算收敛性。其中，作为更简单的算法 GMFS2，可以推荐为 GMFS 的标准算法。若要论不足，GMFS 会比 MFS 计算成本高一些，但考虑到边界解法本身的计算成本很低，因此，在通常的计算中，这种计算成本的增加完全可以忽略不计。

8.8 本章小结

MFS 作为一种非常独特的边界解法，其起源很早[16,53]，而且一直得到较为普遍的应用和关注。MFS 强劲的生命力或许得益于其本身所具有的三大明显优势。其一，MFS 完全依赖边界节点计算，是天然无网格的，恰好符合数值方法无网格化发展的大趋势，而且作为边界解法，自身还有降维计算的优势。其二，这种方法是配点计算的，非常简单、高效和易于数值实施，无须像其他弱式边界无网格法那样执行奇异积分运算。其三，如果 MFS 取到合适的虚边界偏移，其数值计算完全可以是高精度和快速收敛的，其精度和收敛速率甚至要高于传统的边界元法 (BEM)。

众所周知，MFS 存在一个"虚边界选择难题"，而且这个久未攻克的问题已经成为 MFS 进一步应用和发展的障碍。纵观近 20 年来，从事相关研究的科学工作者们为克服这一难题做了大量努力。在我们了解的这些研究工作中，诚然均有效解决了虚边界选择的问题，很多方案不再需要虚边界了。但这些改进的方法中，又或多或少带来了其他一些偏于负面的，或者更为棘手的问题。比如，修正基本解方法 (MMFS) 对任意的问题获得去奇异的基本解尚存疑问，而且一般技术人员很难掌握，实现上有一定复杂性。边界点法 (BKM) 或边界配点法 (BCoM) 改用通解或去奇异解代替奇异基本解。然而，很多非常实际的问题并不存在通解，而采用去奇异解又会影响精度。奇异边界法 (SBM) 引入强度因子来代替奇异性，但需要特别技术先确定强度因子，一个问题需要两次求解。边界分布源方法 (BDS) 在本书中虽然归类为弱式方法，但其执行策略更像 MFS。BDS 虽然有获得解析的奇异积分的可能，但在实际应用中仍然避免不了执行奇异积分运算。

GMFS 也可视为介点原理在边界解法中的一个应用示范。相比于 MFS，GMFS 的稳定性和收敛性均得到明显改善。种种迹象表明，GMFS 不失为一种有效的改进方法。而且这种改进效果也非常简单，并不会带来其他难以处理的问题，也能够完全继承 MFS 的简单、高效、易执行、高精度、快速收敛等诸多优点。其后续的完善和发展，以及在实际问题中的应用潜力仍然值得关注。

参 考 文 献

[1] Mukherjee Y X, Mukherjee S. The boundary node method for potential problems[J]. International Journal for Numerical Methods in Engineering, 1997, 40(5): 797-815.

[2] Kothnur V S, Mukherjee S, Mukherjee Y X. Two-dimensional linear elasticity by the boundary node method[J]. International Journal of Solids and Structures, 1999, 36(8): 1129-1147.

[3] Li G, Aluru N R. Boundary cloud method: A combined scattered point/boundary integral approach for boundary-only analysis[J]. Computer Methods in Applied Mechanics and Engineering, 2002, 191(21): 2337-2370.

[4] Zhang J, Yao Z, Li H. A hybrid boundary node method[J]. International Journal for Numerical Methods in Engineering, 2002, 53(4): 751-763.

[5] Zhang J, Qin X, Han X, et al. A boundary face method for potential problems in three dimensions[J]. International Journal for Numerical Methods in Engineering, 2009, 80(3): 320-337.

[6] 程玉民, 陈美娟. 弹性力学的一种边界无单元法 [J]. 力学学报, 2003, 35(2): 181-186.

[7] Gu Y T, Liu G R. A boundary radial point interpolation method (BRPIM) for 2-D structural analyses[J]. Structural Engineering and Mechanics, 2003, 15(5): 535-550.

[8] Ren H P, Cheng Y M, Zhang W. An interpolating boundary element-free method (IBEFM) for elasticity problems[J]. Science China Physics, Mechanics and Astronomy, 2010, 53(4): 758-766.

[9] Golberg M A. The method of fundamental solutions for Poisson's equation[J]. Engineering Analysis with Boundary Elements, 1995, 16(3): 205-213.

[10] Liu Y J. A new boundary meshfree method with distributed sources[J]. Engineering Analysis with Boundary Elements, 2010, 34(11): 914-919.

[11] Golberg M A. The method of fundamental solutions for Poisson's equation[J]. Engineering Analysis with Boundary Elements, 1995, 16(3): 205-213.

[12] Fairweather G, Karageorghis A. The method of fundamental solutions for elliptic boundary value problems[J]. Advances in Computational Mathematics, 1998, 9(1-2): 69-95.

[13] Fairweather G, Karageorghis A, Martin P A. The method of fundamental solutions for scattering and radiation problems[J]. Engineering Analysis with Boundary Elements, 2003, 27(7): 759-769.

[14] Chen C S, Golberg M A, Hon Y C. The method of fundamental solutions and quasi-Monte-Carlo method for diffusion equations[J]. International Journal for Numerical Methods in Engineering, 1998, 43(8): 1421-1435.

[15] Chen C S, Karageorghis A, Smyrlis Y S. The Method of Fundamental Solutions: A Meshless Method[M]. Atlanta: Dynamic Publishers, 2008.

[16] Kupradze V D, Aleksidze M A. The method of functional equations for the approximate solution of certain boundary value problems[J]. USSR Computational Mathematics & Mathematical Physics, 1964, 4(4): 82-126.

[17] Kupradze V D. A method for the approximate solution of limiting problems in mathematical physics[J]. USSR Computational Mathematics and Mathematical Physics, 1964, 4(6): 199-205.

[18] Kupradze V D. On the approximate solution of problems in mathematical physics[J]. Russian Mathematical Surveys, 1967, 22(2): 58-108.

[19] Chen J T, Chang M H, Chen K H, et al. The boundary collocation method with meshless concept for acoustic eigenanalysis of two-dimensional cavities using radial basis function[J]. Journal of Sound and Vibration, 2002, 257(4): 667-711.

[20] Chen J T, Chang M H, Chen K H, et al. Boundary collocation method for acoustic eigenanalysis of three-dimensional cavities using radial basis function[J]. Computational Mechanics, 2002, 29(4): 392-408.

[21] Chen J T, Hong H K. Review of dual boundary element methods with emphasis on hypersingular integrals and divergent series[J]. Applied Mechanics Reviews, 1999, 52(1): 17-33.

[22] Chen J T, Chen C T, Chen P Y, et al. A semi-analytical approach for radiation and scattering problems with circular boundaries[J]. Computer Methods in Applied

Mechanics & Engineering, 2007, 196(25–28): 2751-2764.

[23] Chen W. Meshfree boundary particle method applied to Helmholtz problems[J]. Engineering Analysis with Boundary Elements, 2002, 26(7): 577-581.

[24] Chen W, Tanaka M. A meshless, integration-free, and boundary-only RBF technique[J]. Computers & Mathematics with Applications, 2002, 43(3): 379-391.

[25] Chen W, Hon Y C. Numerical investigation on convergence of boundary knot method in the analysis of homogeneous Helmholtz, modified Helmholtz, and convection–diffusion problems[J]. Computer Methods in Applied Mechanics and Engineering, 2003, 192(15): 1859-1875.

[26] Young D L, Chen K H, Lee C W. Novel meshless method for solving the potential problems with arbitrary domain[J]. Journal of Computational Physics, 2005, 209(1): 290-321.

[27] Young D L, Chen K H, Chen J T, et al. A modified method of fundamental solutions with source on the boundary for solving Laplace equations with circular and arbitrary domains[J]. Computer Modeling in Engineering and Sciences, 2007, 19(3): 197-221.

[28] Young D L, Chen K H, Liu T Y, et al. Hypersingular meshless method for solving 3D potential problems with arbitrary domain[J]. Computer Modeling in Engineering and Sciences (CMES), 2009, 40(3): 225-269.

[29] Šarler B. Solution of potential flow problems by the modified method of fundamental solutions: Formulations with the single layer and the double layer fundamental solutions[J]. Engineering Analysis with Boundary Elements, 2009, 33(12): 1374-1382.

[30] Hon Y C, Wu Z. A numerical computation for inverse boundary determination problem[J]. Engineering Analysis with Boundary Elements, 2000, 24(7): 599-606.

[31] 陈文. 奇异边界法: 一个新的、简单、无网格、边界配点数值方法 [J]. 固体力学学报, 2009, 30(6): 592-599.

[32] Chen W, Fu Z J. A novel numerical method for infinite domain potential problems[J]. Chinese Sci Bull, 2010, 55: 1598-1603.

[33] Yang J J, Zheng J L, Wen P H. Generalized method of fundamental solution (GMFS) for boundary value problems[J]. Engineering Analysis with Boundary Elements, 2018, 94: 25-33.

[34] Zhang J, Yao Z. Meshless regular hybrid boundary node method[J]. CMES-Computer Modeling in Engineering and Sciences, 2001, 2(3): 307-318.

[35] 陈文, 傅卓佳, 魏星. 科学与工程计算中的径向基函数方法 [M]. 北京: 科学出版社, 2014.

[36] Chen W. Symmetric boundary knot method[J]. Engineering Analysis with Boundary Elements, 2002, 26(6): 489-494.

[37] Hon Y C, Chen W. Boundary knot method for 2D and 3D Helmholtz and convection-diffusion problems under complicated geometry[J]. International Journal for Numerical Methods in Engineering, 2003, 56(13): 1931-1948.

[38] Chen W, Hon Y C. Numerical investigation on convergence of boundary knot method in the analysis of homogeneous Helmholtz, modified Helmholtz, and convection–diffusion problems[J]. Computer Methods in Applied Mechanics and Engineering, 2003, 192(15): 1859-1875.

[39] Jin B, Zheng Y. Boundary knot method for some inverse problems associated with the Helmholtz equation[J]. International Journal for Numerical Methods in Engineering, 2005, 62(12): 1636-1651.

[40] Chen W, Shen L J, Shen Z J, et al. Boundary knot method for Poisson equations[J]. Engineering Analysis with Boundary Elements, 2005, 29(8): 756-760.

[41] Jin B, Zheng Y. Boundary knot method for the Cauchy problem associated with the inhomogeneous Helmholtz equation[J]. Engineering Analysis with Boundary Elements, 2005, 29(10): 925-935.

[42] Fu Z J, Chen W, Qin Q H. Boundary knot method for heat conduction in nonlinear functionally graded material[J]. Engineering Analysis with Boundary Elements, 2011, 35(5): 729-734.

[43] Chen W, Wang F Z. A method of fundamental solutions without fictitious boundary[J]. Engineering Analysis with Boundary Elements, 2010, 34(5): 530-532.

[44] Chen W, Fu Z J, Jin B T. A truly boundary-only meshfree method for inhomogeneous problems based on recursive composite multiple reciprocity technique[J]. Engineering Analysis with Boundary Elements, 2010, 34(3): 196-205.

[45] Gu Y, Chen W, He X Q. Singular boundary method for steady-state heat conduction in three dimensional general anisotropic media[J]. International Journal of Heat & Mass Transfer, 2012, 55(17–18): 4837-4848.

[46] Chen W, Gu Y. Recent advances on singular boundary method[C]. Joint international workshop on Trefftz method VI and method of fundamental solution II, Taiwan China. 2011.

[47] Kim S. An improved boundary distributed source method for two-dimensional Laplace equations[J]. Engineering Analysis with Boundary Elements, 2013, 37(7): 997-1003.

[48] Chen C S, Karageorghis A, Li Y. On choosing the location of the sources in the MFS[J]. Numerical Algorithms, 2016, 72(1): 107-130.

[49] Li M, Chen C S, Karageorghis A. The MFS for the solution of harmonic boundary value problems with non-harmonic boundary conditions[J]. Computers & Mathematics with Applications, 2013, 66(11): 2400-2424.

[50] Karageorghis A, Fairweather G. The method of fundamental solutions for the numerical solution of the biharmonic equation[J]. Journal of Computational Physics, 1987, 69(2): 434-459.

[51] Poullikkas A, Karageorghis A, Georgiou G. Methods of fundamental solutions for harmonic and biharmonic boundary value problems[J]. Computational Mechanics, 1998,

21(4-5): 416-423.

[52] Pei X, Chen C S, Dou F. The MFS and MAFS for solving Laplace and biharmonic equations[J]. Engineering Analysis with Boundary Elements, 2017, 80: 87-93.

[53] 文丕华. 求解弹性地基圆板问题的点源法 [J]. 工程力学, 1987, 4(2): 18-26.

第9章 结合式无网格法

在发展出的无网格方法中,可能某一种方法有其独特的优势,但又存在某种局限性;而另外一种方法的优势和局限性正好与前一种方法相反,那么将两种方法结合应用就可能将各自的优势形成互补,并消除彼此的局限性,这就是结合式无网格法得到发展的基本思想。当前,随着无网格法研究的深入,以及独立方法的丰富,越来越多的结合式无网格法相继被提出,并得到实际应用。本章将重点介绍几种结合式无网格法,或者说是耦合的无网格方法。

9.1 无网格强弱式法

考虑到配点法有计算效率高的特点,但其对导数边界条件的施加不够精确;而 MLPG 法计算效率较低,但可以精确施加导数边界条件。Liu 和 Gu 等[1-5] 提出的无网格强弱式法 (MWS) 即将这两种方法的优势相结合的一种方法,对域内节点和本质边界上的节点采用配点法离散,而对自然边界上的节点采用 MLPG 法离散,如图 9-1 所示。

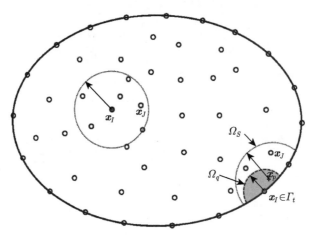

图 9-1 MWS 法图示

Fig. 9-1 Schematics of the MWS method

对应于二维弹性力学问题,MWS 法得到的离散系统方程一般形式为

$$\boldsymbol{K} \cdot \boldsymbol{U} = \boldsymbol{F} \tag{9-1}$$

式中，系统刚度矩阵 K 和系统荷载矩阵 F 的离散形式定义为

$$K_{IJ} = \begin{cases} L^{\mathrm{T}}DB_J, & x_I \in \Omega \\ \int_{\Omega_q} V_p^{\mathrm{T}}DB_J \mathrm{d}\Omega - \int_{\Gamma_q \cup \Gamma_{qu}} w_p n_p DB_J \mathrm{d}\Gamma, & x_I \in \Gamma_t \\ N_J, & x_I \in \Gamma_u \end{cases} \quad (9\text{-}2)$$

$$F_I = \begin{cases} b_I, & x_I \in \Omega \\ \int_{\Omega_q} w_p b_p \mathrm{d}\Omega + \int_{\Gamma_{qt}} w_p \bar{t}_p \mathrm{d}\Gamma, & x_I \in \Gamma_t \\ \bar{u}_I, & x_I \in \Gamma_u \end{cases} \quad (9\text{-}3)$$

式中，有关符号的定义可参见第 6 章 MLPG 法和第 7 章 PIM 法，此处不再赘述。

从方法构造的意义上讲，MWS 法有其值得肯定之处。但应当注意到，配点法对域内节点进行离散，需要执行二阶导数近似，这种高阶导数近似要求是造成数值方法不够稳定的一个重要原因。此外，MLPG 法对不规则边界存在积分的困难，对边界节点使用 MLPG 法离散并没有发挥出其应有的优势。

9.2 杂交有限差分法

差分近似很容易执行导数运算，但其对节点通常有规则布设的要求。Wen 和 Aliabadi[6] 发展了一种杂交有限差分法 (HFDM)。对于可以规则布点的区域，使用差分近似，对应的节点即差分节点。而在不规则边界附近，可以灵活地设置不规则节点，并使用 MLS 近似，对应的节点即无网格节点，如图 9-2 所示，图中的实心点即过渡性无网格节点。

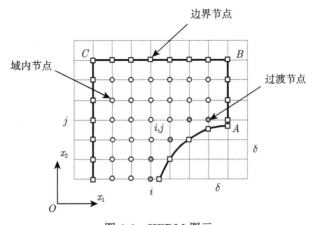

图 9-2　HFDM 图示

Fig. 9-2　Schematics of the HFDM

对问题的规则区域,采用单位长为 δ 的方格来布设差分节点,则域内差分节点的场量导数很容易计算,即

$$\left.\frac{\partial^2 u}{\partial x_1^2}\right|_{i,j} = \frac{1}{\delta^2}\left(u_{i+1,j} - 2u_{i,j} + u_{i-1,j}\right) \tag{9-4}$$

$$\left.\frac{\partial^2 u}{\partial x_2^2}\right|_{i,j} = \frac{1}{\delta^2}\left(u_{i,j+1} - 2u_{i,j} + u_{i,j-1}\right) \tag{9-5}$$

$$\left.\frac{\partial^2 u}{\partial x_1 \partial x_2}\right|_{i,j} = \frac{1}{4\delta^2}\left(u_{i+1,j+1} - u_{i+1,j-1} + u_{i-1,j-1} - u_{i-1,j+1}\right) \tag{9-6}$$

式中,下标 i 和 j 表示对应维度上的点序号索引。对应于二维弹性力学问题,则平衡方程对应写为

$$\begin{aligned}\frac{\partial^2 u_1}{\partial x_1^2} + \frac{1-\nu}{2}\frac{\partial^2 u_1}{\partial x_2^2} + \frac{1+\nu}{2}\frac{\partial^2 u_2}{\partial x_1 \partial x_2} + b_1 = 0 \\ \frac{\partial^2 u_2}{\partial x_2^2} + \frac{1-\nu}{2}\frac{\partial^2 u_2}{\partial x_1^2} + \frac{1+\nu}{2}\frac{\partial^2 u_1}{\partial x_1 \partial x_2} + b_2 = 0\end{aligned} \tag{9-7}$$

将式 (9-4)~ 式 (9-6) 代入上式,则可得到差分节点上的离散方程:

$$\begin{aligned}&u_1^{i+1,j} - (3-\nu)u_1^{i,j} + u_1^{i-1,j} + \frac{1-\nu}{4}\left(u_1^{i,j+1} + u_1^{i,j-1}\right) \\ &+ \frac{1+\nu}{8}\left(u_2^{i+1,j-1} - u_2^{i+1,j-1} + u_2^{i-1,j-1} - u_2^{i-1,j+1}\right) + \delta^2 b_1 = 0 \\ &u_2^{i+1,j} - (3-\nu)u_2^{i,j} + u_2^{i-1,j} + \frac{1-\nu}{4}\left(u_2^{i,j+1} + u_2^{i,j-1}\right) \\ &+ \frac{1+\nu}{8}\left(u_1^{i+1,j-1} - u_1^{i+1,j-1} + u_1^{i-1,j-1} - u_1^{i-1,j+1}\right) + \delta^2 b_2 = 0\end{aligned} \tag{9-8}$$

对于随机的过渡性的无网格节点,其离散方程用 MLS 近似导出,此处不再特别介绍。

显然,HFDM 对于一些具有局部性不规则边界的问题,会非常实用,具有数值实施简单和计算高效的优势。

9.3 无限元无网格法

为了应对无限域问题的求解,Wen 等[7] 提出了无限元与无网格耦合的方法,即无限元无网格法 (IEIMA)。此处,我们忽略具体的数值实现细节,只简要介绍其数值离散思想。

9.3 无限元无网格法

IEIMA 采用有限块法 (FBM)[8-11] 作为基础计算方法。对任意复杂的几何结构，需要对物理域进行分片处理，然后对不规则的物理域进行规范化映射处理。如图 9-3(b) 所示的一个具有开边界的无限域，可以采用 4 节点映射的方法使其成为一个规范化域，如图 9-3(a) 所示。规范域中的 ($\xi=1$) 边界即物理域的无限空间的映射。其映射函数写为

$$N_1 = -\frac{\xi(1-\eta)}{(1-\xi)}, \quad N_2 = \frac{(1+\xi)(1-\eta)}{2(1-\xi)}, \quad N_3 = \frac{(1+\xi)(1+\eta)}{2(1-\xi)}, \quad N_4 = -\frac{\xi(1+\eta)}{(1-\xi)} \tag{9-9}$$

则物理域中任意点 (x_k, y_k) 与规范域中对应点 (ξ_k, η_k) 有如下关系：

$$x_k = \sum_{i=1}^{4} N_i(\xi_k, \eta_k) x_i, \quad y_k = \sum_{i=1}^{4} N_i(\xi_k, \eta_k) y_i \tag{9-10}$$

则规范域上的偏导数计算形式为

$$\frac{\partial N_1}{\partial \xi} = -\frac{1-\eta}{(1-\xi)^2}, \quad \frac{\partial N_2}{\partial \xi} = \frac{1-\eta}{(1-\xi)^2}, \quad \frac{\partial N_3}{\partial \xi} = \frac{1+\eta}{(1-\xi)^2}, \quad \frac{\partial N_4}{\partial \xi} = -\frac{1+\eta}{(1-\xi)^2} \tag{9-11}$$

$$\frac{\partial N_1}{\partial \eta} = \frac{\xi}{1-\xi}, \quad \frac{\partial N_2}{\partial \eta} = -\frac{1+\xi}{2(1-\xi)}, \quad \frac{\partial N_3}{\partial \eta} = \frac{1+\xi}{2(1-\xi)}, \quad \frac{\partial N_4}{\partial \eta} = -\frac{\xi}{1-\xi} \tag{9-12}$$

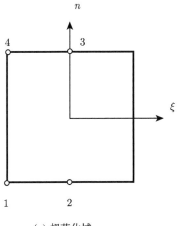
(a) 规范化域

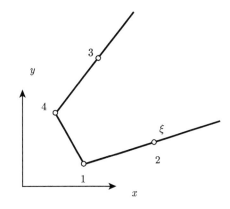
(b) 物理域

图 9-3 4 节点映射

Fig. 9-3 Four nodes mapping

分片后的域总可以映射为一个规范域，只不过所需的映射节点可能要更多一些。比如，对一个包含圆孔的无限长板，其 1/4 结构被分成 3 片，其中，I 和 II 为

有限块域，Ⅲ为无限元域，如图 9-4 所示。最终，其物理域上的节点离散如图 9-5 所示。

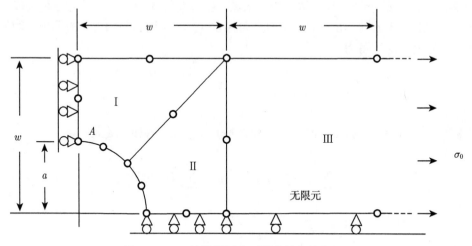

图 9-4　1/4 结构带圆孔无限条的分片处理

Fig. 9-4　Slicing for a quarter of infinite strip containing circular hole

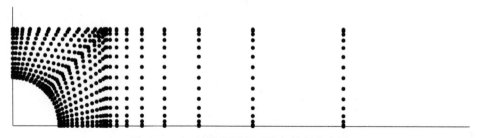

图 9-5　1/4 结构带圆孔无限条的节点布置

Fig. 9-5　Nodal distribution for the quarter of infinite strip containing circular hole

需提及的是，对于多个有限块或分片域的交点处，如图 9-6 所示，则交点处的位移和应力需平衡处理，即

$$u_\beta^{\mathrm{I}}(\boldsymbol{x}) = u_\beta^{\mathrm{II}}(\boldsymbol{x}) = \cdots = u_\beta^{\mathrm{X}}(\boldsymbol{x}) \tag{9-13}$$

$$\begin{aligned}
\sum_{q=\mathrm{I}}^{\mathrm{X}} \left(\sigma_x^{(q)}[\sin\theta_2^{(q)} - \sin\theta_1^{(q)}] - \tau_{xy}^{(q)}[\cos\theta_2^{(q)} - \cos\theta_1^{(q)}] \right) = 0 \\
\sum_{q=\mathrm{I}}^{\mathrm{X}} \left(\tau_{xy}^{(q)}[\sin\theta_2^{(q)} - \sin\theta_1^{(q)}] - \sigma_y^{(q)}[\cos\theta_2^{(q)} - \cos\theta_1^{(q)}] \right) = 0
\end{aligned} \tag{9-14}$$

9.3 无限元无网格法

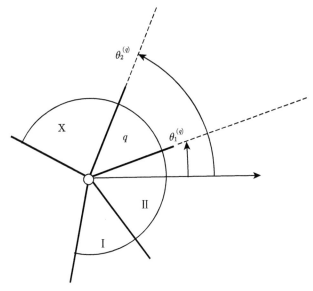

图 9-6 多个有限块的交点

Fig. 9-6 Joint with blocks

在映射得到的规范域上，FBM 对规范域采用规则节点离散，如图 9-7 所示。该域上任意点的场量 $u(\boldsymbol{\xi})$ 采用 Lagrange 插值多项式近似：

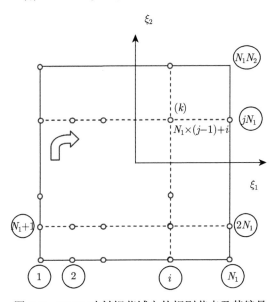

图 9-7 FBM 映射规范域上的规则节点及其编号

Fig. 9-7 Uniformly distributed nodes in mapped domain and numbering system for FBM

$$u(\boldsymbol{\xi}) = \sum_{i=1}^{N_1} \sum_{j=1}^{N_2} F(\xi, \xi_i) G(\eta, \eta_j) u^{(k)} \tag{9-15}$$

式中，

$$F(\xi, \xi_i) = \prod_{\substack{m=1 \\ m \neq i}}^{N_1} \frac{\xi - \xi_m}{\xi_i - \xi_m}, \quad G(\eta, \eta_j) = \prod_{\substack{n=1 \\ n \neq j}}^{N_2} \frac{\eta - \eta_n}{\eta_j - \eta_n} \tag{9-16}$$

则 FBM 的形函数表示为

$$\phi_k = F(\xi, \xi_i) G(\eta, \eta_j) = \prod_{\substack{m=1 \\ m \neq i}}^{N_\xi} \frac{\xi - \xi_m}{\xi_i - \xi_m} \prod_{\substack{n=1 \\ n \neq j}}^{N_\eta} \frac{\eta - \eta_n}{\eta_j - \eta_n} \tag{9-17}$$

则其一阶偏导数写为

$$\frac{\partial \phi_k}{\partial \xi} = \frac{\partial F(\xi, \xi_i)}{\partial \xi} G(\eta, \eta_j), \quad \frac{\partial \phi_k}{\partial \eta} = F(\xi, \xi_i) \frac{\partial G(\eta, \eta_j)}{\partial \eta} \tag{9-18}$$

式中，

$$\frac{\partial F}{\partial \xi} = \sum_{l=1}^{N_\xi} \prod_{i=1, i \neq i, i \neq l}^{N_\xi} (\xi - \xi_i) \Big/ \prod_{m=1, m \neq i}^{N_\xi} (\xi_i - \xi_m) \tag{9-19}$$

$$\frac{\partial G}{\partial \eta} = \sum_{l=1}^{N_\eta} \prod_{j=1, j \neq i, j \neq l}^{N_\eta} (\eta - \eta_j) \Big/ \prod_{n=1, n \neq j}^{N_\eta} (\eta_j - \eta_n) \tag{9-20}$$

须提及的是，FBM 形函数具有 Kronecker Delta 性质。

最终，IEIMA 采用配点方法构建离散系统方程，此处不予详述。IEIMA 采用了 FBM 作为基础计算方法，很容易实现有限域与无限域的嵌套，是求解无限域问题的一种非常实用而有效的方法。

9.4 最小二乘序列函数法

通常的方法对求解高阶偏微分方程问题，通常是不适应或精度和稳定性是难以得到保证的。为了应对高阶偏微分方程的求解，Huang 等 [12] 提出了一种非常简单易用的最小二乘序列函数法 (MLSS)。此处，主要结合一些常见的四阶偏微分方程问题对其进行介绍。

一个泛适的四阶 PDEs 定义为

$$Lu(x, y) = f(x, y), \quad (x, y) \in \Omega \tag{9-21}$$

9.4 最小二乘序列函数法

$$Bu(x,y) = g(x,y), \quad (x,y) \in \partial\Omega \tag{9-22}$$

式中，L 为一个任意的四阶偏微分算子；B 为一个任意的不高于四阶的边界条件偏微分算子。假定序列函数

$$M = a \cdot A = \sum_{j=0}^{n}\sum_{i=0}^{j} a_{ij}x^{j-i}y^{i} \tag{9-23}$$

是满足边界控制方程和边界条件的一个序列函数表示的近似解。式中，$a = [a_1, a_2, \cdots, a_m]$ 为系数向量；$A = \begin{bmatrix} 1, x, y, x^2, xy, y^2, \cdots, x^n y^n \end{bmatrix}^{\mathrm{T}}$ 为序列多项式基（共 m 项）。则有

$$LM = a \cdot L \cdot A = f(x,y), \quad (x,y) \in \Omega \tag{9-24}$$

$$BM = a \cdot B \cdot A = g(x,y), \quad (x,y) \in \partial\Omega \tag{9-25}$$

MLSS 对问题域和边界用 $N \geqslant m$ 个节点离散，其中域内布设 h 个节点，则边界上布设 $N-h$ 个节点，如图 9-8 所示。则由此 N 个节点，可以得到系统方程：

$$K_{N \times m} \cdot a_{m \times 1} = B_{N \times 1} \tag{9-26}$$

式中，

$$K = [\mathcal{L}A_1, \mathcal{L}A_2, \cdots, \mathcal{L}A_h, \mathcal{B}A_{h+1}, \mathcal{B}A_{h+2}, \cdots, \mathcal{B}A_N]_{m \times N}^{\mathrm{T}} \tag{9-27}$$

$$B = [f_1, f_2, \cdots, f_h, g_1, g_2, \cdots, g_{N-h}]_{1 \times N}^{\mathrm{T}} \tag{9-28}$$

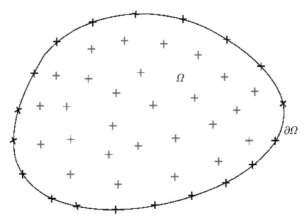

图 9-8 MLSS 图示

Fig. 9-8 Shematics of the MLSS

为了求解系数向量，需构造如下最小二乘范数：

$$J(\boldsymbol{a}) = \sum_{i=1}^{N} (\boldsymbol{B}_i - \boldsymbol{K}_i \boldsymbol{a})^2 \tag{9-29}$$

令 $\partial J/\partial a = 0$，则可解得

$$\boldsymbol{a} = \left(\boldsymbol{K}^{\mathrm{T}} \boldsymbol{K}\right)^{-1} \boldsymbol{K}^{\mathrm{T}} \boldsymbol{B} \tag{9-30}$$

由式 (9-23)，则问题域上任意点的场量及其导数可以计算得到

$$u(x, y) = \boldsymbol{a} \cdot \boldsymbol{A} \tag{9-31}$$

$$\begin{cases} u_{,i} = \boldsymbol{a} \cdot \boldsymbol{A}_{,i} \\ u_{,ij} = \boldsymbol{a} \cdot \boldsymbol{A}_{,ij} \end{cases} \tag{9-32}$$

接下来我们给出几个算例对方法进行检验。

算例 1　非齐次双调和方程问题

一个非齐次双调和方程问题定义为

$$\Delta^2 u(x,y) = \bar{f}(x,y), \quad (x,y) \in \Omega \tag{9-33}$$

给定 Robin BCs 为

$$u(x,y) = \bar{u}(x,y), \quad \Delta u(x,y) = \bar{g}(x,y), \quad (x,y) \in \partial\Omega \tag{9-34}$$

问题的解析解设为

$$u(x,y) = \sin(y^2 + x) - \cos(y - x^2) \tag{9-35}$$

问题域定义为

$$\Omega = \{(x,y) \,|\, x = \rho\cos\theta, y = \rho\sin\theta, 0 \leqslant \theta \leqslant 2\pi\} \tag{9-36}$$

式中，

$$\rho = 1 + \cos^2 4\theta \tag{9-37}$$

域上的节点布置如图 9-9(a) 所示，域上的解析解场量分布如图 9-9(b) 所示。

9.4 最小二乘序列函数法

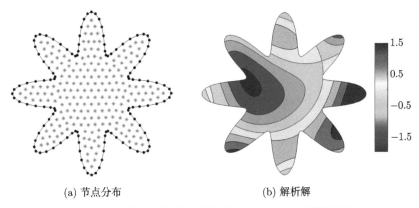

(a) 节点分布　　　　　(b) 解析解

图 9-9　非齐次双调和方程问题的节点分布及其解析解

Fig. 9-9　Nodal distributation and analytical solution for the inhomogeneous biharmonic problem

接下来，我们考察 MLSS 求解该问题的收敛性。首先定义两种误差范数：

$$\text{RMSE} = \sqrt{\frac{1}{n_t}\sum_{j=1}^{n_t}\left(\widehat{u}_j - u_j\right)^2} \tag{9-38}$$

$$\text{RMSEx} = \sqrt{\frac{1}{n_t}\sum_{j=1}^{n_t}\left(\frac{\partial \widehat{u}_j}{\partial x} - \frac{\partial u_j}{\partial x}\right)^2} \tag{9-39}$$

式中，用 $\widehat{u}_j$ 和 u_j 分别表示第 j 个测量点的数值解和解析解；n_t 表示域内随机分布的测量点数量，计算中直接取域内布置的场节点。

多项式序列函数的项数分别取为 $m=20, 25, 30$，则对应的求解误差如图 9-10

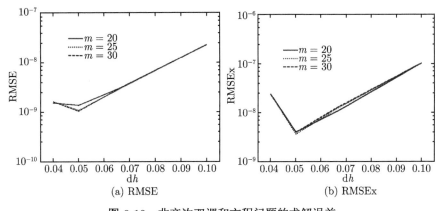

(a) RMSE　　　　　(b) RMSEx

图 9-10　非齐次双调和方程问题的求解误差

Fig. 9-10　Numerical errors for the inhomogeneous biharmonic problem

所示，图中 dh 表示平均节点间距。可以看出，MLSS 求解该问题具有很高的计算精度，并须注意到，场节点密度达到一定密度时，如图中 dh < 0.05 时，误差反而呈随节点加密而增大的趋势，表明节点密度超过一定限度时，系统矩阵将会呈现病态性。

算例 2 薄板振动方程问题

一个薄板振动方程问题定义为

$$(\Delta - 10)^2 u(x,y) = \bar{f}(x,y), \quad (x,y) \in \Omega \tag{9-40}$$

给定 Robin BCs 为

$$u(x,y) = \bar{u}(x,y), \quad \Delta u(x,y) = \bar{g}(x,y), \quad (x,y) \in \partial\Omega \tag{9-41}$$

问题的解析解设为

$$u(x,y) = \sin \pi x \cosh y - \cos \pi x \sinh y \tag{9-42}$$

问题域定义为

$$\Omega = \{(x,y) | x = \rho\cos\theta, y = \rho\sin\theta, 0 \leqslant \theta \leqslant 2\pi\} \tag{9-43}$$

式中，

$$\rho = 3 + \sin\theta\cos 5\theta \tag{9-44}$$

域上的节点分布如图 9-11(a) 所示，域上的解析解场量分布如图 9-11(b) 所示。

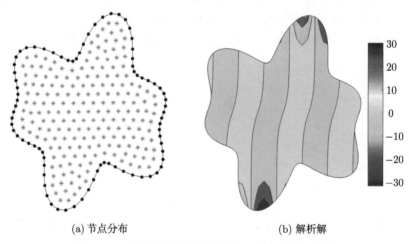

(a) 节点分布　　　　　　　　(b) 解析解

图 9-11　薄板振动方程问题的节点分布及其解析解

Fig. 9-11　Nodal distributation and analytical solution for the plate vibration equation problem

9.4 最小二乘序列函数法

接下来，我们考察 MLSS 求解该问题的收敛性。多项式序列函数的项数分别取为 $m=10, 15, 20$，则对应的求解误差如图 9-12 所示。可以看出，MLSS 求解该问题同样具有很高的计算精度。同时，也可观察到场节点密度达到一定密度时，误差呈随节点加密而增大的反常趋势。

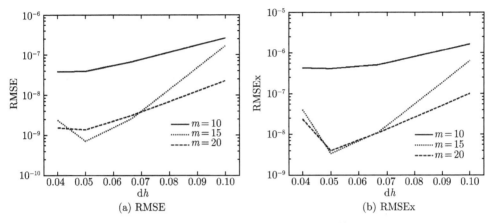

(a) RMSE (b) RMSEx

图 9-12 薄板振动方程问题的求解误差

Fig. 9-12 Numerical errors for the plate vibration equation problem

算例 3 多参数四阶微分方程问题

考虑一个更为一般化的四阶微分方程问题，一个多参数四阶微分方程定义为

$$\alpha \frac{\partial^4 u}{\partial x^4} + (a + bx + cy) \frac{\partial^4 u}{\partial x^2 \partial y^2} + \beta \frac{\partial^4 u}{\partial y^4} = \bar{f}(x,y), \quad (x,y) \in \Omega \quad (9\text{-}45)$$

此处，参数 α, β, a, b, c 等均取为 1。给定 Robin BCs 为

$$u(x,y) = \bar{u}(x,y), \quad \Delta u(x,y) = \bar{g}(x,y), \quad (x,y) \in \partial\Omega \quad (9\text{-}46)$$

问题的解析解设为

$$u(x,y) = \sin x \cos y \quad (9\text{-}47)$$

问题域定义为

$$\Omega = \{(x,y) | x = \rho \cos\theta, y = \rho \sin\theta, 0 \leqslant \theta \leqslant 2\pi\} \quad (9\text{-}48)$$

式中，

$$\rho = \sin\theta + \sqrt{1 - 0.998 \cdot \cos^2\theta} + \sqrt{1 - 0.125 \cdot \cos^2\theta} \quad (9\text{-}49)$$

域上的节点分布如图 9-13(a) 所示，域上的解析解场量分布如图 9-13(b) 所示。

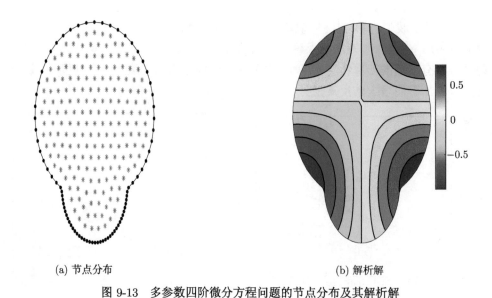

(a) 节点分布　　　　　　　　　　　(b) 解析解

图 9-13　多参数四阶微分方程问题的节点分布及其解析解

Fig. 9-13　Nodal distributation and analytical solution for the 4th-order partial differential equation with variable coefficients

接下来，我们考察 MLSS 求解该问题的收敛性。多项式序列函数的项数分别取为 $m=15, 20, 25$，则对应的求解误差如图 9-14 所示。可以看出，MLSS 求解该问题同样具有很高的计算精度。在这一问题中，可看出 $m=15$ 时表现出最佳的收敛性，由此说明项数 m 并不总是越大越好。

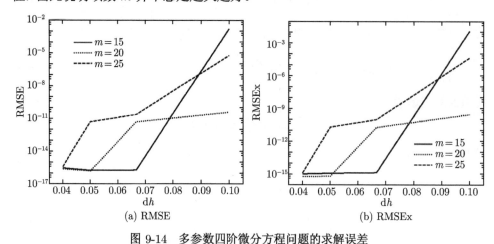

(a) RMSE　　　　　　　　　　　(b) RMSEx

图 9-14　多参数四阶微分方程问题的求解误差

Fig. 9-14　Numerical errors for the 4th-order partial differential equation with variable coefficients

9.5 无网格局部强弱法

总而言之，MLSS 本质上是一种最小二乘全局逼近的配点求解方法，是求解高阶偏微分方程问题的一种简单、灵活、易用的无网格法。

9.5 无网格局部强弱法

在弱式无网格法中，Atluri 等提出的无网格局部彼得罗夫–伽辽金 (MLPG) 法是一种重要且得到广泛应用的方法。然而，MLPG 法在实际应用中也面临一些严峻的挑战。该方法对边界条件的施加需要执行边界积分运算，对于复杂而不规则的边界，不准确的数值积分将导致数值解的失真，而精确的数值积分需要对边界进行足够精确的刻画和度量，这在实际数值执行中会面临多重困难。无网格强弱式法 (MWS)[1] 是一种基于求解效率考虑的结合式方法，对于复杂而不规则的边界，仍然存在数值实施困难的问题。由此，本书作者提出了无网格局部强弱 (MLSW) 法 [13]，其数值执行策略是对域内节点和边界上的节点采用不同的数值离散方式。MLSW 对域内节点采用 MLPG5 离散，对本质边界上的节点采用配点离散，对自然边界 (导数边界) 采用 MIP 法离散，如图 9-15 所示。

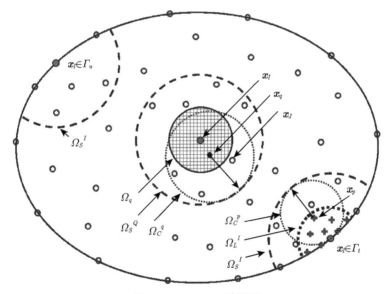

图 9-15 MLSW 法图示

Fig. 9-15 Schematics of the MLSW method

对应于式 (2-1)～ 式 (2-3) 定义的二维弹性力学问题，MLSW 法得到的离散系统方程写为

$$\boldsymbol{K} \cdot \boldsymbol{U} = \boldsymbol{F} \tag{9-50}$$

其系统刚度矩阵 K 和系统荷载矩阵 F 定义为

$$K_{IJ} = \begin{cases} \int_{\partial \Omega_q} nDB_J \mathrm{d}\Gamma_q, & \begin{cases} x_I \in \Omega \\ \Omega_q \cap \Gamma = \varnothing \end{cases} \\ \sum_{p=1}^{N_p} nDB_p \phi_J(x, x_p), & x_I \in \Gamma_t \\ I_J, & x_I \in \Gamma_u \end{cases} \quad (9\text{-}51)$$

$$F_I = \begin{cases} \int_{\Omega_q} b_q(x) \mathrm{d}\Omega, & \begin{cases} x_I \in \Omega \\ \Omega_q \cap \Gamma = \varnothing \end{cases} \\ \bar{t}_I(x), & x_I \in \Gamma_t \\ \bar{u}_I(x), & x_I \in \Gamma_u \end{cases} \quad (9\text{-}52)$$

式中,

$$n = \begin{bmatrix} n_x & 0 & n_y \\ 0 & n_y & n_x \end{bmatrix}, \quad I_J = \begin{bmatrix} \phi_J & 0 \\ 0 & \phi_J \end{bmatrix}, \quad D = \frac{E}{1-\nu^2} \begin{bmatrix} 1 & \nu & 0 \\ \nu & 1 & 0 \\ 0 & 0 & (1-\nu)/2 \end{bmatrix} \quad (9\text{-}53)$$

$$B_{J|p} = \begin{bmatrix} \dfrac{\partial \phi_{J|p}}{\partial x} & 0 & \dfrac{\partial \phi_{J|p}}{\partial y} \\ 0 & \dfrac{\partial \phi_{J|p}}{\partial y} & \dfrac{\partial \phi_{J|p}}{\partial x} \end{bmatrix}^{\mathrm{T}} \quad (9\text{-}54)$$

显然,MLSW 法在边界上不再需要执行积分计算。需注意,在数值实施中,对于域内节点,应保证其积分子域不与全局边界相交,即 $\Omega_q \cap \Gamma = \varnothing$,这在数值实施中容易实现,而且子域的偏移不会对数值结果带来明显影响。接下来,结合几个数值算例对 MLSW 法的计算效果进行检验。

算例 1 Poisson 方程问题

一个 Poisson 方程定义为

$$\frac{\partial^2 u(x)}{\partial x^2} + \frac{\partial^2 u(x)}{\partial y^2} = -\sin x - \cos y \quad (9\text{-}55)$$

其自然边界 $\rho(\theta : \pi \leqslant \theta \leqslant 2\pi)$ 条件取为

$$\frac{\partial u(x)}{\partial x} + \frac{\partial u(x)}{\partial y} = \cos x - \sin y \quad (9\text{-}56)$$

并将 $\rho(\theta : 0 \leqslant \theta \leqslant \pi)$ 作为本质边界条件 (将解析解方程代入)。该问题的解析解为

$$u(x) = \sin x + \cos y \quad (9\text{-}57)$$

9.5 无网格局部强弱法

此处取 $a=3$, $b=1$。问题域上的节点离散及对应的数值求解结果如图 9-16 所示。

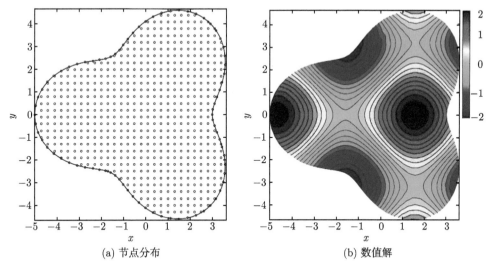

(a) 节点分布　　　　　　　　　(b) 数值解

图 9-16　Poisson 方程问题的节点分布及其数值解

Fig. 9-16　Nodal distributation and numerical solution for the Poisson's equation problem

在对比计算中,把配点法也列入比较对象。为了便于直观考察数值方法的求解精度,将数值解和精确解映射到问题域内 $r=2$ 的圆上,求解结果如图 9-17 所示。可见,MLSW 方法能够得到非常精确的数值解,而且其求解精度明显高于配点法。

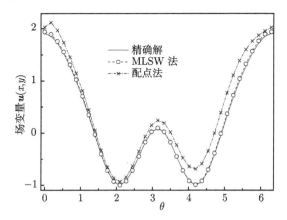

图 9-17　Poisson 方程问题的数值解和精确解比较

Fig. 9-17　Comparing the exact solution and numerical solution for the Poisson's equation problem

问题域的边界由外旋轮线函数确定：

$$\rho(\theta) = \sqrt{(a+b)^2 + 1 - 2(a+b)\cos(a\theta/b)} \tag{9-58}$$

算例 2　Laplace 方程问题

一个 Laplace 方程定义为

$$\frac{\partial^2 u(\boldsymbol{x})}{\partial x^2} + \frac{\partial^2 u(\boldsymbol{x})}{\partial y^2} = 0 \tag{9-59}$$

其自然边界条件 (NBC) 取为

$$n_x \frac{\partial u(\boldsymbol{x})}{\partial x} + n_y \frac{\partial u(\boldsymbol{x})}{\partial y} = \frac{n_x(x-0.5) + n_y(y-0.5)}{(x-0.5)^2 + (y-0.5)^2} \tag{9-60}$$

该问题的解析解为

$$u(\boldsymbol{x}) = \ln\sqrt{(x-0.5)^2 + (y-0.5)^2} \tag{9-61}$$

问题域的外边界定义为

$$\rho(\theta) = \exp(\sin\theta)\sin^2 2\theta + \exp(\cos\theta)\cos^2 2\theta \tag{9-62}$$

域内包含一个半径 $r=0.2$ 的圆孔，圆心位于 ($x_0=0.5$, $y_0=0.5$)。在数值计算中，将不规则外边界取为自然边界，将内含圆孔取为本质边界。问题域上的节点离散及对应的数值求解结果如图 9-18 所示。

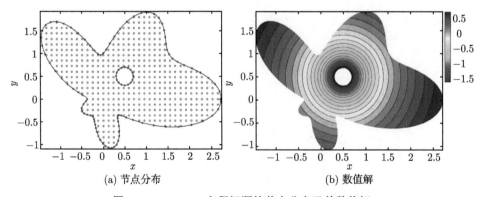

(a) 节点分布　　(b) 数值解

图 9-18　Laplace 方程问题的节点分布及其数值解

Fig. 9-18　Nodal distribution and numerical solution for the Laplace's equation problem

9.5 无网格局部强弱法

数值方法中,不精确的边界条件施加会导致系统方程丧失协调性,从而可能严重影响求解精度。本书方法中用 MIP 法施加 NBC 是一个重要的数值实施策略。为了验证其必要性,结合本算例,在 MLPG5 方法基础上,改用配点法施加 NBC 和本质边界条件的数值结果也列入比较对象。将数值解和精确解结果映射到问题域内 $\rho' = \rho/2$ 的闭曲线上,如图 9-19 所示。可见,MLSW 方法能对该问题给出精确的解答,且求解精度明显优于其他两种数值方法。显然,用配点法施加 NBC 的 MLPG5 方法,其求解精度也不够理想,由此说明,推荐的方法用 MIP 法施加 NBC 是非常必要的。

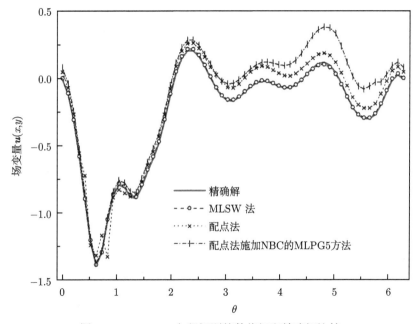

图 9-19 Laplace 方程问题的数值解和精确解比较

Fig. 9-19 Comparing the exact solution and numerical solution for the Laplace's equation problem

算例 3 内含圆孔的弹性薄板问题

考虑一个内含圆孔的不规则弹性薄板。其不规则外边界由式 (9-58) 定义,边界曲线参数取 $a=4$, $b=1$。利用结构的对称性,取 1/4 结构计算。结合 7.4 节中算例 1 给出的带圆孔无限板问题。在求解域的 x 轴和 y 轴上,根据带圆孔无限板问题的位移解,施加对称位移边界条件。在求解域的外边界上,根据带圆孔无限板问题的应力解,施加表面力荷载。计算域上的节点离散,及全求解域上对应的 x 方向应力数值求解结果如图 9-20 所示。

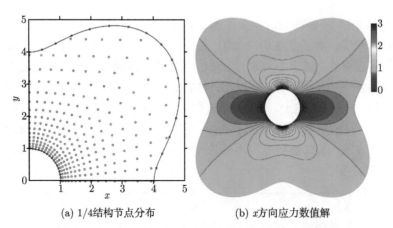

(a) 1/4结构节点分布　　(b) x方向应力数值解

图 9-20　内含圆孔弹性薄板问题的节点分布及其数值解

Fig. 9-20　Nodal distributation and numerical solution for the elasticity plate with a hole

在比较计算中，将求解域 y 轴上的 x 方向应力解结果进行分析。MLSW 法、配点法和采用配点技术施加 NBC 的 MLPG5 方法的数值解结果，及其与精确解的对照如图 9-21 所示。可见，MLSW 法能够给出非常精确的数值解，而另外两种方法在内圆孔附件应力集中区域的解均有明显偏差。此外，由配点技术施加 NBC 的 MLPG5 方法，其数值解曲线光滑性较差，在区间求解稳定性方面反而不及完全的配点法，由此可进一步说明推荐方法所采用算法的合理性。

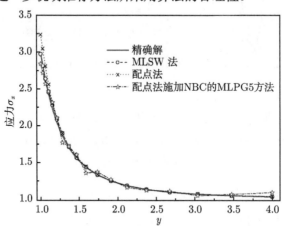

图 9-21　含圆孔弹性薄板问题的数值解和精确解比较

Fig. 9-21　Comparing the exact solution and numerical solution for the elasticity plate with a hole

所建议的 MLSW 法，可以视为是对 MLPG 法实用性的一个扩展，采用配点法施加本质边界条件，采用 MIP 法施加自然边界条件，完全避免了局部弱式方法的边界积分运算。因此，推荐方法能够将弱式方法和强式方法的各自优势结合起来，并消减了各自的局限性。通过初步数值检验，MLSW 法容易处理不规则域问题，而且具有精确、稳定、可靠的计算性能。

9.6 本章小结

本章介绍了几种结合式的无网格方法。其中，最小二乘序列函数法 (MLSS) 也可归类为配点类的方法，鉴于其采用了全局逼近的技术，因此列入本章予以介绍。与 MLSS 相类似的一种方法被称为特解法 (MPS)[14-17]，该法采用特解函数作为试函数，而非通常的多项式序列函数。相比较而言，MLSS 具有更为简单、灵活和普适的优点。

实际上，结合式方法的研究内容非常丰富。比如，有限元与 EFG 的结合法 (FE-EFG)[18]，EFG 与 BEM 的结合法 (EFG-HBEM)[19]，MLPG 与 FEM 的结合法 (MLPG-FE)[20]，最小二乘配点与 EFG 相结合的无网格伽辽金最小二乘法 (MGLS)[21]，SPH 与 FEM 的结合法 (SPH-FE)[22]，MLPG 与 FDM 的结合法 (MLPG-FD)[23]，MLPG 与配点结合法 (MLPG-CM)[24]，EFG 和 BEM 结合的方法 (EFG-BEM)[25] 等。

结合式方法将多种离散技术进行耦合，提供了一种将不同方法的优势相结合，并弥补各自局限性的解决思路。这种求解技术在解决一些特定的实际问题中，往往会具有某种应用上的优势。

参 考 文 献

[1] Liu G R, Gu Y T. A meshfree method: meshfree weak-strong (MWS) form method, for 2-D solids[J]. Computational Mechanics, 2003, 33(1): 2-14.

[2] Liu G R, Wu Y L, Ding H. Meshfree weak-strong (MWS) form method and its application to incompressible flow problems[J]. International Journal for Numerical Methods in Fluids, 2004, 46(10): 1025-1047.

[3] Gu Y T, Liu G R. A meshfree weak-strong (MWS) form method for time dependent problems[J]. Computational Mechanics, 2005, 35(2): 134-145.

[4] Liu G R, Gu Y T. An Introduction to Meshfree Methods and Their Programming[M]. Netherlands: Springer Science & Business Media, 2005.

[5] Dehghan M, Salehi R. A meshfree weak-strong (MWS) form method for the unsteady magnetohydrodynamic (MHD) flow in pipe with arbitrary wall conductivity[J]. Computational Mechanics, 2013, 52(6): 1445-1462.

[6] Wen P H, Aliabadi M H. A hybrid finite difference and moving least square method

for elasticity problems[J]. Engineering Analysis With Boundary Elements, 2012, 36(4): 600-605.

[7] Wen P H, Yang J J, Huang T, et al. Infinite element in meshless approaches[J]. European Journal of Mechanics-A/Solids, 2018, 72: 175-185.

[8] Wen P H, Cao P, Korakianitis T. Finite block method in elasticity[J]. Engineering Analysis With Boundary Elements, 2014, 46: 116-125.

[9] Li M, Wen P H. Finite block method for transient heat conduction analysis in functionally graded media[J]. International Journal For Numerical Methods in Engineering, 2014, 99(5): 372-390.

[10] Li M, Lei M, Munjiza A, et al. Frictional contact analysis of functionally graded materials with Lagrange finite block method. International Journal For Numerical Methods in Engineering, 2015, 103(6): 391-412.

[11] Li M, Meng L X, Hinneh P, et al. Finite block method for interface cracks[J]. Engineering Fracture Mechanics, 2016, 156: 25-40.

[12] Huang W, Ban Y, Sulaj D. Solving 4th-order PDEs with the method of particular solutions and the method of least square series[R]. Hunan: Changsha University of Science and Technology, 2018.

[13] 杨建军, 郑健龙. 无网格局部强弱 (MLSW) 法求解不规则域问题 [J]. 力学学报, 2017, 49(3): 659-666.

[14] Wen P H, Chen C S. The method of particular solutions for solving scalar wave equations[J]. International Journal for Numerical Methods in Biomedical Engineering, 2010, 26(12): 1878-1889.

[15] Chen C S, Fan C M, Wen P H. The method of approximate particular solutions for solving certain partial differential equations[J]. Numerical Methods for Partial Differential Equations, 2012, 28(2): 506-522.

[16] Chen C S, Fan C M, Wen P H. The method of approximate particular solutions for solving elliptic problems with variable coefficients[J]. International Journal of Computational Methods, 2011, 8(03): 545-559.

[17] Muleshkov A S, Golberg M A, Chen C S. Particular solutions of Helmholtz-type operators using higher order polyhrmonic splines[J]. Computational Mechanics, 1999, 23(5-6): 411-419.

[18] Belytschko T, Organ D, Krongauz Y. A coupled finite element-element-free Galerkin method[J]. Computational Mechanics, 1995, 17(3): 186-195.

[19] Liu G R, Gu Y T. Coupling of element free Galerkin and hybrid boundary element methods using modified variational formulation[J]. Computational Mechanics, 2000, 26(2): 166-173.

[20] Liu G R, Gu Y T. Meshless local Petrov–Galerkin (MLPG) method in combination with finite element and boundary element approaches[J]. Computational Mechanics,

2000, 26(6): 536-546.

[21] Pan X F, Zhang X, Lu M W. Meshless Galerkin least-squares method[J]. Computational Mechanics, 2005, 35(3): 182-189.

[22] De Vuyst T, Vignjevic R, Campbell J C. Coupling between meshless and finite element methods[J]. International Journal of Impact Engineering, 2005, 31(8): 1054-1064.

[23] Atluri S N, Liu H T, Han Z D. Meshless local Petrov-Galerkin (MLPG) mixed finite difference method for solid mechanics[J]. Computer Modeling in Engineering & Sciences, 2006, 15(1): 1-16.

[24] Atluri S N, Liu H T, Han Z D. Meshless local Petrov-Galerkin (MLPG) mixed collocation method for elasticity problems[J]. Computer Modeling in Engineering & Sciences, 2006, 4(3): 141-152.

[25] Zhang Z, Liew K M, Cheng Y. Coupling of the improved element-free Galerkin and boundary element methods for two-dimensional elasticity problems[J]. Engineering Analysis with Boundary Elements, 2008, 32(2): 100-107.

第10章 无网格法应用

当前,无网格法已经逐步被应用于解决一些实际的工程问题。从理论上讲,经典方法可求解的问题,比如固体力学问题、流体力学问题、热力学问题、电磁力学问题等,无网格法对同样的问题并不会存在求解障碍。相反,在一些特殊的问题中,可能传统方法存在求解困难。而无网格法凭借其独特的优势,反而更适合于求解这些特殊问题。本章对无网格法求解这些特殊问题,或者是一些前沿应用,作一些简单扼要的介绍,以帮助读者更全面地了解无网格法的应用趋势。

10.1 大变形问题中的应用

有限元法求解大变形问题时,由于网格畸变可能导致求解失败。而无网格法不严格依赖于网格,甚至是对网格完全免除的,所以也就不必担心这种"网格畸变障碍",可以有效地求解大变形问题。

当前,无网格法已被用于分析非线性结构[1-4]、薄壳结构[5]、结构成型[6-8]、超弹性材料[9-12]、工程滑坡[13-19]等大变形问题。图10-1给出一个圆锥壳的翻转过程的无网格法模拟[5],重现了结构在实现内面外翻过程中的大变形现象。

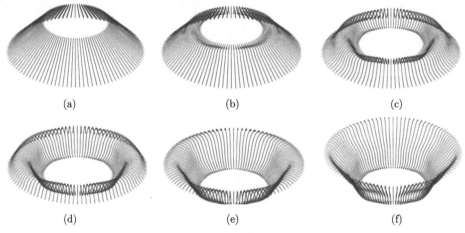

图 10-1 圆锥壳的翻转过程[5]

Fig. 10-1 The snap-through of a conic shell[5]

10.1 大变形问题中的应用

图 10-2 给出一个受集中荷载挤压的圆筒变形过程 [5]，可见筒壳结构在对偶集中力的作用下，表现为大变形过程。图 10-3 给出的是一个地震触发的山体滑坡过程模拟 [20]，其大变形过程已经超越连续介质力学的范畴。实际上，类似于离散元法 (discrete element method, DEM)[21-26]，无网格法不仅适用于连续介质力学问题，也可以适用于非连续介质力学问题，应用范畴上也比传统数值方法更为广泛。

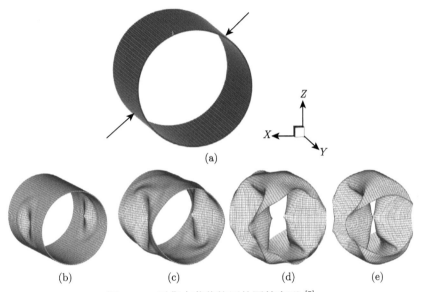

图 10-2 受集中荷载挤压的圆筒变形 [5]

(a) 结构受力模型; (b)∼(e) 变形过程

Fig. 10-2 The deformation sequence of a pinched cylinder[5]

(a) numerical model; (b)∼(e) deformation process

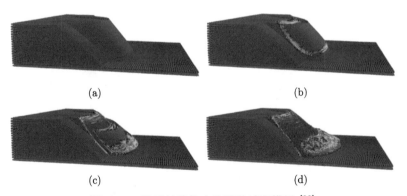

图 10-3 地震触发的山体滑坡过程模拟 [20]

Fig. 10-3 Simulation of a landslide triggered earthquake[20]

10.2 断裂与破坏问题中的应用

无网格法基于散点执行计算,对计算单元没有连通性的要求,很容易实现自适应计算,因此天然适用于处理开裂和裂纹扩展此类动态边界问题。求解断裂破坏和裂纹扩展问题,是近代无网格法最成功的应用专题之一。目前发展的很多无网格法已被应用于求解断裂问题,比如 EFG 法[27-33]、MLPG 法[34-38]、RKPM[39,40]、Hp 云法[41,42]、XFEM[43-51] 等。

图 10-4 给出一个带初始裂纹的钢板在冲击荷载作用下,其脆性裂纹扩展及其应力场的模拟结果[52],研究表明数值计算能够精确地模拟试验结果。

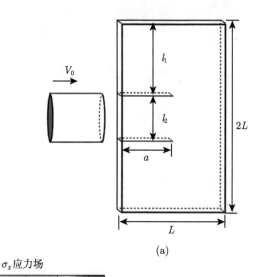

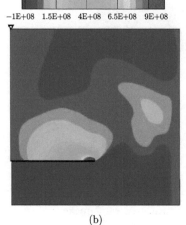

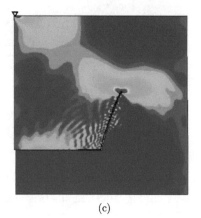

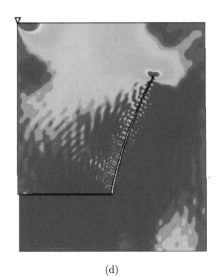 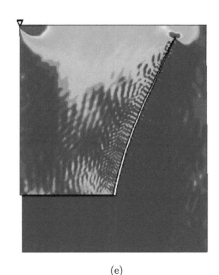

(d)   (e)

图 10-4   裂纹扩展及其应力场[52]

(a) 数值模型；(b)～(e) 裂纹扩展

Fig. 10-4   Stress field with the crack propagation[52]

(a) numerical model; (b)～(e) crack growth

Zhao 等[53]对材料界面裂纹及其附近的应力场进行了模拟分析，其数值模型如图 10-5 所示。裂纹附近的应力场在不同材料中，其分布可能具有明显的差异性，图 10-6 给出其应力场的时程变化模拟结果。

Yang 等[54]对一个合金构件的断裂破坏演化进行了数值分析，图 10-7 给出的是数值计算结果和试验结果的对比；图 10-8 则给出了开裂与破坏的演化过程，上部图为实体试验的影像，下部图为对应的数值模拟结果。

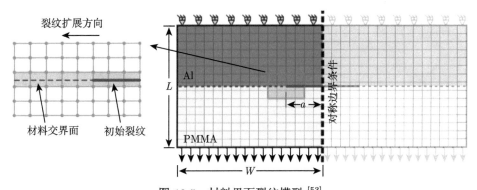

图 10-5   材料界面裂纹模型[53]

Fig. 10-5   Interface crack model of a biomaterial structure[53]

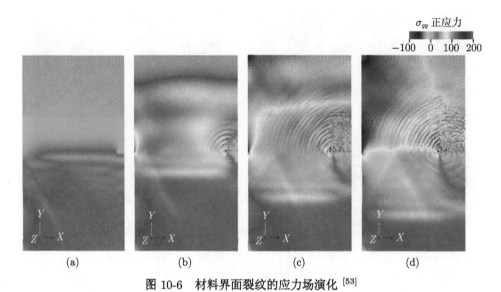

图 10-6 材料界面裂纹的应力场演化[53]

Fig. 10-6 Normal stress evolution within the bimaterial specimen[53]

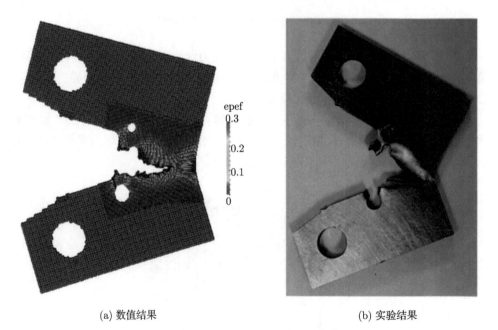

图 10-7 合金构件的破坏试验[54]

Fig. 10-7 Comparison of cracking patterns for a metal-alloy structure[54]

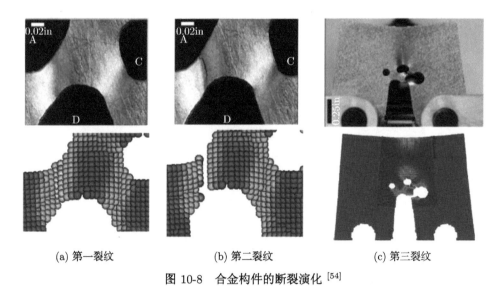

(a) 第一裂纹　　　(b) 第二裂纹　　　(c) 第三裂纹

图 10-8　合金构件的断裂演化 [54]

Fig. 10-8　Crack evolution for the metal-alloy structure[54]

10.3　冲击与爆炸问题中的应用

冲击与爆炸问题涉及结构的大变形、碎裂与解体。无网格法的散点计算特性，再辅以特殊的形函数构造方法，可以有效地对此类问题进行数值分析和模拟。Rabczuk 和 Belytschko[55] 对混凝土厚板的板心位置受一个小直径圆柱刚体冲击时，结构发生脆裂的破坏过程进行了分析，如图 10-9 所示，给出了结构冲击破坏的时程演化模拟结果。

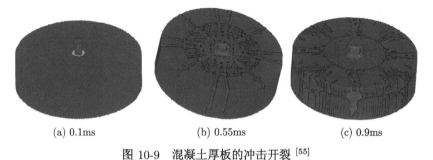

(a) 0.1ms　　　(b) 0.55ms　　　(c) 0.9ms

图 10-9　混凝土厚板的冲击开裂 [55]

Fig. 10-9　Impact cracks on concrete slabs[55]

Yreux 和 Chen[56] 对铝棒冲击问题进行了数值模拟。一段铝棒 (Taylor 杆) 以一定初始速度射向一面刚体墙，图 10-10 给出了最终铝棒冲击端发生塑性变形的结果，给出的三种数值结果是对使用不同基函数的 RKPM 计算结果的比较。

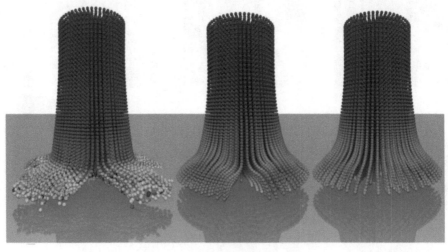

(a) 常数基　　　　　　(b) 自适应基　　　　　　(c) 准线性基

图 10-10　冲击作用后的 Taylor 杆变形[56]

Fig. 10-10　Deformed shape after impact for a Taylor bar[56]

Yreux 和 Chen[56] 对子弹击穿混凝土板的过程进行了数值模拟，能够生动而真实地演示子弹击穿时那种物质碎屑崩裂四射的场景，如图 10-11 所示。

图 10-11　子弹击穿混凝土板[56]

Fig. 10-11　Bullet penetrating a concrete panel[56]

为了应对天体小行星撞击地球的潜在威胁，一种尚在论证的方案是通过空天火箭将核弹射向撞击而来的不速之客，以核爆的方式来免除危险。Kaplinge 和

Wie[57,58] 采用 SPH 法对核爆摧毁小行星并改变其运动轨迹的问题进行了模拟分析，如图 10-12 所示。

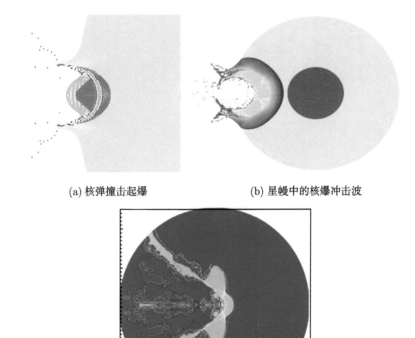

(a) 核弹撞击起爆   (b) 星幔中的核爆冲击波

(c) 裂解星体碎片分布

图 10-12　小行星的核爆 [57,58]

Fig. 10-12　Nuclear fragmentation of a crisis asteroid[57,58]

10.4　微细观力学问题中的应用

材料的微观尺度上表现为粒子属性，而无网格法的散点计算特性很容易对材料的微观结构进行直接描述，因此在微细观力学问题中得到广泛应用。石墨烯具有优异的光学、电学、力学特性，在材料学、微纳加工、能源、生物医学和药物传递等方面具有重要的应用前景，被认为是未来一种革命性的材料。其中，石墨烯的热传导性是一个研究热点问题 [59-61]，Hu 等 [62] 使用分子动力学方法对石墨烯的热力学性质进行了数值模拟，如图 10-13 所示，结果表明石墨烯不仅具有超常的热传导性，而且表现为一种在不同方向上传导能力的显著差别性，即异向性。

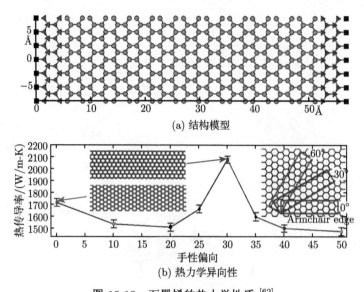

(a) 结构模型

(b) 热力学异向性

图 10-13　石墨烯的热力学性质[62]

Fig. 10-13　Thermal properties of graphene[62]

碳纳米管材料凭借其优异的物理性能，自问世以来受到极大关注[63]。在微观尺度上，很难对其力学性能进行直接测试，Liew 等[64-67]采用无网格法对其进行了数值模拟。图 10-14 是碳纳米管的屈曲变形数值模拟结果[66]，基于此类数值模拟，并辅以显微观测，便可间接测定其基本力学常数。

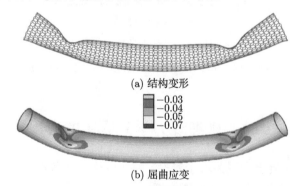

(a) 结构变形

(b) 屈曲应变

图 10-14　碳纳米管的屈曲变形[66]

Fig. 10-14　Buckling of carbon nanotubes[66]

声子晶体是由弹性固体周期排列在另一种固体或流体介质中形成的一种新型功能材料。弹性波在声子晶体中传播时，受其内部结构的作用，在一定频率范围 (带隙) 内被阻止传播, 而在其他频率范围 (通带) 可以无损耗地传播。Zheng 等 [68,69] 对声子晶体的能带结构进行了数值分析, 如图 10-15 所示。

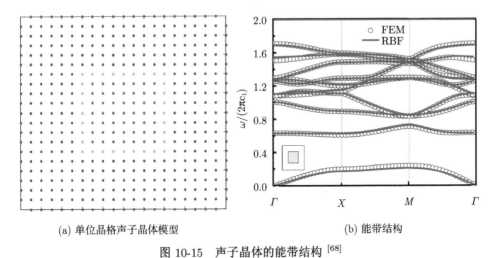

(a) 单位晶格声子晶体模型　　　　　(b) 能带结构

图 10-15　声子晶体的能带结构 [68]

Fig. 10-15　Band structure of phononic crystal[68]

10.5　流体力学问题中的应用

流体是气体和液体的总称。在人们的生活和生产活动中随时随地都可遇到流体, 所以流体力学是与人类日常生活和生产事业密切相关的。大气运动、海水运动 (包括波浪、潮汐、中尺度涡旋、环流等) 乃至地球深处熔浆的流动都是流体力学的研究内容。20 世纪初, 随着航空技术的发展, 人类的活动范围扩展到太空。航空航天事业的蓬勃发展是同流体力学的分支学科——空气动力学和气体动力学的发展紧密相连的。

无网格法采用散点近似, 可以彻底或部分地消除网格, 不需要网格的初始划分和重构, 不仅可以保证计算的精度, 而且可以大大减小计算的难度。传统的流体分析软件在解决自由液面、高速碰撞、移动边界或是移动物体等工程问题时, 需要一套独特的网格重划分或是网格自适应算法, 由于在网格质量、算法求解、收敛性等方面存在一系列问题, 所以传统方法受到了很大的限制。而快速发展的无网格法在分析自由液面、多相流、流固耦合等问题中逐步取得显著进步。

圆柱绕流是观察流体旋涡的一个经典试验。Bouscasse 等对这一试验现象进行了无网格法数值分析 [70,71]。图 10-16 给出了 SPH 法的数值计算结果, 并比较了圆柱浸没深度 h 和圆柱直径尺度 d 对涡街分布形态的影响。结果表明对自由表面流而言，圆柱浸没深度对涡流分布的影响比较显著。

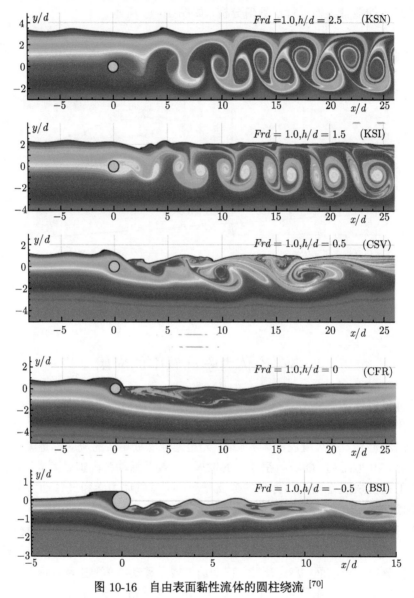

图 10-16 自由表面黏性流体的圆柱绕流 [70]

Fig. 10-16 Viscous flow past a circular cylinder interacting with a free surface [70]

在工程实际问题中，更多可能面临流固耦合的多物理场分析。比如，图 10-17 是对直升机在水上的迫降模拟 [72,73]，其流场分析采用 SPH 法，固体结构 (直升机) 则采用有限元法建模。另如，图 10-18 给出汽车轮胎的水滑行为模拟 [74]，也采用了类似的分析方法。

10.5 流体力学问题中的应用　　　　　　　　　　　　　　　　　　　　　　　　· 235 ·

(a) 水压分布

(b) 流速分布

图 10-17　直升机的水上迫降 [73]

Fig. 10-17　The ditching simulation of a helicopter on water[73]

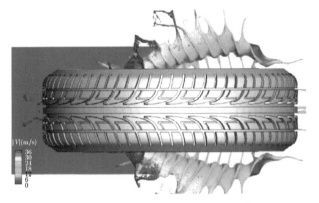

图 10-18　汽车轮胎的水滑行为模拟 [74]

Fig. 10-18　Tire aquaplaning simulation[74]

10.6 生物医学中的应用

生物医学工程 (biomedical-engineering) 是一门新兴学科, 它综合工程学、物理学、生物学和医学的理论和方法, 在各层次上研究人体系统的状态变化, 并解决医学中的有关问题, 保障人类健康, 为疾病的预防、诊断、治疗和康复服务。生物医学工程学是在电子学、微电子学、现代计算机技术、化学、高分子化学、力学、近代物理学、光学、射线技术、精密机械和近代高新技术发展的基础上, 与医学结合的条件下发展起来的。其中, 数值计算科学对生物医学工程学的发展也起到了重要的支撑作用, 并在生物力学和医学影像学方面扮演着重要角色。生物力学是运用力学的理论和方法, 研究生物组织和器官的力学特性, 研究机体力学特征与其功能的关系。生物力学的研究成果对了解人体伤病机理, 确定治疗方法有着重大意义, 同时可为人工器官和组织的设计提供依据。而医学影像学是临床诊断疾病的主要手段之一, 也是世界上科研开发的重点课题。医用影像设备主要采用 X 射线、超声、放射性核素磁共振等进行成像。

Chen 等 [75,76] 采用 RKPM 对骨骼肌的收缩应力进行了数值分析, 如图 10-19 所示。对于生命体内的这种力学行为, 是很难直接探测的, 而采用数值方法, 却能够给出理性的分析结果。

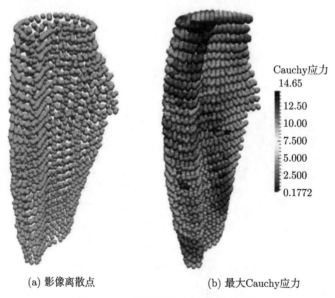

(a) 影像离散点　　　　(b) 最大Cauchy应力

图 10-19　骨骼肌的力学模拟 [75,76]

Fig. 10-19　Skeletal muscle simulation[75,76]

10.6 生物医学中的应用

物质点法 (material point method,MPM) 将物质域离散为一组粒子,这些粒子具有物质变量属性,比如密度、速度和应力等,而粒子的空间和运动信息伴随背景网格定义[77,78]。Zhou 等[79] 采用 MPM 对人脑的受冲击行为进行了力学模拟分析,如图 10-20 所示,其中 (c)~(e) 的上下图分别为侧视图和俯视图。

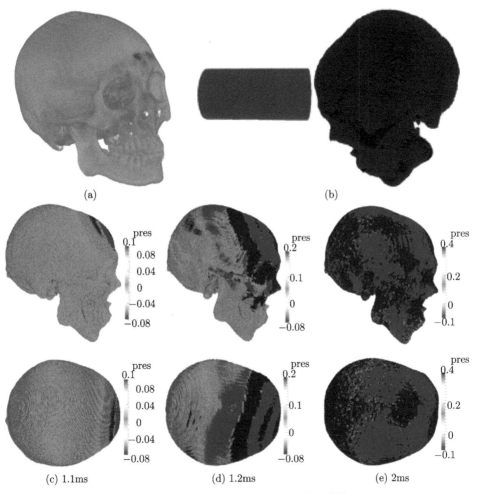

图 10-20 人脑受冲击作用的力学模拟[79]

(a) 人脑模型; (b) 人脑冲击模型; (c)~(e) 人脑骨骼在不同时间步的压力场

Fig. 10-20 Human head impact simulation[79]

(a) head model; (b) head impact model; (c)~(e) pressure contour of the skull bone at different time steps

心电图扫描反成像技术是一种将心电扫描数据进行逆向重构并生成影像的数值模拟方法。基于 MFS 的无网格法计算优势,Wang 和 Rudy[80] 提出了基于 MFS

的心电图扫描反成像技术。图 10-21 给出的是同步右心室心内膜和左心室心外膜起搏的心电图扫描反成像结果，(a) 为电势图 (起搏 40ms 时)，(b) 为搏动等时差图，每一类图均给出 3 个不同的成像视角，图中标有 * 号的位置为起搏点位置。

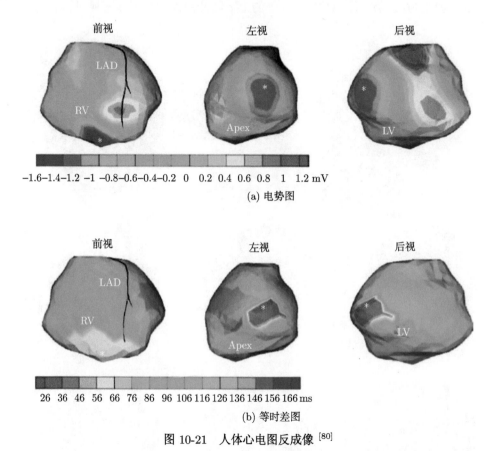

图 10-21　人体心电图反成像 [80]

Fig. 10-21　Human inverse electrocardiography[80]

随着无网格法的逐步完善和成熟，其数值求解技术在生物医学工程中的应用也逐步推进。当前，无网格方法在微创手术模拟 [81-83]、生物力学 [84]、医学影像 [85-87]、肿瘤检测 [88-93] 等医学诊疗中已经得到实际应用。

10.7　无网格法商业软件开发

科学计算软件的商业开发是将科学计算理论向实际应用转化的重要途径。比如应用非常广泛的 Ansys、Abaqus、Nastran、Adina 等有限元仿真软件已经成为全世界工程设计和工业制造领域的基本工具。由于有限元法仿真软件起步较早，目前已发展得非常成熟，所以无网格法商业软件开发的竞争压力较大。即便如此，无网格法凭借其先进的计算概念和强劲的求解能力，正逐步渗透到商业软件领域。近年来，无网格法的商业软件开发也逐步取得一些进展。

较早前，因为传统的有限元商业软件很难处理一些冲击和破坏问题。光滑粒子流体动力学法 (SPH) 作为早期表现出这种求解能力的一种无网格法，已经被集成到一些有限元商业软件中，如 Ls-Dyna、Pam-Crash、Abaqus/Explict，扩展了此类传统软件的求解功能。

XFlow 是总部位于西班牙马德里的 Next Limit Technologies 公司研发的一款流体力学无网格法软件。XFlow 在微观尺度使用分子运动理论 (DSMC) 求解动力学方程，在介观尺度使用格子玻尔兹曼法求解玻尔兹曼传输方程，在宏观尺度使用光滑粒子法和涡粒子法求解纳维 – 斯托克斯方程。2016 年末，Next Limit Technologies 公司被法国著名的飞机制造商和三维工程仿真软件供应商达索系统公司收购。

shonDy 流体仿真软件是由德国 shonDynamics 公司研发的一款无网格法流体分析软件。软件的主要模拟功能包括：黏性流体的流动，流固耦合，对流传热与导热，以及弹性体形变等。

NoGrid Points 软件是基于有限点方法 (finite pointset method, FPM) 的新一代无网格计算流体力学软件，代表了流体数值仿真的最新发展方向。该软件由德国 NoGrid 公司研发。

MPM3D 是清华大学张雄课题组自 2004 年开始研发的开源三维显式物质点法并行数值仿真软件 (计算机软件著作权登记号 2009SRBJ4761, 2009)，可用于模拟超高速碰撞、冲击、侵彻和爆炸等强冲击载荷作用下材料与结构的力学行为[94]。

Particleworks 流体仿真软件是由日本 Prometech software 公司开发的一种无网格法流体分析与模拟软件。该软件由日本东京大学的 Seiichi Koshizuka 基于自主算法 MPS[95] 而研发。

麦栗 (Maili) 无网格法仿真分析软件由本书作者基于无网格介点法 (MIP) 而研发 (计算机软件著作权登记号 2016SR327219, 2016)，并由我国"麦粒软件"公司[96] 进行商业开发和推广。Maili 的初期产品主要专注于固体结构分析功能开发，也是国际上首个无网格法固体结构仿真分析商业化软件。

10.8　本 章 小 结

无网格法作为一种强大而富于生命力的数值计算方法，近年来其应用领域和应用范围正在迅速拓展。本章介绍的一些内容，可帮助读者从更广阔的视野了解其应用潜力和优势。即便如此，所介绍的内容相比于实际的应用广度而言，仍然是比较狭窄的一部分。我们同时也注意到，近年来无网格法在量子力学[97-100]、天文物理学[101-104]、分子生物学[105-107]、纳米材料[108-112]等前沿科学领域也逐步得到应用，在更广泛的工程科学领域，一些新颖的应用更是难以穷举。我们有理由相信，随着无网格法的进一步发展和完善，这种独特的计算方法在科学研究、工程制造和工业生产中必将发挥更广泛、更深远的应用价值。

参 考 文 献

[1] Li S, Hao W, Liu W K. Numerical simulations of large deformation of thin shell structures using meshfree methods[J]. Computational Mechanics, 2000, 25(2-3): 102-116.

[2] Chen J S, Hillman M, Chi S W. Meshfree methods: Progress made after 20 years[J]. Journal of Engineering Mechanics, 2017, 143(4): 04017001.

[3] Chen J S, Pan C, Wu C T, et al. Reproducing kernel particle methods for large deformation analysis of non-linear structures[J]. Computer Methods in Applied Mechanics and Engineering, 1996, 139(1-4): 195-227.

[4] Liu W K, Jun S. Multiple-scale reproducing kernel particle methods for large deformation problems[J]. International Journal for Numerical Methods in Engineering, 1998, 41(7): 1339-1362.

[5] Li S, Hao W, Liu W K. Numerical simulations of large deformation of thin shell structures using meshfree methods[J]. Computational Mechanics, 2000, 25(2-3): 102-116.

[6] Chen J S, Roque C M O L, Pan C, et al. Analysis of metal forming process based on meshless method[J]. Journal of Materials Processing Technology, 1998, 80: 642-646.

[7] Bonet J, Kulasegaram S. Correction and stabilization of smooth particle hydrodynamics methods with applications in metal forming simulations[J]. International Journal for Numerical Methods in Engineering, 2000, 47(6): 1189-1214.

[8] Gu Y T. An adaptive local meshfree updated Lagrangian approach for large deformation analysis of metal forming[C]. Advanced Materials Research. Trans Tech Publications, 2010, 97: 2664-2667.

[9] Chen J S, Pan C, Wu C T. Large deformation analysis of rubber based on a reproducing kernel particle method[J]. Computational Mechanics, 1997, 19(3): 211-227.

[10] Calvo B, Martinez M A, Doblare M. On solving large strain hyperelastic problems with the natural element method[J]. International Journal for Numerical Methods in Engineering, 2005, 62(2): 159-185.

[11] Grindeanu I, Chang K H, Choi K K, et al. Design sensitivity analysis of hyperelastic structures using a meshless method[J]. AIAA Journal, 1998, 36(4): 618-627.

[12] Gu Y T, Wang Q X, Lam K Y. A meshless local Kriging method for large deformation analyses[J]. Computer Methods in Applied Mechanics and Engineering, 2007, 196(9-12): 1673-1684.

[13] Soga K, Alonso E, Yerro A, et al. Trends in large-deformation analysis of landslide mass movements with particular emphasis on the material point method[J]. Géotechnique, 2015, 66(3): 248-273.

[14] Pastor M, Haddad B, Sorbino G, et al. A depth-integrated, coupled SPH model for flow-like landslides and related phenomena[J]. International Journal for Numerical and Analytical Methods in Geomechanics, 2009, 33(2): 143-172.

[15] McDougall S, Hungr O. A model for the analysis of rapid landslide motion across three-dimensional terrain[J]. Canadian Geotechnical Journal, 2004, 41(6): 1084-1097.

[16] Huang Y, Zhang W, Xu Q, et al. Run-out analysis of flow-like landslides triggered by the Ms 8.0 2008 Wenchuan earthquake using smoothed particle hydrodynamics[J]. Landslides, 2012, 9(2): 275-283.

[17] McDougall S, Hungr O. Dynamic modelling of entrainment in rapid landslides[J]. Canadian Geotechnical Journal, 2005, 42(5): 1437-1448.

[18] Huang Y, Dai Z. Large deformation and failure simulations for geo-disasters using smoothed particle hydrodynamics method[J]. Engineering Geology, 2014, 168: 86-97.

[19] Ataie-Ashtiani B, Shobeyri G. Numerical simulation of landslide impulsive waves by incompressible smoothed particle hydrodynamics[J]. International Journal for Numerical Methods in Fluids, 2008, 56(2): 209-232.

[20] Chen J S, Hillman M, Chi S W. Meshfree methods: progress made after 20 years[J]. Journal of Engineering Mechanics, 2017, 143(4): 04017001.

[21] Moresi L, Dufour F, Mühlhaus H B. A Lagrangian integration point finite element method for large deformation modeling of viscoelastic geomaterials[J]. Journal of Computational Physics, 2003, 184(2): 476-497.

[22] Munjiza A A. The Combined Finite-Discrete Element Method[M]. Hoboken: John Wiley & Sons, 2004.

[23] Munjiza A, Owen D R J, Bicanic N. A combined finite-discrete element method in transient dynamics of fracturing solids[J]. Engineering computations, 1995, 12(2): 145-174.

[24] Kruggel-Emden H, Simsek E, Rickelt S, et al. Review and extension of normal force models for the discrete element method[J]. Powder Technology, 2007, 171(3): 157-173.

[25] Mishra B K, Rajamani R K. The discrete element method for the simulation of ball mills[J]. Applied Mathematical Modelling, 1992, 16(11): 598-604.

[26] McDowell G R, Harireche O, Konietzky H, et al. Discrete element modelling of geogrid-reinforced aggregates[J]. Proceedings of the Institution of Civil Engineers-Geotechnical Engineering, 2006, 159(1): 35-48.

[27] Belytschko T, Lu Y Y, Gu L. Element-free Galerkin methods[J]. International Journal for Numerical Methods in Engineering, 1994, 37(2): 229-256.

[28] Belytschko T, Gu L, Lu Y Y. Fracture and crack growth by element free Galerkin methods[J]. Modelling and Simulation in Materials Science and Engineering, 1994, 2(3A): 519-534.

[29] Belytschko T, Lu Y Y, Gu L. Crack propagation by element-free Galerkin methods[J]. Engineering Fracture Mechanics, 1995, 51(2): 295-315.

[30] Belytschko T, Lu Y Y, Gu L, et al. Element-free Galerkin methods for static and dynamic fracture[J]. International Journal of Solids and Structures, 1995, 32(17-18): 2547-2570.

[31] Belytschko T, Tabbara M. Dynamic Fracture using element-free Galerkin methods[J]. International Journal for Numerical Methods in Engineering, 1996, 39(6): 923-938.

[32] Fleming M, Chu Y A, Moran B, et al. Enriched element-free Galerkin methods for crack tip fields[J]. International Journal for Numerical Methods in Engineering, 1997, 40(8): 1483-1504.

[33] Sukumar N, Moran B, Black T, et al. An element-free Galerkin method for three-dimensional fracture mechanics[J]. Computational Mechanics, 1997, 20(1-2): 170-175.

[34] Ching H K, Batra R C. Determination of crack tip fields in linear elastostatics by the meshless local Petrov-Galerkin (MLPG) method[J]. Computer Modeling in Engineering and Sciences, 2001, 2(2): 273-289.

[35] Qian L F, Batra R C, Chen L M. Static and dynamic deformations of thick functionally graded elastic plates by using higher-order shear and normal deformable plate theory and meshless local Petrov–Galerkin method[J]. Composites Part B: Engineering, 2004, 35(6-8): 685-697.

[36] Sladek J, Sladek V, Krivacek J, et al. Meshless local Petrov-Galerkin method for stress and crack analysis in 3-D axisymmetric FGM bodies[J]. Computer Modeling in Engineering & Sciences, 2005, 8(3): 259-270.

[37] Kaiyuan L, Shuyao L, Guangyao L. A simple and less-costly meshless local Petrov-Galerkin (MLPG) method for the dynamic fracture problem[J]. Engineering Analysis with Boundary Elements, 2006, 30(1): 72-76.

[38] Rabczuk T, Areias P. A meshfree thin shell for arbitrary evolving cracks based on an extrinsic basis[J]. Computer Modeling in Engineering and Sciences, 2006, 16(2): 115-130.

[39] Liu W K, Hao S, Belytschko T, et al. Multiple scale meshfree methods for damage fracture and localization[J]. Computational Materials Science, 1999, 16(1-4): 197-205.

[40] Hao S, Liu W K, Chang C T. Computer implementation of damage models by finite element and meshfree methods[J]. Computer Methods in Applied Mechanics and Engineering, 2000, 187(3-4): 401-440.

[41] Pereira J P, Duarte C A, Guoy D, et al. Hp-Generalized FEM and crack surface representation for non-planar 3-D cracks[J]. International Journal for Numerical Methods in

Engineering, 2009, 77(5): 601-633.

[42] Pereira J P, Duarte C A, Jiao X. Three-dimensional crack growth with hp-generalized finite element and face offsetting methods[J]. Computational Mechanics, 2010, 46(3): 431-453.

[43] Duarte C A, Hamzeh O N, Liszka T J, et al. A generalized finite element method for the simulation of three-dimensional dynamic crack propagation[J]. Computer Methods in Applied Mechanics and Engineering, 2001, 190(15): 2227-2262.

[44] Sukumar N, Moës N, Moran B, et al. Extended finite element method for three-dimensional crack modelling[J]. International Journal for Numerical Methods in Engineering, 2000, 48(11): 1549-1570.

[45] Moës N, Dolbow J, Belytschko T. A finite element method for crack growth without remeshing[J]. International Journal for Numerical Methods in Engineering, 1999, 46(1): 131-150.

[46] Fleming M, Chu Y A, Moran B, et al. Enriched element-free Galerkin methods for crack tip fields[J]. International Journal for Numerical Methods in Engineering, 1997, 40(8): 1483-1504.

[47] Garzon J, O'Hara P, Duarte C A, et al. Improvements of explicit crack surface representation and update within the generalized finite element method with application to three-dimensional crack coalescence[J]. International Journal for Numerical Methods in Engineering, 2014, 97(4): 231-273.

[48] Dolbow J, Moës N, Belytschko T. An extended finite element method for modeling crack growth with frictional contact[J]. Computer Methods in Applied Mechanics and Engineering, 2001, 190(51-52): 6825-6846.

[49] Laborde P, Pommier J, Renard Y, et al. High-order extended finite element method for cracked domains[J]. International Journal for Numerical Methods in Engineering, 2005, 64(3): 354-381.

[50] Areias P, Belytschko T. Analysis of three-dimensional crack initiation and propagation using the extended finite element method[J]. International Journal for Numerical Methods in Engineering, 2005, 63(5): 760-788.

[51] Sukumar N, Chopp D L, Moran B. Extended finite element method and fast marching method for three-dimensional fatigue crack propagation[J]. Engineering Fracture Mechanics, 2003, 70(1): 29-48.

[52] Wen L, Tian R. Improved XFEM: Accurate and robust dynamic crack growth simulation[J]. Computer Methods in Applied Mechanics and Engineering, 2016, 308: 256-285.

[53] Zhao J, Bessa M A, Oswald J, et al. A method for modeling the transition of weak discontinuities to strong discontinuities: From interfaces to cracks[J]. International Journal for Numerical Methods in Engineering, 2016, 105(11): 834-854.

[54] Yang P, Gan Y, Zhang X, et al. Improved decohesion modeling with the material

point method for simulating crack evolution[J]. International Journal of Fracture, 2014, 186(1-2): 177-184.

[55] Rabczuk T, Belytschko T. A three-dimensional large deformation meshfree method for arbitrary evolving cracks[J]. Computer Methods in Applied Mechanics and Engineering, 2007, 196(29-30): 2777-2799.

[56] Yreux E, Chen J S. A quasi-linear reproducing kernel particle method[J]. International Journal for Numerical Methods in Engineering, 2017, 109(7): 1045-1064.

[57] Kaplinger B, Wie B, Dearborn D. Nuclear fragmentation/dispersion modeling and simulation of hazardous near-Earth objects[J]. Acta Astronautica, 2013, 90(1): 156-164.

[58] Kaplinger B, Wie B. Optimized GPU simulation of a disrupted near-earth object including self gravity[C]. Presented as paper AAS-11-266 at the 21st AAS/AIAA Space Flight Mechanics Meeting, New Orleans, LA. 2011.

[59] Novoselov K S, Geim A K, Morozov S V, et al. Electric field effect in atomically thin carbon films[J]. Science, 2004, 306(5696): 666-669.

[60] Balandin A A, Ghosh S, Bao W, et al. Superior thermal conductivity of single-layer graphene[J]. Nano Letters, 2008, 8(3): 902-907.

[61] Balandin A A. Thermal properties of graphene and nanostructured carbon materials[J]. Nature Materials, 2011, 10(8): 569.

[62] Hu J, Ruan X, Chen Y P. Thermal conductivity and thermal rectification in graphene nanoribbons: A molecular dynamics study[J]. Nano Letters, 2009, 9(7): 2730-2735.

[63] Iijima S. Helical microtubules of graphitic carbon[J]. Nature, 1991, 354(6348): 56-58.

[64] Liew K M, He X Q, Wong C H. On the study of elastic and plastic properties of multiwalled carbon nanotubes under axial tension using molecular dynamics simulation[J]. Acta Materialia, 2004, 52(9): 2521-2527.

[65] Liew K M, Wong C H, He X Q, et al. Nanomechanics of single and multiwalled carbon nanotubes[J]. Physical Review B, 2004, 69(11): 115429.

[66] Sun Y, Liew K M. The buckling of single-walled carbon nanotubes upon bending: The higher order gradient continuum and mesh-free method[J]. Computer Methods in Applied Mechanics and Engineering, 2008, 197(33-40): 3001-3013.

[67] Yan J W, Liew K M, He L H. Analysis of single-walled carbon nanotubes using the moving Kriging interpolation[J]. Computer Methods in Applied Mechanics and Engineering, 2012, 229: 56-67.

[68] Zheng H, Zhang C, Wang Y, et al. A meshfree local RBF collocation method for antiplane transverse elastic wave propagation analysis in 2D phononic crystals[J]. Journal of Computational Physics, 2016, 305: 997-1014.

[69] Zheng H, Zhang C, Wang Y, et al. A local RBF collocation method for band structure computations of 2D solid/fluid and fluid/solid phononic crystals[J]. International Journal for Numerical Methods in Engineering, 2017, 110(5): 467-500.

[70] Bouscasse B, Colagrossi A, Marrone S, et al. SPH modelling of viscous flow past a circular cylinder inter-acting with a free surface[J]. Computers & Fluids, 2017, 146: 190-212.

[71] Dehghan M, Abbaszadeh M. Proper orthogonal decomposition variational multiscale element free Galerkin (POD-VMEFG) meshless method for solving incompressible Navier-Stokes equation[J]. Computer Methods in Applied Mechanics and Engineering, 2016, 311: 856-888.

[72] Shadloo M S, Oger G, Le Touzé D. Smoothed particle hydrodynamics method for fluid flows, towards industrial applications: Motivations, current state, and challenges[J]. Computers & Fluids, 2016, 136: 11-34.

[73] Guibert D, De Leffe M, Oger G, et al. Efficient parallelisation of 3D SPH schemes[C]. 7th International SPHERIC Workshop, Prato, Italy. 2012: 259-265.

[74] Barcarolo D, Candelier J, Guibert D, et al. Hydrodynamic performance simulations using sph for automotive applications[C]. 9th international SPHERIC workshop, Paris, France. 2014: 321-326.

[75] Basava R R, Chen J S, Zhang Y, et al. Pixel based meshfree modeling of skeletal muscles[C]. International Symposium Computational Modeling of Objects Represented in Images. Springer, Cham, 2014: 316-327.

[76] Chen J S, Basava R R, Zhang Y, et al. Pixel-based meshfree modelling of skeletal muscles[J]. Computer Methods in Biomechanics and Biomedical Engineering: Imaging & Visualization, 2016, 4(2): 73-85.

[77] 廉艳平, 张帆, 刘岩, 等. 物质点法的理论和应用 [J]. 力学进展, 2013, 43 (2): 237-264.

[78] 张雄, 廉艳平, 刘岩, 等. 物质点法 [M]. 北京: 清华大学出版社, 2013.

[79] Zhou S, Zhang X, Ma H. Numerical simulation of human head impact using the material point method[J]. International Journal of Computational Methods, 2013, 10(04): 1350014.

[80] Wang Y, Rudy Y. Application of the method of fundamental solutions to potential-based inverse electrocardiography[J]. Annals of Biomedical Engineering, 2006, 34(8): 1272-1288.

[81] Basdogan C, De S, Kim J, et al. Haptics in minimally invasive surgical simulation and training[J]. IEEE Computer Graphics and Applications, 2004, 24(2): 56-64.

[82] Miller K, Wittek A, Joldes G, et al. Modelling brain deformations for computer-integrated neurosurgery[J]. International Journal for Numerical Methods in Biomedical Engineering, 2010, 26(1): 117-138.

[83] Delingette H, Pennec X, Soler L, et al. Computational models for image-guided robot-assisted and simulated medical interventions[J]. Proceedings of the IEEE, 2006, 94(9): 1678-1688.

[84] Belinha J. Meshless Methods in Biomechanics[M]. Switzerland: Springer International Publishing, 2014.

[85] Guilkey J E, Hoying J B, Weiss J A. Computational modeling of multicellular constructs with the material point method[J]. Journal of Biomechanics, 2006, 39(11): 2074-2086.

[86] Chen T, Wang X, Chung S, et al. Automated 3D motion tracking using Gabor filter bank, robust point matching, and deformable models[J]. IEEE Transactions on Medical Imaging, 2010, 29(1): 1-11.

[87] Wang H, Amini A A. Cardiac motion and deformation recovery from MRI: A review[J]. IEEE Transactions on Medical Imaging, 2012, 31(2): 487-503.

[88] Opfer R, Wiemker R. A new general tumor segmentation framework based on radial basis function energy minimization with a validation study on LIDC lung nodules[C]. Medical Imaging 2007: Image Processing. International Society for Optics and Photonics, 2007, 6512: 651217.

[89] Li M, Miller K, Joldes G R, et al. Biomechanical model for computing deformations for whole-body image registration: A meshless approach[J]. International Journal for Numerical Methods in Biomedical Engineering, 2016, 32(12): eo2771.

[90] Diaz I, Boulanger P. Atlas to patient registration with brain tumor based on a mesh-free method[C]. Engineering in Medicine and Biology Society (EMBC), 2015 37th Annual International Conference of the IEEE. IEEE, 2015: 2924-2927.

[91] Horton A, Wittek A, Miller K. Towards meshless methods for surgical simulation[C]. Proceedings of MICCAI Workshop. Copenhagen. 2006: 34-43.

[92] Fumeaux C, Kaufmann T, Shaterian Z, et al. Conformal and multi-scale time-domain methods: From unstructured meshes to meshless discretisations[M]//Ahmed I, Chen Z. Computational Electromagnetics—Retrospective and Outlook. Singapore: Springer, 2015: 139-165.

[93] Doyle B, Miller K, Wittek A, et al. Computational Biomechanics for Medicine: New Approaches and New Applications[M]. Switzerland: Springer, 2015.

[94] 刘岩, 张雄, 刘平, 等. 空间碎片防护问题的物质点无网格法与软件系统 [J]. 载人航天, 2015 (5): 503-509.

[95] Koshizuka S, Oka Y. Moving-particle semi-implicit method for fragmentation of incompressible fluid[J]. Nuclear Science and Engineering, 1996, 123(3): 421-434.

[96] 麦粒软件：www.mailisoft.com.

[97] Mohebbi A, Abbaszadeh M, Dehghan M. The use of a meshless technique based on collocation and radial basis functions for solving the time fractional nonlinear Schrödinger equation arising in quantum mechanics[J]. Engineering Analysis with Boundary Elements, 2013, 37(2): 475-485.

[98] Chen J S, Hu W, Puso M. Orbital HP-clouds for solving Schrödinger equation in quantum mechanics[J]. Computer Methods in Applied Mechanics and Engineering, 2007,

196(37-40): 3693-3705.

[99] Dehghan M, Mirzaei D. Numerical solution to the unsteady two-dimensional Schrödinger equation using meshless local boundary integral equation method[J]. International Journal for Numerical Methods in Engineering, 2008, 76(4): 501-520.

[100] Dehghan M, Abbaszadeh M, Mohebbi A. The numerical solution of nonlinear high dimensional generalized Benjamin–Bona–Mahony–Burgers equation via the meshless method of radial basis functions[J]. Computers & Mathematics with Applications, 2014, 68(3): 212-237.

[101] Davé R, Thompson R, Hopkins P F. MUFASA: Galaxy formation simulations with meshless hydrodynamics[J]. Monthly Notices of the Royal Astronomical Society, 2016, 462(3): 3265-3284.

[102] Zhu Q, Li Y. The Formation of a Milky way-sized disk galaxy 1. A Comparison of numerical methods[J]. The Astrophysical Journal, 2016, 831(1): 52.

[103] Few C G, Dobbs C, Pettitt A, et al. Testing hydrodynamics schemes in galaxy disc simulations[J]. Monthly Notices of the Royal Astronomical Society, 2016, 460(4): 4382-4396.

[104] Lupi A, Volonteri M, Silk J. Simplified galaxy formation with mesh-less hydrodynamics[J]. Monthly Notices of the Royal Astronomical Society, 2017, 470(2): 1673-1686.

[105] Qin C, Tian J, Yang X, et al. Galerkin-based meshless methods for photon transport in the biological tissue[J]. Optics Express, 2008, 16(25): 20317-20333.

[106] Koumoutsakos P. Multiscale flow simulations using particles[J]. Annu. Rev. Fluid Mech., 2005, 37: 457-487.

[107] Filipovic N, Ivanovic M, Kojic M. A comparative numerical study between dissipative particle dynamics and smoothed particle hydrodynamics when applied to simple unsteady flows in microfluidics[J]. Microfluidics and Nanofluidics, 2009, 7(2): 227-235.

[108] Kiani K, Ghaffari H, Mehri B. Application of elastically supported single-walled carbon nanotubes for sensing arbitrarily attached nano-objects[J]. Current Applied Physics, 2013, 13(1): 107-120.

[109] Parand K, Hemami M. Application of meshfree method based on compactly supported radial basis function for solving unsteady isothermal gas through a micro–nano porous medium[J]. Iranian Journal of Science and Technology, Transactions A: Science, 2017, 41(3): 677-684.

[110] Kiani K. A meshless approach for free transverse vibration of embedded single-walled nanotubes with arbitrary boundary conditions accounting for nonlocal effect[J]. International Journal of Mechanical Sciences, 2010, 52(10): 1343-1356.

[111] Zhang L W, Lei Z X, Liew K M. Free vibration analysis of functionally graded carbon nanotube-reinforced composite triangular plates using the FSDT and element-free

IMLS-Ritz method[J]. Composite Structures, 2015, 120: 189-199.

[112] Liew K M, Lei Z X, Zhang L W. Mechanical analysis of functionally graded carbon nanotube reinforced composites: A review[J]. Composite Structures, 2015, 120: 90-97.